Industry 4.0 Key Technological Advances and Design Principles in Engineering, Education, Business, and Social Applications

This book offers an in-depth look at Industry 4.0's applications and provides a conceptual framework for design principles and easy implementation. The book touches on the impact of Industry 4.0 and also examines the key technological advances and potential economic and technical benefits through case studies featuring real-world implementations.

Industry 4.0 Key Technological Advances and Design Principles in Engineering, Education, Business, and Social Applications discusses the impact of Industry 4.0 and workforce transformation. The book examines the key technological advances that support Industry 4.0 and examines their potential economic and technical benefits through case studies. It covers the connection Industrial 4.0 has with IT and communication technologies and demonstrates the technological advancements and how to use their benefits towards and through examples of real-world applications.

This book offers a conceptual framework and road map for those making decisions as they go through the next stage of transformation. This book mainly targets academicians, professionals, business professionals, start-up companies, and researchers at undergraduate, postgraduate, and doctoral levels.

Industry 4.0 Key Technological Advances and Design Principles in Engineering, Education, Business, and Social Applications

Edited by
Sagaya Aurelia and Ossama Embarak

CRC Press
Taylor & Francis Group
Boca Raton London New York

CRC Press is an imprint of the
Taylor & Francis Group, an **informa** business

Designed cover image: Shutterstock—metamorworks

MATLAB® and Simulink® are trademarks of The MathWorks, Inc. and are used with permission. The MathWorks does not warrant the accuracy of the text or exercises in this book. This book's use or discussion of MATLAB® or Simulink® software or related products does not constitute endorsement or sponsorship by The MathWorks of a particular pedagogical approach or particular use of the MATLAB® and Simulink® software.

First edition published 2025
by CRC Press
2385 NW Executive Center Drive, Suite 320, Boca Raton FL 33431

and by CRC Press
4 Park Square, Milton Park, Abingdon, Oxon, OX14 4RN

CRC Press is an imprint of Taylor & Francis Group, LLC

© 2025 selection and editorial matter, Sagaya Aurelia and Ossama Embarak; individual chapters, the contributors

ISBN: 978-1-032-37687-5 (hbk)
ISBN: 978-1-032-38064-3 (pbk)
ISBN: 978-1-003-34333-2 (ebk)

DOI: 10.1201/9781003343332

Typeset in Times
by Apex CoVantage, LLC

Contents

Contents

About the Editors

Over her 20-year career in the teaching profession, Dr. Sagaya Aurelia has worked with leading engineering colleges and universities in India and overseas. She is currently working at CHRIST University. She has completed her bachelor's degree in electronics and communication, masters in information technology, and doctorate in computer science and engineering. She has completed a postgraduate diploma in business administration and a diploma in e-commerce application development. She has published several papers in international and national journals and conferences and won the best paper award for many of them. She serves as a member of IEEE, ISTE, the International Economics Development Research Center, and the International Association of Computer Science and Information Technology. She was awarded the Peace Leader and Peace Educator Award from the United Nation's global compact's global mission.

Dr. Ossama Embarak is currently employed as a university professor. He received his PhD in computer and information technology (artificial intelligence, machine learning, and data mining) from Heriot-Watt University in Scotland, UK, in 2012. He has over 15 years of experience in teaching, research, research project supervision, and industry partnerships. Ossama works as a journal reviewer, Springer book editor, and has numerous publications in conferences and books on computing. He was awarded a 200k grant to investigate the fragmentation of students' knowledge and skills using machine learning and to create a student radar system for live predictions and recommendations. He oversaw numerous industry collaboration research projects funded by the Fujairah Chamber of Commerce, the Ministry of Interior, the Ministry of Environment, and the Ministry of Tourism.

Ossama is an Apple-certified trainer. He is also the author of numerous books and book chapters on data science, recommendation systems, and smart city technologies. He has led numerous workshops and seminars on AI, machine learning, privacy, mobile applications, data science, data analysis and visualization, emerging technologies, etc. He is also passionate about new technologies, particularly artificial intelligence, data science, machine learning, and bioinformatics. Besides his academic credentials, he is a member of social and academic communities such as Oracle Academy, IBM Middle East Skills Academy, IEEE, Fellow of the Higher Education Academy (FHEA), and Toastmaster.

List of Contributors

Rasha Abousamra
Higher Colleges of Technology
Seyouh, Sharjah, UAE

Maryam Alameeri
Higher Colleges of Technology
Fujairah, UAE

Fatima Aldarmaki
Higher Colleges of Technology
Fujairah, UAE

Alnaqbi Ali
Higher Colleges of Technology
Fujairah, UAE

Maryam Almesmari
Higher Colleges of Technology
Fujairah, UAE

Rula Azzawi
Higher Colleges of Technology
Sharjah, Sharjah, UAE

Afsaneh Behforouz
Ball State University
Muncie, Indiana, USA

Chanyalew Belachew
Higher Colleges of Technology
Fujairah, UAE

Abdulla Desmal
Higher Colleges of Technology
Ras Al-Khaimah, UAE

Shailja Dixit
Amity University
Lucknow, India

M Durairaj
Bharathidasan University
Tiruchirappalli, Tamilnadu, India

Ossama Embarak
Higher Colleges of Technology
Fujairah, UAE

Dilip Kumar Gayen
College of Engineering and
    Management
Kolaghat, KTPP Township,
    West Bengal, India

Hamad Hareb
Higher Colleges of Technology
Fujairah, UAE

Mohammed Hassouna
Higher Colleges of Technology
Sharjah, Sharjah, UAE

Osama Hosam
Higher Colleges of Technology
Sharjah City, Sharjah
    Province, UAE

Osama Hossameldeen
Higher Colleges of Technology
Muwaileh, Sharjah, UAE

David Hua
Ball State University
Muncie, Indiana, USA

Zakea Ilagure
Higher Colleges of Technology
Ras Al-Khaimah, UAE

Shamoona Imtiaz
Mälardalen University
Västerås, Sweden

Yaseen Ishfaq
Prince Sattam Bin Abdul Aziz
    University, Alkharj
Saudi Arabia

Hicham Itani
Higher Colleges of Technology
Ras Al-Khaimah, UAE

Dan Ivanov
Higher Colleges of Technology
Muwaileh, Sharjah, UAE

Anju Kalluvelil Janardhanan
Southern Cross Institute
Parramatta, NSW, Australia

Deepa V Jose
CHRIST (Deemed to be University)
Bangalore, Karnataka, India

Manu K S
CHRIST University
Bangalore, Karnataka, India

Natarajan K
Department of Computer Science and
    Engineering
School of Engineering and
    Technology
CHRIST (Deemed to be University)
Kengeri Campus, Bengaluru
Karnataka, India

Faisal Kalota
Ball State University
Muncie, Indiana, USA

Jeevanantham Lenin
Alliance University
Bangalore, Karnataka, India

Lina Momani
Higher Colleges of Technology
Fujairah, UAE

P Nagamalleswrarao
Chalapathi Institute of Technology
Guntur, Andhra Pradesh, India

Pulicherla Padmaja
Hyderabad Institute of Technology
    and Management
Hyderabad, India

Oliver Popov
Stockholm University
Stockholm, Sweden

Shaikha Al Qassemi
Higher Colleges of Technology
Ras Al-Khaimah, UAE

Kavitha Rajamohan
CHRIST University
Bangalore, Karnataka, India

Sangeetha Rangasamy
CHRIST University
Bangalore, Karnataka, India

Jaume Rius i Riu
Telenor Sverige AB
Solna, Sweden

Reeja S R
VIT-AP University
Amaravati, Andhra Pradesh, India

J Sandeep
CHRIST (Deemed to be University)
Bengaluru, Karnataka, India

Charles Savarimuthu
University of Technology and Applied
    Sciences
Al Musanna, Muladdah, Mussanah
Sultanate of Oman

Monikka Reshmi Sethurajan
Department of Computer Science and
    Engineering
School of Engineering and Technology
CHRIST (Deemed to be University),
    Kengeri Campus
Bengaluru, Karnataka, India

Tamanna Jena Singhdeo
FDU Vancouver Campus
Vancouver, Canada

K Srinivasarao
ICFAI University
Hyderabad, Telangana, India

Akansha Abhi Srivastava
Institute of Hotel Management (IHM)
Lucknow, India

Amit Kumar Srivastava
Invertis University
Bareilly, India

Libin Thomas
CHRIST (Deemed to be
    University)
Bengaluru, Karnataka, India

Priya Thomas
CHRIST (Deemed to be
    University)
Bangalore, Karnataka, India

Ashwin Kumar V
CHRIST University
Bengaluru, Karnataka, India

Faizal N Vasaya
CHRIST (Deemed to be
    University)
Bengaluru, Karnataka, India

Jasim Yousif
Higher Colleges of Technology
Fujairah, UAE

Preface

In the rapidly evolving landscape of technology, the emergence of Industry 4.0 has brought forth a revolution that transcends the boundaries of traditional industries, transforming every aspect of our lives. The convergence of advanced technologies, connectivity, and data-driven insights has paved the way for innovative solutions in engineering, education, business, and social applications. This book, *Industry 4.0 Key Technological Advances and Design Principles in Engineering, Education, Business, and Social Applications*, delves into the myriad dimensions of this transformative era.

Chapter 1: The Fourth Industrial Revolution: An Overview of Technologies and Socio-Economic Implications

The journey begins with exploring Industry 4.0, providing a foundational understanding of its core principles, significance, and impact across various sectors. The reader is introduced to the technological underpinnings that set the stage for the subsequent chapters.

Chapter 2: An Insight into the Consensus Protocols and Data Security in IoT Applications Using Blockchain

With the proliferation of interconnected devices, securing data becomes paramount. This chapter explores how blockchain technology provides a robust framework for ensuring data integrity and security in IoT applications.

Chapter 3: Technological Advances in Smart Monitoring Imaging Systems Based on Machine Learning

Building upon the foundation of IoT, this chapter delves into developing smart monitoring systems. These systems offer insights into various domains, including infrastructure, health care, and environment, by leveraging real-time data collection and analysis.

Chapter 4: IoT-Based Seaweed Cultivation for a Sustainable Economy

Highlighting the breadth of Industry 4.0's impact, this chapter showcases how IoT is pivotal in promoting sustainable practices, as demonstrated by seaweed cultivation. This innovative approach not only supports economic growth but also fosters ecological balance.

Chapter 5: Security Analysis and Planning for Enterprise Networks: Incorporating Modern Security Design Principles

In an age where digital threats are ever-present, this chapter investigates security strategies for safeguarding enterprise networks. It addresses the evolving challenges of cybersecurity and presents methodologies for effective defense.

Chapter 6: A Comparative Study of Structureless Data Aggregation Protocols in Wireless Sensor Networks (WSNs)

Wireless Sensor Networks hold immense potential in Industry 4.0 applications. This chapter offers a comparative analysis of data aggregation protocols, highlighting their suitability for diverse scenarios.

Chapter 7: HR Analytics: A Model of Modern Human Resource Practices

Moving into the business realm, this chapter explores how Industry 4.0 principles reshape human resource management. Integrating data analytics and human-centric insights is pivotal in modern HR practices.

Chapter 8: Technology and Design Advancements: Implementation of Cloud Computing Using a Service-Oriented Architecture (SOA) Model

Integrating cloud computing and service-oriented architecture has transformed business operations. This chapter illustrates how this integration enhances scalability, flexibility, and efficiency in business processes.

Chapter 9: Redefining Traditional Education Using Augmented Reality and Virtual Reality

The educational landscape evolves as AR and VR redefine traditional learning methods. This chapter illustrates how immersive technologies enhance engagement and knowledge retention.

Chapter 10: AI-Based Automatic Detection of IP Network Performance in Telecommunication

Telecommunications undergoes a paradigm shift with AI-based network performance analysis. The chapter examines how AI algorithms streamline network monitoring and maintenance.

Chapter 11: AI-Based Movie Recommendation System

Entertainment and technology converge in this chapter, exploring AI's role in curating personalized movie recommendations. This intersection showcases how Industry 4.0 enriches user experiences.

Chapter 12: Criminal Investigation Using Deep Learning and Image Processing

The fusion of deep learning and image processing accelerates criminal investigations. This chapter outlines how Industry 4.0 empowers law enforcement agencies with advanced tools for evidence analysis.

Chapter 13: Machine Learning Validation for Project Success Multivariate Modeling

Project management receives a boost from machine learning as this chapter explores its role in predicting and enhancing project success rates.

Chapter 14: Deep Learning Method for Anomaly Detection in Cyber Physical Systems (CPS) and Social Networks

As cyber-physical systems proliferate, anomaly detection becomes vital. The chapter examines how deep learning techniques offer real-time insights into potential malfunctions.

Chapter 15: Industry 4.0: Intelligent Routing and Scheduling Algorithm for EV Charging Based on EV Battery Management System

Green transportation gains impetus as this chapter unveils routing and scheduling algorithms for electric vehicle charging. The integration of battery management systems optimizes the charging process.

Chapter 16: Design of Eco-Friendly Batteries Using PCM

Integrating Phase Change Material (PCM) batteries revolutionizes domestic energy management. This chapter explores their role in enhancing energy efficiency and storage.

Chapter 17: Synthesizing User Comments to Prospective Viral Interests on New Media Using Sentiment Analysis for Effective Data Visualization Premeditated for Journalism 4.0

In journalism, Industry 4.0 brings forth innovative ways of analyzing user comments and sentiments for effective data visualization. This chapter delves into how journalism is transformed in the digital era.

Chapter 18: A Data-Driven Model-Based Application to Interpret Anomalies Utilizing Mutual Information in Satellite Data

Drawing upon the power of data-driven insights, this chapter introduces a model-based approach for interpreting anomalies in satellite data, showcasing the far-reaching applications of Industry 4.0.

As you journey through the pages of this book, you will traverse the vast expanse of Industry 4.0, exploring its multidisciplinary impact on engineering, education, business, and social realms. Each chapter offers a unique perspective on the ever-evolving landscape of technology, shedding light on how innovation and integration can shape a future defined by connectivity, efficiency, and sustainability.

Dr. Sagaya Aurelia
Bengaluru, India

Dr. Ossama Embarak
Fujairah, United Arab Emirates

Acknowledgments

"Every good and perfect gift is from above."

Writing a book is a journey; this one has been no exception. It's with immense gratitude that we extend our heartfelt recognition to the individuals and entities who have played a pivotal role in bringing this work to fruition.

First and foremost, we want to express our most profound appreciation to our families. Their unwavering support, encouragement, and understanding during the long hours of writing and research have been our pillars of strength.

We are indebted to our mentors and advisors, whose guidance and expertise have been instrumental in shaping this book. Their wisdom and insights have enriched this work and our understanding of the subject matter.

We want to thank our colleagues and peers, including Dr. Fr. Joseph CC, vice chancellor and director of the Center for Research Projects at CHRIST (Deemed to be University). We also want to extend our appreciation to the Department of Computer Science at CHRIST (Deemed to be University), India and the Higher Colleges of Technology, United Arab Emirates, as well as all staff members, seniors, and colleagues from India, and the UAE, who have provided invaluable support throughout this publication journey. Their engaging discussions, constructive feedback, and camaraderie have enriched this experience.

Our gratitude extends to the reviewers who provided constructive feedback and helped refine the contents of this book. Their expertise and attention to detail have undoubtedly improved the quality of this work.

We would also like to acknowledge the support of CRC Press for believing in this project and their professionalism throughout the publishing process. A special thanks to Cindy Carelli.

To our friends and well-wishers who offered encouragement and cheered us on, thank you for being a constant source of motivation.

Last but not least, to the readers of this book, we hope you find it insightful and valuable. Your interest in this subject makes authors like us continue exploring and sharing knowledge.

Ultimately, any success this book achieves results from collective effort and the support of an incredible network of people. We are truly grateful to each one of you.

With heartfelt thanks,

Dr. Sagaya Aurelia
Dr. Ossama Embarak

1 The Fourth Industrial Revolution

An Overview of Technologies and Socio-Economic Implications

*Faisal Kalota, David Hua,
and Afsaneh Behforouz*

1.1 INTRODUCTION

The Terminator movies paint a future in which machines are at war with human beings, and the machines are trying to take over the world. While this idea seems far-fetched, the current technological advancements in artificial intelligence (AI), big data, cloud computing, edge computing, the Internet of Things, and relevant technologies paint a picture that machines are supporting humans and possibly slowly replacing them in various domains. For example, AI has been utilized in finance, health care, marketing, and recruitment (Pyo, Hwang, and Yoon 2021).

The world is witnessing the Fourth Industrial Revolution (FIR), also known as Industry 4.0 (I4.0). Some of the technologies stated earlier are at the cornerstone of Industry 4.0. While it may seem that I4.0 is only about the manufacturing industries, it should be noted that I4.0 also impacts many other industries and has various socio-economic implications. This chapter provides a brief historical perspective of the different industrial revolutions, followed by an overview of some of the Industry 4.0 technologies and, lastly, some of the socio-economic implications of Industry 4.0.

1.2 AN INDUSTRIAL REVOLUTION

According to the *Merriam-Webster Dictionary*, a revolution can be defined as something that is (a) sudden, (b) radical, (c) complete change, or (d) "activity or movement designed to effect fundamental changes in the socio-economic situation" (Meriam Webster n.d.). Throughout history, there have been many revolutions around the globe, each leaving a long-lasting mark in history and impacting the direction of history; for example, American Revolution, French Revolution, Scientific Revolution, etc. Over the last 250 years, the world has been transformed through three previous industrial revolutions, and the world is currently going through the fourth one. This section briefly overviews the different industrial revolutions and their impact.

DOI: 10.1201/9781003343332-1

1

1.2.1 FIRST INDUSTRIAL REVOLUTION

Defined by the Munich Personal RePEc Archive, the First Industrial Revolution kicked off in the middle of the 1750s and concluded around the 1840s (Mohajan 2021). It is often referred to as a turning point in human history because it transformed human and animal labor. The inventions and innovations of this revolution promoted the design and development of advanced manufacturing processes in the United States and other European Nations. Samuel Slater and Francis Cabot Lowell significantly impacted the Industrial Revolution in the United States because they introduced textile manufacturing (Brooks 2018; Nicholas and Guilford 2016). The introduction of machinery also impacted the mining and manufacturing industries (Dima n.d.).

Since many of the factories and mills were located in the cities, people left the farms in search of economic opportunities in cities, and this became a catalyst for urbanization. This boom in population also resulted in some socio-economic issues. For example, due to little or no regulations, people worked in poor factory conditions and lived in small, overcrowded quarters; this created an unhygienic environment that spread diseases and resulted in higher death rates.

1.2.2 SECOND INDUSTRIAL REVOLUTION

The Second Industrial Revolution, which roughly lasted between the late 1800s and early 1900s, was not limited to Britain only (Mokyr 1998). This period, driven by electricity and steel production, was a global transformation with the largest reach of any period. The key advancements during this period included electricity, steel, micro-inventions, and production. Some of the major inventions of this revolution include air brakes, light bulbs, wireless communication, petroleum refining, QWERTY typewriter keyboard, skyscrapers, tractors, and safety razors (Kiger 2021).

Although electricity was discovered in the early 1800s, it became a catalyst for innovation when scientists Joseph Henry and Michael Faraday created the first electric motor (Mokyr 1998). Steel was also a material that was invented. Innovations by Henry Bessemer lowered the cost of manufacturing high-quality steel that was stronger, more malleable, and longer lasting (Agarwal and Agarwal 2017). The improvement in steel manufacturing positively impacted transportation because it became more efficient to develop railroads. This, in turn, positively impacted commerce and the transportation of goods. The development and improvement in steel production also impacted the shipping, automobile, manufacturing, and construction industries. With the introduction of steel and electricity, the industrial world went through a period of hard work and efficiency.

Unfortunately, with the positive advancements, there were also some negative consequences. For example, because of all the new jobs and tasks, children were pushed into factories to work to accomplish simple tasks. Often these children were underpaid (Johnston 2020). Due to the lack of appropriate rules and regulations, there were several other issues, such as inhumane work conditions, poor sanitation due to overcrowding, and the rise of corporate monopolies (Richmond Vale Academy 2022).

1.2.3 THIRD INDUSTRIAL REVOLUTION

The late 1900s and the early 2000s marked the era of the Third Industrial Revolution. The world saw advancements in computers, automobiles, and antibiotics. The main drivers of this impactful time of change are computerization, lean production, the rise of the Internet, and biotechnology (Taalbi 2019). The exponential growth of technology led to the digitization of many processes in different industries. This also had a direct impact on the skills of the workforce. In addition to mechanical skills, the new jobs also required people to have technical knowledge (Liu and Grusky 2013). This created opportunities for people to develop new skills and transition out of lower-paying jobs; hence, it also improved the socio-economic conditions for many people.

Continuous technological innovation resulted in better quality products and a positive impact on life. For example, there was an evolution in medical technologies to support doctors (Rafferty n.d.). While advancements in medical sciences positively impacted life expectancy, there were a few drawbacks as well. The major population shift to the cities and industrial towns created sewer and sanitation issues (Rafferty n.d.). Additionally, the pollution from factories (Mohajan 2021) created a new set of environmental problems.

1.2.4 FOURTH INDUSTRIAL REVOLUTION

The root of the term Industry 4.0 (I4.0) can be traced back to 2011 in Germany (Puriwat and Tripopsakul 2020). It was an initiative launched by the German Government (European Commission 2017). Industry 4.0 is built on the foundations of previous industrial revolutions (Carvalho and Cazarini, 2020). Industry 4.0 also represents an integration of various domains such as additive manufacturing, artificial intelligence (AI), big data, cloud computing, the Internet of Things (IoT), and edge computing. The tightknit integration of various hardware and software technologies led to the development of cyber-physical systems. The next section describes some of the technologies associated with Industry 4.0.

1.3 INDUSTRY 4.0 TECHNOLOGIES

"Industry 4.0 is based on the digitalization and automation of factories" (Ortiz, Marroquin, and Cifuentes 2020, 15) using various technologies, including but not limited to the Internet of Things (IoT), cloud computing, artificial intelligence, machine learning, edge computing, cybersecurity, digital twins (IBM n.d.a.), and additive manufacturing. This section describes some of the technologies in the Industry 4.0 ecosystem.

1.3.1 ARTIFICIAL INTELLIGENCE AND MACHINE LEARNING

"Intelligence is the capacity to learn from experience, using metacognitive processes to enhance learning, and the ability to adapt to the surrounding environment" (Sternberg and Sternberg 2012, 17). Artificial intelligence (AI) "is the study of how to make computers do things which, at the moment, people do better" (Rich and Knight 1991, 3). Although Rich and Knight acknowledge that this is not a universally accepted definition, it may still be valid. Developers are trying to make machines do things that humans have historically done better. For example, replacing humans in factories, supporting doctors, playing chess, playing Go, etc. Joshi classifies it as "a science of finding theories and methodologies that can help machines understand the world and accordingly react to situations in the same way that humans do" (2017, 8). So among other things, AI aims to mimic human intelligence and behavior.

Machines and software must be programmed to mimic human intelligence and behavior, which is one of the goals of machine learning (ML). Machine Learning is a collection of algorithms and techniques used to design systems that learn from data (Lee 2019). This is accomplished by implementing a set of methods for the automated analysis of structure in data (Fisher, Breckon, and Dawson-Howe et al. 2016). There are many different ML techniques. Few of the ML techniques include the following: decision trees, K-means, K-nearest neighbor, naïve bayes, regressions analysis, and support vector machines.

ML is broadly categorized into supervised machine learning (SML) and unsupervised machine learning (UsML). Supervised machine learning is utilized with labeled data for classification and regression. Classification is used to predict discrete outcomes, whereas regression predicts continuous output. Unsupervised machine learning is intended to discover patterns in unlabeled data without human intervention. These algorithms involve clustering the data based on differences or similarities within the data and finding relationships between them. Such algorithms would also perform the task of dimension reduction so that a model isn't overfitted.

Another branch of AI is deep learning (DL). Deep learning is a subset of machine learning, and it utilizes an artificial neural network (ANN) as a learning process. Neural networks mimic the functionality of a human brain. A fairly recent application of deep learning is generative AI, which is used to generate content such as text and images. An example of generative AI is ChatGPT.

AI relies on various machine learning and deep learning techniques and algorithms. The output produced by these algorithms is only as good as the data that is fed into these algorithms. Hence, it's important that data is valid and reliable. This also brings us to the next topic: big data.

1.3.2 BIG DATA

As of 2022, over 90% of the data in the world was created in the previous two years (Baltzan and Philips 2022). The significance of data cannot be emphasized enough. It has been called the "new oil," "natural resources" (Shahul 2018), and the "new soil" (McCandless 2010). At the same time, some may argue against the analogy of data being the new oil because, unlike oil, data are not a finite resource and can be reused (Virnes 2021). However, no one can argue the significance of data. If utilized properly, data can provide significant benefits by allowing us to generate information that enables business intelligence to generate knowledge (Baltzan and Phillips 2022, 7).

Over the last decade, big data has gained significant traction and attention. Big data does not have a universal definition. While there are different definitions of big data, it is often characterized by various features. These features include variety, velocity, veracity, and volume (Qamar and Raza 2020). Variety, volume, and velocity imply that the data comes in different forms in very large quantities at a very high speed. Such data cannot be processed with traditional methods (SAS n.d.). Some researchers (Pyo, Hwang, and Yoon 2021) also consider "value" as one of the features of big data. Whether big or not, data can allow organizations to perform various descriptive, diagnostic, predictive, or prescriptive analytics (Baltzan and Philips 2022). Data is also utilized to generate real-time dynamic reports in dashboards.

1.3.3 CLOUD COMPUTING

The practice of cloud computing (CC) dates back to the mainframe era of the 1950s, in which users used dumb terminals to access mainframe servers (IBM 2017). From a more contemporary perspective, it provides the user with different products and services, such as computing power, computing infrastructure, software, etc., whenever and wherever the user needs it (Hurwitz, Bloor, Kaufman et al. 2009) through the Internet (O'Leary, O'Leary, and O'Leary 2023). It is a pool of configurable computing resources rapidly provisioned to the end user (NIST n.d.a).

Cloud computing provides three common models to provide various services: Infrastructure as a Service (IaaS), Platform as a Service (PaaS), and Software as a Service (SaaS). IaaS provides access to the infrastructure, such as servers, storage, operating systems, etc. PaaS caters to the developer community and provides them with access to a framework of tools and technologies to develop applications. Most everyday people are likely to be familiar with SaaS. They probably have utilized it as well because it provides them access to applications on the cloud, such as emails and productivity applications.

Cloud computing platforms have evolved significantly and provide advanced technologies such as artificial intelligence, machine learning, and IoT (SAP n.d.). Hence, it plays an important role in smart manufacturing and supporting small or medium-sized manufacturers in cost reduction (IBM n.d.a). With the availability of 5G networks and the advanced technologies in IoT devices, it has become easier to collect and share massive amounts of data. However, sending data over the network to the data centers for processing can increase network traffic and add latency. Hence, it is important to consider how modern IoT devices and edge computing can alleviate some of these challenges.

1.3.4 INTERNET OF THINGS

The Internet of Things (IoT) is "the continuing development of the Internet that allows everyday objects embedded with electronic devices to send and receive data over the Internet" (O'Leary, O'Leary, and O'Leary 2023, 45). IoT devices are Internet-enabled devices connected to the Internet and communicate with other IoT devices. These include anything from high-tech devices to mundane items such as baby diapers. When these things are connected and communicate with one another, they can provide very sophisticated services.

IoT devices play an important role in Smart factories. Smart factories are characterized by four important features (a) sensors, (b) interoperability, (c) integration, and (d) virtual reality (VR) techniques (Kalsoom, Ramzan, and Ahmed et al. 2020). Sensors can be either passive or smart (Kalsoom, Ramzan, and Ahmed et al. 2020). They play a key role in smart factories by collecting various data from the environment for additional processing. Some of these data can be sent to the cloud, while other data may be processed locally at the edge.

1.3.5   EDGE COMPUTING

In the cloud computing paradigm, the computing resources are centrally located away from the data origination point. The massive amount of generated data can also cause network congestion due to data transmission to the cloud, which is why edge computing becomes important. In the edge computing framework, most computations are done close to the data's origination point, also known as the edge of the network.

Edge computing can provide agility and autonomy in smart manufacturing. In the edge computing paradigm, "substantial computing and storage resources are placed at the Internet's edge in close proximity to mobile devices or sensors" (Satyanarayanan 2017, 30). Edge computing and cloud computing are not mutually exclusive, but these technologies are intended to complement each other. The nodes at the Edge can collect and process some of the raw data, and if needed, they can also upload the data to the cloud for additional processing. This can reduce network congestion and latency. Edge computing provides certain benefits over the traditional cloud computing model, such as fast data processing and analysis, security, lower cost due to low energy consumption, and low bandwidth cost (Cao, Liu, and Meng et al. 2020).

1.3.6   AUTOMATION AND ROBOTICS

Automation is the use of scripting to allow technology to conduct repetitive or routine tasks and processes with little or no human intervention. This is done to optimize business processes, increase productivity, and increase revenue. Related to this topic is the concept of robotic process automation (RPA), which is defined as "the use of software with artificial intelligence (AI) and machine learning capabilities to handle high-volume, repeatable tasks that previously required a human to perform" (Baltzan and Philips 2022, 82). Traditional MIS automation differs from RPA because RPA can adapt to the environment (Baltzan and Philips 2022, 82). Hence autonomous robots can react to the environment and make a decision accordingly.

Robots have long been utilized in industrial sectors. Incorporating embedded intelligence and communications directly into the robots has introduced them to what is known as the Internet of Robotic Things (IoRT), which allows the robots to share data around them and take various actions based on the data (Javaid, Haleem, and Singh et al. 2021). In addition to IoRT, collaborative robots are utilized in manufacturing and industrial settings. Collaborative robots, also known as Cobots, can interact with humans (Javaid, Haleem, and Singh et al. 2021). Such machines with a certain level of artificial intelligence built into them can be used to automate various processes, especially the ones that are routine and repetitive.

1.3.7 Augmented Reality, Virtual Reality, and Mixed Reality

Virtual reality (VR) is a "computer-simulated environment that can be a simulation of the real world or imaginary world" (Baltzan and Philips 2022, 72). Augmented reality (AR), on the other hand, is the "viewing of the physical world with computer-generated layers of information added to it" (Baltzan and Philips 2022, 72). There are numerous applications of both of these technologies; for example, augmented reality is used by retailers to allow customers to see how a piece of furniture would look in their homes.

Mixed reality (MR) integrates real-world and digital elements that allow you to interact with physical and virtual objects using advanced sensing and imaging technologies (Intel n.d.). A prime example of such technology is the Microsoft HoloLens, an "ergonomic, untethered self-contained holographic device with enterprise-ready applications to increase user accuracy and output" (Microsoft n.d.). Such a device benefits various industry sectors, including manufacturing, engineering and construction, health care, and education. A study conducted by the Forrester Group, which Microsoft commissioned, found that over a three-year period, the organizations that implemented Microsoft HoloLens saw a collective 177% return on investment (ROI), $11.9 million in present value (PV) benefits, $7.6 million in net present value (NPV), and a payback (cost recovery) in 13 months (Forrester 2021). AR, MR, and VR have the potential to transform many processes in different industries, and one prime example is the creation of digital twins.

1.3.8 Digital Twins

"A digital twin is a virtual model designed to accurately reflect a physical object" (IBM n.d.b). For example, various components of an automobile can be outfitted with sensors. Based on the data and information from these sensors, a virtual (digital) twin is created, which is a replica of the actual physical model. Digital twins (DT) provide a bi-direction communication and sharing of various types of data between the physical object and virtual representation; this allows the DT to perform various tasks such as (a) simulating the health condition of the physical object, (b) increasing the safety and reliability of physical object, (c) product or process optimization, (d) design validation, and (e) track and predict the performance of the physical object (Singh, Fuenmayor, and Hinchy et al. 2021). For example, the digital twin can be tested with various settings and optimization algorithms, which can then be implemented on the actual physical object.

Depending on the level of magnification, from the micro-level to the macro-level, digital twins are classified into four categories: (a) component or part twins, (b) asset twins, (c) system or unit twins, and (d) process twins (IBM n.d.b; Woods 2018). As the name implies, a component or part twin is a component of a larger asset, while the asset is the entire physical object. A system is a collection of assets. A process twin provides a business-level view of the system (Woods 2018). For example, a turbine can be classified as an asset, whereas a blade in the turbine would be considered a component. A collection of turbines would be classified as the system, and a business-level view to optimize them would be part of the process twin.

Digital twins can also utilize a combination of augmented reality, virtual reality, and mixed reality. For example, Tu, Autiosalo, and Jadid et al. (2021) developed a proof of concept for a digital twin for a crane that utilized mixed reality running on the Microsoft HoloLens. In another study, a proof of concept of a digital twin using mixed reality for robotic construction was developed (Ravi, Ng, and Salinas et al. 2021). Based on their model of DT for human-robot interaction (HRI), Gallala, Kumar, and Hichri et al. (2022) concluded that it achieved efficient HRI. DTs have great potential in many different sectors, including but not limited to aerospace and defense, automotive, construction, health care, the industrial sector, and more.

1.3.9 ADDITIVE MANUFACTURING

In additive manufacturing (AM), "digital designs guide the fabrication of complex, three-dimensional products that are built up, layer by layer" (NIST n.d.b); 3D printing, also known as additive manufacturing, "refers to the process of constructing a three-dimensional object from a digital, or computer-based, file" (Coe 2021, 1). This allows the customer to design and develop prototypes more efficiently before mass production. It has various applications in different domains, including but not limited to medical sciences (Doucleff 2015; US-FDA 2017), the automotive industry (Loganathan, Krishnamoorthi, and Jospeh et al. 2023), and consumer goods (Dunham, Mosadegh, and Romito et al. 2018), to name a few.

Haleem and Javid (2019) identified 13 AM applications in Industry 4.0; among them are customization, prototyping, design and development, customer satisfaction, and improved supply chain performance. However, it should also be noted that it does have some drawbacks, due to which it may not be suitable for the mass production of regular parts (Dilberoglu, Gharehpapagh, and Yaman et al. 2017; Ngo, Kashani, and Imbalzano et al. 2018). These drawbacks may include the limited or low strength of the product (Haleem and Javid 2019; Ngo, Kashani, and Imbalzano et al. 2018), initial start-up costs, and additional post-processing requirements (Padasak 2022).

1.3.10 BLOCKCHAIN

Security should be a concern for each of the Industry 4.0 technologies presented. These technologies depend on the use or transmission of data. It is incumbent upon those developing these technologies to secure the integrity of this data. The final Industry 4.0 technology to be presented in this chapter may serve that purpose. A blockchain is a form of distributed computing. It is a "distributed ledger, consisting of blocks of data that maintain a permanent and tamperproof record of transactional data" (Baltzan and Philips 2022, 250). Blockchains can be broadly categorized as permissioned or permissionless. A permissionless blockchain, also known as a public blockchain, allows anyone to participate by submitting and confirming a transaction, whereas permissioned blockchains, also known as a private blockchain, do not permit open participation either for submission or transaction verification (Solat, Calvez, and Nait-Abdesselam et al. 2021).

Blockchain is utilized by industries such as banking, health care, legal and smart contracts to authenticate payments and detect fraud (Baltzan and Philips 2022). Two of the most common applications of blockchains are cryptocurrencies and non-fungible tokens (NFTs). Due to the complexity of blockchain, it may not be a viable option for all organizations or transactions. A blockchain may be appropriate in instances where the participants and workflows are highly distributed, but it may not be appropriate for applications with just a few participants who know and trust each other (US-GAO 2022).

1.4 SOCIO-ECONOMIC IMPLICATIONS

The previous section described some of the Industry 4.0 technologies and their applications. This section discusses some socio-economic implications associated with advancing Industry 4.0 technologies. Some of these socio-economic implications are related to organizational impact, workforce development, safety, security and ethical concerns, and legal issues.

1.4.1 ORGANIZATIONAL IMPACT

Change management is a key component of any successful organization. Change management "is a comprehensive, cyclic, and structured approach for transitioning individuals, groups, and organizations from a current state to a future state with intended business benefits. It helps organizations to integrate and align people, processes, structures, culture, and strategy" (PMI 2013, 7). Hence, the execution of the change management plan requires it to be continuously monitored and updated as necessary. As part of the strategic plan, organizations must address challenges or concerns associated with adopting new technologies and processes. These challenges could be either internal or external to the organization. Some of the internal challenges are technology adoption, talent development, legal procedures, and safety procedures. External challenges could be the client or partner-facing processes.

1.4.2 WORKFORCE DEVELOPMENT AND ECONOMIC IMPLICATIONS

The global epidemic of vacating jobs, known as the Great Resignation, represented a time when citizens across the country were abandoning their duties and resigning from their jobs. At the same time, massive layoffs by organizations also burden the economy. Such events greatly impact the supply chain process (Kalota and Trinkle 2022). Integrating AI and automation necessitates an intelligent workforce (Trinkle, Cyrus, and Nouhan et al. 2021).

Automation will also impact jobs and vary by the type of job and the industry. By mid-2030, up to 30% of the jobs could be automatable (PwC n.d.) According to Manyika, Lund, and Chui et al. (2017), automation could displace between 400–800 million individuals from their jobs by 2030, and these people would have to seek out new employment opportunities; additionally, roughly 375 million may need to switch occupational categories. In the long run, automation can have a bigger impact on people with lower educational levels (PwC n.d.).

Keynes (1963), a British economist from the early 1900s, talked about technological unemployment, which is the loss of jobs due to technological advancements and innovations. Computerization creates a cultural shift and changes the labor supply and is described as hollowing out middle-income jobs (Frey and Osborne 2013); the manufacturing sector may also feel a similar impact of Industry 4.0 technologies and processes. This may result in many people losing their jobs and increasing unemployment. The transition to unemployment benefits will significantly reduce discretionary spending (Ganong and Pascal 2019), causing a negative effect throughout the economy.

There will be an urgent need to develop and recruit talent and workforce (Deloitte n.d.; Zaouini 2017) for organizations to sustain growth with the changing landscape of technology. This workforce must be able to manage and work alongside the technology (Kalota and Trinkle 2022). Like any profession, jobs associated with Industry 4.0 will require certain skill sets. These skills can be classified into two broad categories: (a) technical skills and (b) nontechnical skills. The technical skills could include various technology domains such as IoT, autonomous robots, 3D printing/additive manufacturing, simulation, data analytics (Jegnathan, Khan, and Raju et al. 2018; Piñol, Porta, and Arévalo et al. 2017), artificial intelligence, cloud computing and cyber security (Jegnathan, Khan, and Raju et al. 2018), software development, programming, data science, data management, and security (Lorenz, Küpper, and Rüßmann et al. 2016). There are additional technical skills that are needed for Industry 4.0, such as an understanding of computer hardware and software design, enterprise resource planning (ERP), edge computing, computer networking, etc.

The nontechnical skills would include skills like emotional intelligence (Forbes Expert Panel 2021; Paschou, Rapaccini, and Adrodegari et al. 2018; WEF 2016), critical thinking, problem-solving, negotiation skills (Paschou, Rapaccini, and Adrodegari et al. 2018; WEF 2016); collaboration skills (Paschou, Rapaccini, and Adrodegari et al. 2018); and time management, people management, oral and written expression (WEF 2016). Financial management, leadership skills, and project coordination are additional skills that will benefit job seekers in Industry 4.0. Therefore, governments, academic institutions, and businesses need to invest in the professional development of their workforce. This would allow machines to perform repetitive tasks while humans focus on innovations.

1.4.3 SAFETY, SECURITY, AND ETHICAL CONCERNS

Manufacturing and warehouse environments are not free of issues associated with occupational health and hazards (OHH); therefore, it becomes important for industrial organizations to maintain a safe workplace (Forbes 2022). Working around or with industrial robots may pose safety risks (OSHA n.d.). In 2021, the combined number of fatal work injuries in the construction, logistics and warehousing, and manufacturing industries was 2,345 in the USA (US-BLS, n.d.). Globally, there are 313 million workplace accidents and 2.3 million deaths (Rosenberg, Minc, and Falkoff et al.

2016). Organizations and government agencies must work collaboratively to continuously develop and update procedures to ensure the safety of employees.

Communication is a key component of Industry 4.0, providing powerful and pervasive connectivity between people, equipment, and devices (Forbes 2022). The need for the various IoT devices to communicate and share information with each other and share information with humans cannot be emphasized enough. Security is also an important issue associated with Industry 4.0 (Bezdicek 2019; Forbes 2022). Hence, cybersecurity risks automatically increase with so many devices connected to the Internet.

Likewise, blockchains also pose security threats. Permissionless blockchains can have issues associated with scalability and throughput, whereas permissioned blockchain may not be tamperproof in some instances (Solat, Calvez, and Nait-Abdesselam 2021). There are concerns for data reliability, long-term data security, quantum computing, and energy consumption with blockchains (US-GAO 2022). Blockchain may also pose privacy concerns since there may be sensitive data on public blockchains (US-GAO 2022).

In a capitalist society, things often boil down to the bottom line: money. It is said that when a product is free, then the consumer becomes the product. However, with the pervasiveness of intelligent technologies, there is an ongoing collection of data about human activity, even for the products that have been paid for. While it is often touted that this is to provide better service to consumers, there is a greater risk of misusing this data. Therefore, it becomes imperative that laws are put in place to prevent illegal or unethical data use.

1.4.4   LEGAL CHALLENGES

Industry 4.0 is not just one technology. Instead, it is comprised of many technologies. Whether these technologies work in a standalone mode or integrated mode, they pose different legal and security challenges. For example, the level of legal challenges associated with cloud computing and big data analytics is considered medium, and the level of legal challenges associated with IoT and cyber-physical systems (CPS) is considered high (Harbart 2020). With integrated or networked technologies, without well-documented traceability procedures or contractual agreements, it will be difficult to determine which party would be liable for faulty products (Schröder 2016).

The preservation of the environment and human health has resulted in stricter regulations regarding industrial automation (Bezdicek 2019). With more humans working alongside robots, it is important to have legal regulations and safety procedures in place. As stated earlier, the interconnected nature of technologies makes it challenging to understand who or what is at fault. One must also understand that human-robot interaction (HRI) goes beyond the factories and warehouses. Robots are used for different purposes in different settings, such as agriculture, health care, and law enforcement. The pervasiveness of robots and automation will continue to grow, making it paramount that explicit considerations are given to the ethics of HRI (Riek and Howard 2014).

1.5   FUTURE DIRECTIONS AND CONCLUSION

The Fourth Industrial Revolution is the natural evolution of the technological advancements made in previous industrial revolutions, and it will continue to impact different industries, businesses, and smart cities. Therefore, businesses and organizations must develop internal and external strategies to embrace business and technological changes to be successful.

Industry 4.0 technologies, such as AI, big data, cloud computing, etc., will significantly impact the workforce. In some instances, these technologies will eliminate certain jobs, whereas in other cases, these technologies will change the workers' roles. Therefore, it is recommended that businesses, government agencies, and institutions develop agile and responsive professional development programs to support workers in keeping up with the latest technology trends.

Industry 4.0 technologies and business processes may also have varying impacts on society and businesses in different forms, including but not limited to ethical, legal, and security concerns. As has been seen throughout history, technological advancements can have negative consequences if not used properly. Therefore, it is recommended that government agencies actively support businesses to develop sound policies and procedures that protect consumers and society from unethical practices.

Through education, awareness, standards, and regulation, the technologies that underlie Industry 4.0 will positively transform how individuals, industry, and society will operate. At the same time, it is equally important for businesses, government agencies, community partners, and institutions to work together for a better future continuously.

REFERENCES

Agarwal, H., and R. Agarwal. 2017. "First Industrial Revolution and Second Industrial Revolution: Technological Differences and the Differences in Banking and Financing of the Firms." *Saudi Journal of Humanities and Social Sciences* 2 (11).

Baltzan, Paige, and Amy Philips. 2022. *Business Driven Information System* (8th ed.). New York, NY: McGraw Hill.

Bezdicek, Jan. 2019. "Industry 4.0: The Challenges and Risks | Rockwell Automation." *Rockwell Automation.* www.rockwellautomation.com/en-us/company/news/blogs/industry-4-0—the-challenges-and-risks.html.

Brooks, Rebecca Beatrice. 2018. "The Industrial Revolution in America—History of Massachusetts Blog." *History of Massachusetts Blog.* https://historyofmassachusetts.org/industrial-revolution-america/.

Cao, Keyan, Yefan Liu, Gongjie Meng, and Qimeng Sun. 2020. "An Overview on Edge Computing Research." *IEEE Access* 8 (January). Institute of Electrical and Electronics Engineers: 85714–85728. doi:10.1109/access.2020.2991734.

Carvalho, Núbia Gabriela Pereira, and Edson Walmir Cazarini. 2020. "Industry 4.0—What Is It?" In: Ortiz, J. H., editor. *Industry 4.0—Current Status and Future Trends* (pp. 3–11). London, UK: IntechOpen.

Coe, Diana C. 2021. "Advances in 3D Printing." *Advances in 3D Printing*, April 1. https://search-ebscohost-com.proxy.bsu.edu/login.aspx?direct=true&db=sch&AN=149778563&site=ehost-live&scope=site.

Deloitte. n.d. *Industry 4.0 Challenges and Solutions for the Digital Transformation and Use of Exponential Technologies*. https://www2.deloitte.com/cn/en/pages/consumer-industrial-products/articles/industry-4-0-challenges-and-solutions.html.

Dilberoglu, Ugur M., Bahar Gharehpapagh, Ulas Yaman, and Melik Dolen. 2017. "The Role of Additive Manufacturing in the Era of Industry 4.0." *Procedia Manufacturing* 11 (January). Elsevier BV: 545–554. doi:10.1016/j.promfg.2017.07.148.

Dima, Adrian. n.d. "Short History of Manufacturing: From Industry 1.0 to Industry 4.0." *KFactory*, Accessed June 17, 2023. https://kfactory.eu/short-history-of-manufacturing-from-industry-1–0-to-industry-4–0/.

Doucleff, Michaeleen. 2015. "Engineers Create A Titanium Rib Cage Worthy of Wolverine." *NPR*, September 15. www.npr.org/sections/health-shots/2015/09/15/440361621/engineers-create-a-titanium-rib-cage-worthy-of-wolverine.

Dunham, Simon, Bobak Mosadegh, Eva Romito, and Mohamed Zgaren. 2018. "Applications of 3D Printing." In *Elsevier EBooks* (pp. 61–78). doi:10.1016/b978-0-12-803917-5.00004-3.

European Commission. 2017. "Germany—Industry 4.0 | Advanced Technologies for Industry." *Advanced Technologies for Industry*. https://ati.ec.europa.eu/reports/policy-briefs/germany-industry-40#:~:text=%E2%80%9Cindustrie%204.0%E2%80%9D(Industry %204.0,Affairs%20and%20Energy%20(BMWI).

Fisher, Robert N., Toby P. Breckon, Kenneth M. Dawson-Howe, Andrew Fitzgibbon, Craig M. Robertson, Emanuele Trucco, and Christopher Williams. 2016. *Dictionary of Computer Vision and Image Processing*. Wiley Online Library. doi:10.1002/9781119286462.

Forbes. 2022. "Preparing for Industry 4.0? These 4 Challenges Should Inform Your Plan." *Forbes*, August 16. www.forbes.com/sites/tmobile/2022/08/16/preparing-for-industry-40-these-4-challenges-should-inform-your-plan/?sh=14918a524975.

Forbes Expert Panel. 2021. "How Automation Will Impact Future Job Markets: 10 Predictions." *Forbes*, March 4. www.forbes.com/sites/forbescoachescouncil/2021/03/04/how-automation-will-impact-future-job-markets-10-predictions/?sh=52cf9d2a2ed7.

Forrester. 2021. *The Total Economic Impact of Mixed Reality Using Microsoft HoloLens 2*. https://tools.totaleconomicimpact.com/go/microsoft/HoloLens2/.

Frey, Carl Benedikt, and Michael A. Osborne. 2013. "The Future of Employment: How Susceptible Are Jobs to Computerisation?" *Oxford Martin School* (June). www.oxfordmartin.ox.ac.uk/publications/the-future-of-employment/.

Gallala, A., Atal Anil Kumar, Bassem Hichri, and Peter Plapper. 2022. "Digital Twin for Human–Robot Interactions by Means of Industry 4.0 Enabling Technologies." *Sensors* 22 (13). MDPI: 4950. doi:10.3390/s22134950.

Ganong, Peter, and Noel J. Pascal. 2019. "Consumer Spending During Unemployment: Positive and Normative Implications." *American Economic Review* 109 (7), 2383–2424.

Haleem, Abid, and Mohd Javid. 2019. "Additive Manufacturing Applications in Industry 4.0: A Review." *Journal of Industrial Integration and Management* 4 (4). World Scientific: 1930001. https://doi.org/10.1142/S2424862219300011.

Harbart, D. 2020. "Legal Challenges of Digitalization and Automation in the Context of Industry 4.0." *Procedia Manufacturing* 51, 938–942.

Hurwitz, Judith S., Robin Bloor, Marcia Kaufman, and Fern Halper. 2009. *Cloud Computing for Dummies*. Hoboken, NJ: Wiley Publishing Inc.

IBM. n.d.a. *What Is Industry 4.0?* Accessed January 24, 2023. www.ibm.com/topics/industry-4-0.

IBM. n.d.b. *What Is a Digital Twin?* Accessed January 24, 2023. www.ibm.com/topics/what-is-a-digital-twin.

IBM. 2017. *A Brief History of Cloud Computing*. Accessed February 05, 2023. www.ibm.com/cloud/blog/cloud-computing-history.

Intel. n.d. *Demystifying the Virtual Reality Landscape*. Accessed February 1, 2023. www.intel.com/content/www/us/en/tech-tips-and-tricks/virtual-reality-vs-augmented-reality.html.

Javaid, Mohd, Abid Haleem, Ravi P. Singh, and Rajiv Suman. 2021. "Substantial Capabilities of Robotics in Enhancing Industry 4.0 Implementation." *Cognitive Robotics* 1 (January). Elsevier BV: 58–75. doi:10.1016/j.cogr.2021.06.001.

Jeganathan, L., A. Nayeemulla Khan, Jagadeesh Kannan Raju, and Sambandam Narayanasamy. 2018. *On a Frame Work of Curriculum for Engineering Education 4.0. 2018 World Engineering Education Forum—Global Engineering Deans Council (WEEF-GEDC).* doi:10.1109/weef-gedc.2018.8629629.

Johnston, Madeleine. 2020. "The Role and Regulation of Child Factory Labour During the Industrial Revolution in Australia, 1873–1885." *International Review of Social History* 65 (3). Cambridge University Press: 433–463. doi:10.1017/s0020859020000322.

Joshi, Prateek. 2017. *Artificial Intelligence with Python.* Birmingham, UK: Packt Publishing, Limited.

Kalota, Faisal, and Dennis Trinkle. 2022. "Managing Alongside of Technology." *Association of Telecom, Mobility, and IT Management Professionals [AOTMP] Insights.* Accessed June 17, 2023. https://aotmp.com/insights-article/managing-alongside-of-technology.

Kalsoom, Tahera, Naeem Ramzan, Shehzad Ahmed, and Masood Ur Rehman. 2020. "Advances in Sensor Technologies in the Era of Smart Factory and Industry 4.0." *Sensors* 20 (23). MDPI: 6783. doi:10.3390/s20236783.

Keynes, John M. 1963. *Essays in Persuasion.* New York: W.W. Norton & Company.

Kiger, Patrick. J. 2021. *8 Groundbreaking Inventions from the Second Industrial Revolution.* www.history.com/news/second-industrial-revolution-inventions.

Lee, Wei-Meng. 2019. *Python Machine Learning.* Indianapolis, IN: John Wiley & Sons, Inc.

Liu, Yujia, and David B. Grusky. 2013. "The Payoff to Skill in the Third Industrial Revolution." *American Journal of Sociology* 118 (5), 1330–1374. https://doi.org/10.1086/669498.

Longanthan, Prabhu, Sangeetha Krishnamoorthi, Jackson Joseph, Jerish K. George, and None Naveen P. 2023. "Development of Advanced 3D Printing Technology in Automotive Industry. Nucleation and Atmospheric Aerosols." *American Institute of Physics.* doi:10.1063/5.0110513.

Lorenz, Markus, Daniel Küpper, Michael Rüßmann, Ailke Heidemann, and Alexandra Bause. 2016. "Time to Accelerate in the Race Toward Industry 4.0." *BCG Global.* Accessed June 17, 2023. www.bcg.com/publications/2016/lean-manufacturing-operations-time-accelerate-race-toward-industry-4.

Manyika, James, Susan Lund, Michael Chui, Jacques Bughin, Jonathan Woetzel, Parul Batra, Ryan Ko, and Saurabh Sanghvi. 2017. *Jobs Lost, Jobs Gained: What the Future of Work Will Mean for Jobs, Skills, and Wages.* Accessed June 16, 2023. www.mckinsey.com/featured-insights/future-of-work/jobs-lost-jobs-gained-what-the-future-of-work-will-mean-for-jobs-skills-and-wages.

McCandless, David. 2010. *The Beauty of Data Visualization.* Accessed June 16, 2023. www.ted.com/talks/david_mccandless_the_beauty_of_data_visualization?language=en.

Meriam Webster. n.d. *Revolution.* Accessed January 24, 2023. www.merriam-webster.com/dictionary/revolution.

Microsoft. n.d. "Microsoft HoloLens 2: For Precise, Efficient Hands-Free Work." *Microsoft.* Accessed February 1, 2023. www.microsoft.com/en-us/hololens.

Mohajan, Haradhan K. 2021. "Third Industrial Revolution Brings Global Development." *Journal of Social Sciences and Humanities* 7 (4), 239–251.

Mokyr, Joel. 1998. "The Second Industrial Revolution, 1870–1914." Northwestern University. Accessed June 16, 2023. https://faculty.wcas.northwestern.edu/~jmokyr/castronovo.pdf.

National Institute of Standards and Technology [NIST]. n.d.a. *NIST Cloud Computing Program—NCCP.* Accessed January 24, 2023. www.nist.gov/programs-projects/nist-cloud-computing-program-nccp.

National Institute of Standards and Technology [NIST]. n.d.b. *Additive Manufacturing.* Accessed January 24, 2023. www.nist.gov/additive-manufacturing.

Nicholas, Tom, and Matthew Guilford. 2016. *Samuel Slater & Francis Cabot Lowell: The Factory System in US Cotton Manufacturing*. Accessed January 24, 2023 www.hbs.edu/faculty/Pages/item.aspx?num=46048.

Ngo, Tuan, Alireza Kashani, Gabriele Imbalzano, Kate Nguyen, and David S. C. Hui. 2018. "Additive Manufacturing (3D Printing): A Review of Materials, Methods, Applications and Challenges." *Composites Part B-Engineering* 143 (June). Elsevier BV: 172–196. doi:10.1016/j.compositesb.2018.02.012.

Occupational Safety and Health Administration [OSHA]. n.d. *Warehousing*. Accessed June 17, 2023. www.osha.gov/warehousing/hazards-solutions.

O'Leary, Timothy, Linda O'Leary, and Daniel O'Leary. 2023. *Computing Essentials 2023: Making IT Work for You*. New York, NY: McGraw Hill.

Ortiz, Jesus Hamilton, William Gutierrez Marroquin, and Leonardo Zambrano Cifuentes. 2020. "Industry 4.0: Current Status and Future Trends." In: Ortiz, J. H., editor. *Industry 4.0—Current Status and Future Trends* (pp. 13–28). London, UK: IntechOpen.

Padasak, Zachery. 2022. "Top 10 Advantages and Disadvantages of Using Additive Manufacturing." *Alpha Precision Group*. Accessed June 17, 2023. www.alphaprecisionpm.com/blog/top-10-advantages-and-disadvantages-of-using-additive-manufacturing

Paschou, Theoni, Mario Rapaccini, Federico Adrodegari, and Nicola Saccani. 2018. "Competences in Digital Servitization: A New Framework." *ResearchGaate*. Accessed June 17, 2023. www.researchgate.net/publication/326266410_Competences_in_digital_servitization_a_new_framework.

Piñol, T. Curià, S. Artigas Porta, M. C. Rodríguez Arévalo, and Joaquim Minguella-Canela. 2017. "Study of the Training Needs of Industrial Companies in the Barcelona Area and Proposal of Training Courses and Methodologies to Enhance Further Competitiveness." *Procedia Manufacturing* 13 (January). Elsevier BV: 1426–1431.

PricewaterhouseCoopers [PwC]. n.d. "How Will Automation Impact Jobs?" *PwC*. Accessed January 24, 2023. www.pwc.co.uk/services/economics/insights/the-impact-of-automation-on-jobs.html.

Project Management Institute [PMI]. 2013. *Managing Change in Organization: A Practice Guide*. Newton Square, PA: Project Management Institute, Inc.

Puriwat, Wilert, and Suchart Tripopsakul. 2020. "Preparing for Industry 4.0—Will Youths Have Enough Essential Skills?: An Evidence from Thailand." *International Journal of Instruction* 13 (3). Osmangazi University: 89–104. doi:10.29333/iji.2020.1337a.

Pyo, Dongjin, Jaejin Hwang, and Youngjin Yoon. 2021. *Tech Trend of the 4th Industrial Revolution*. Dulles, VA: Mercury Learning and Information, LLC.

Qamar, Usman, and Muhammad Summair Raza. 2020. *Data Science Concepts and Techniques With Applications*. Singapore: Springer. https://doi.org/10.1007/978-981-15-6133-7.

Rafferty, John P. n.d. "The Rise of the Machines: Pros and Cons of the Industrial Revolution." *Britannica*. Accessed June 10, 2022. www.britannica.com/story/the-rise-of-the-machines-pros-and-cons-of-the-industrial-revolution.

Ravi, Kaushik, Ming-Yen Ng, Jesús Salinas, and Daniel L. Hall. 2021. "Real-Time Digital Twin of Robotic Construction Processes in Mixed Reality." In *ISARC Proceedings of the International Symposium on Automation and Robotics in Construction* (Vol. 38, pp. 451–458). IAARC Publications. doi:10.22260/isarc2021/0062.

Rich, Elaine, and Kevin Knight. 1991. *Artificial Intelligence* (2nd ed.). New York, NY: McGraw-Hill, Inc.

Richmond Vale Academy. 2022. *The Second Industrial Revolution: The Technological Revolution*. Accessed June 16, 2023. https://richmondvale.org/blog/second-industrial-revolution/.

Riek, Laurel, and Don Howard. 2014. *A Code of Ethics for the Human-Robot Interaction Profession*. In SSRN. Proceedings of We Robot. https://ssrn.com/abstract=2757805.

Rosenberg, Minc, Falkoff & Wolf, Law Firm LLP [RMFW]. 2016. *Common Factory Accidents and Injuries*. Accessed February 07, 2023. https://rmfwlaw.com/blog/personal-injury/common-factory-accidents-and-injuries.

SAP. n.d. *What Is Industry 4.0?* Accessed February 07, 2023. www.sap.com/insights/what-is-industry-4-0.html.

SAS. n.d. *Big Data: What Is It and Why It Matters*. Accessed February 07, 2023. www.sas.com/en_us/insights/big-data/what-is-big-data.html.

Satyanarayanan, Mahadev. 2017. "The Emergence of Edge Computing." *IEEE Computer* 50 (1). IEEE Computer Society: 30–39. doi:10.1109/mc.2017.9.

Schröder, Christian. 2016. "The Challenges of Industry 4.0 for Small and Medium-Sized Enterprises." *Friedrich-Ebert-Stiftung*. Accessed June 17, 2023. https://library.fes.de/pdf-files/wiso/12683.pdf.

Shahul, Ameer. 2018. "Data Is the New Oil." *IBM*. Accessed June 17, 2023. www.ibm.com/blogs/digital-transformation/in-en/blog/data-is-the-new-oil-right/.

Singh, Maulshree, Evert Fuenmayor, Eoin P. Hinchy, Yuansong Qiao, Niall Murray, and Declan M. Devine. 2021. "Digital Twin: Origin to Future." *Applied System Innovation* 4 (2). MDPI: 36. doi:10.3390/asi4020036.

Solat, Siamak, Philippe Calvez, and Farid Naït-Abdesselam. 2021. "Permissioned Vs. Permissionless Blockchain: How and Why There Is Only One Right Choice." *Journal of Software* 16 (3), 95–106.

Sternberg, Robert J., and Karin Sternberg. 2012. *Cognitive Psychology* (6th ed.). Wadsworth: Cengage Learning.

Taalbi, Josef. 2019. "Origins and Pathways of Innovation in the Third Industrial Revolution1." *Industrial and Corporate Change* 28 (5), 1125–1148. doi:10.1093/icc/dty053. https://academic.oup.com/icc/article/28/5/1125/5193703?login=false

Trinkle, Dennis, Cyrus Green, Christopher Nouhan, and Paul Faria. 2021. *The CICS 2021 Horizon Report: Future Disrupters in Technology, Leadership, and Business*. Muncie, IN: Centre for Information and Communications Sciences, Ball State University.

Tu, Xinyi, Juuso Autiosalo, Adnane Jadid, Kari Tammi, and Gudrun Klinker. 2021. "A Mixed Reality Interface for a Digital Twin Based Crane." *Applied Sciences* 11 (20). MDPI: 9480. doi:10.3390/app11209480.

United States Bureau of Labor and Statistics [US-BLS]. n.d. *Number and Rate of Fatal Work Injuries, by Private Industry Sector*. Accessed January 21, 2023. www.bls.gov/charts/census-of-fatal-occupational-injuries/number-and-rate-of-fatal-work-injuries-by-industry.htm.

United States Food and Drug Administration [US-FDA]. 2017. *Medical Applications of 3D Printing*. Accessed February 2, 2023. www.fda.gov/medical-devices/3d-printing-medical-devices/medical-applications-3d-printing.

United States Government Accountability Office [US-GAO]. 2022. *Blockchain: Emerging Technology Offers Benefits for Some Applications but Faces Challenges*. Report to congressional requesters. United States Government Accountability Office.

Virnes, Francis Adrain. 2021. "Stop Saying 'Data Is the New Oil' Why the Catchphrase of the Fourth Industrial Revolution May Be a Misnomer." *Medium.com*. Accessed January 24, 2023. https://medium.com/geekculture/stop-saying-data-is-the-new-oil-a2422727218c.

Woods, Dan. 2018. "Why Digital Twins Should Be the CEO's Best Friend." *Forbes*. Accessed June 16, 2023. www.forbes.com/sites/danwoods/2018/07/18/why-digital-twins-should-be-the-ceos-best-friend/?sh=2c2447233c75.

World Economic Forum [WEF]. 2016. *The Future of Jobs: Employment, Skills and Workforce Strategy for the Fourth Industrial Revolution*. Accessed January 20, 2023. www.weforum.org/reports/the-future-of-jobs.

Zaouini, Mustapha. 2017. "Nine Challenges of Industry 4.0. IIoT World." *IIoT World*. Accessed June 06, 2023. https://iiot-world.com/industrial-iot/connected-industry/nine-challenges-of-industry-4–0/.

2 An Insight into the Consensus Protocols and Data Security in IoT applications Using Blockchain

Priya Thomas and Deepa V Jose

2.1 INTRODUCTION

Blockchain technology is one of the leading technologies which have achieved significant importance due to its immutable nature. The concept was introduced with the introduction of bitcoin, which has gained wide popularity. The blockchain uses a distributed ledger where multiple entities record and verify each transaction to guarantee trust. Different applications rely on the blockchain concept to enhance the security aspects. One of the main domains is the Internet of Things (IoT) applications. IoT-enabled applications are subjected to different security attacks which are difficult to resist due to resource constraints. The changes made by attackers in a large network are difficult to detect properly due to their diverse nature. So incorporating the immutable feature of blockchain will make the IoT system more reliable and fault tolerant. IoT devices also suffer from single point of failure as the entire system is centrally controlled by the cloud server. The communication overhead increases as the number of sensor-enabled devices increases. Blockchain overcomes the problem of a single trusted server by its distributed nature. The trust level can be increased as the consensus protocols in blockchain technology enable verifying the transactions happening in the system in a systematic tamperproof way.

The consensus algorithms form the basis of blockchain technologies (Afzaal et al., 2022). The selection of appropriate consensus protocol is very important as it can affect directly the performance of the entire system. Different factors, such as scalability, computational cost, miner incentives, etc. play an important role in deciding the proper protocol for each specific application. The protocols also differ based on the type of blockchain involved. Blockchain can be public like popular bitcoin cryptocurrencies, which can be verified by anyone, or it can also be permissioned blockchain, which restricts the verification process to limited authorized groups of entities. Different enterprises implement permissioned blockchain for enhancing the security and trust. Most of the public blockchains use Ethereum framework for implementation, whereas permissioned blockchains mostly use Hyper Ledger Fabric.

DOI: 10.1201/9781003343332-2

The smart contracts coded in blockchain are responsible for providing the required functionalities to support various applications. Section 2 provides literature review of existing methods. Section 3 discusses the blockchain concepts in detail. Section 4 elaborates on different consensus protocols and applications. Different blockchain use cases are mentioned in section 5 and a proposed blockchain implementation on the smart home use case is discussed in section 6 followed by conclusion.

2.2 LITERATURE REVIEW

Blockchain technology when incorporated into different applications helps to implement distributed control as well as improves security features. Different blockchain applications follow different consensus algorithms. IoT applications greatly suffer from security issues, which make them more dependent on technologies like blockchain, which offer transparency and security. Blockchain technology cannot be directly applied to and needs to be modified accordingly to satisfy the requirements of IoT applications. This greatly depends on which consensus protocol is chosen for implementation. The protocols designed should be lightweight and energy efficient to meet the demands of IoT applications. Numerous design approaches have been proposed.

W. Zheng et al. proposed a platform named NutBaaS which utilizes the blockchain as a service framework. This concept is designed to provide the services of blockchain using the incorporation of a cloud computing platform. The platform supports monitoring services, network deployment services, analysis of smart contracts, and testing over the cloud computing platform (Zheng et al., 2019). The system offers less overhead to business community in maintaining and implementing the blockchain network.

Blockchain technology being newer technology is highly researched. Different papers are available in the literature which explore the various features and open issues in the technology. A recent survey on the technology shows that blockchain when incorporated with technologies such as edge, cloud, and fog will greatly increases its acceptability and can improve the adoption rate (Berdik et al., 2021).

Another recent survey explores the potential benefits of incorporating blockchain with supply chain sector. The paper studies the possibilities of the integration and shows that the performance, automation of the processes, security features, and distributed level of governance in supply chain can be highly improved on incorporation of blockchain to supply chain. The paper also highlights the possibilities of technical adoption (Chang & Chen, 2020), block-supply chain integration, and their social impacts.

Electronic health records (EHRs) are always subject to various privacy issues. The integrity and security of EHRs can be significantly improved if technology like blockchain can be incorporated along with it. A. Shahnaz et al. have proposed a novel framework that implements EHR using blockchain technology. The framework security is improved by defining different access rules for users (Shahnaz et al., 2019). The paper also proposes a solution for the scalability issues faced by blockchain integration.

An efficient sliding window strategy that works on Proof of Work (PoW) consensus algorithm can be used to enhance IoT security and minimize overheads. While using the sliding window approach as the size of the window is not known for an attacker, computation of hash for all possible window sizes is required to initiate any attack (Koshy et al., 2020), thereby increasing the difficulty level of invoking an attack. The method is computationally complex as PoW consumes more resources. The proposed method uses smart home scenario to test the efficiency. The analysis shows that security can be enhanced, and memory overhead can be considerably reduced with this approach.

Another approach proposes a security framework for edge-cloud integration using blockchain enabled distributed framework, which helps security attack detection in the cloud layer and faster attack detection at the edge layer of the IoT network (Medhane et al., 2020). The model uses gateway enabled using SDN to manage the network traffic. The model also tracks malicious flows and blocks them with the help of blockchain principles.

The Proof of Authority (PoA) is a hybrid consensus protocol which combines features of Proof of Work (PoW) and Proof of Stake (PoS). PoA approach includes a committee to which considers the participants PoW power "w" and PoS capability "s" to conduct an election based on a hybrid weight (Liu et al., 2019). The paper designs this new consensus algorithm by combining the features of PoW and PoS in a flexible manner. The performance evaluations shows that the hybrid algorithm eliminates many issues associated with individual algorithms.

Blockchain data keeps on increasing with addition of newer nodes to the framework. To overcome this issue, another approach called LiTiChain, which is comparatively lightweight (Pyoung & Baek, 2020) was proposed. In LiTiChain, only active blocks will be kept in the chain, thereby reducing the storage requirements. This blockchain was successfully implemented in the medical field for secure maintenance of transactions. The proposed work is also capable to address scalability issues in blockchain.

MedChain (Daraghmi et al., 2019) is a scheme which employs Proof of Activity consensus protocol for managing medical records. The system maintains large datasets of patient records in a secure way using the blockchain technology. MedChain uses time-oriented smart contracts for managing the transactions and for authorizing the access to the electronic medical records. The method also uses encryption principles to further enhance the security.

Most of the consensus protocols possess higher overhead. To make them scalable for IoT, a novel lightweight Proof of Block and Trade (PoBT) consensus algorithm was proposed (Biswas et al., 2020). This approach considerably reduces the complexity of common consensus algorithms. The computation time for data validation is very less in the proposed algorithm. The paper also proposes a distributed ledger with fewer memory requirements. The overall evaluations shows that the algorithm is better than the existing algorithms.

To reduce time delay resulting in incorrect executions and for improved system scalability, another protocol called extended Practical Byzantine Fault Tolerance (PBFT) protocol (Xu & Viriyasitavat, 2019) can be used, which offers very little confirmation time to business process operations thereby reducing the time delay. The paper mainly focuses on establishing trust in the Internet of Things environment. For achieving this, the paper proposes a new consensus approach, which is an extended version of the PBFT protocol.

Index Support Vector Machine (SVM) based training algorithm (M. Shen et al., 2019) can effectively address the issues related to data privacy and integrity by employing the blockchain techniques. The proposed system uses blockchain techniques for creating a secure and trustworthy data sharing platform among different data providers, by encrypting the IoT data before recording it to a distributed ledger.

The work proposed by H. Shen et al. (2020) addresses the security attacks such as man in the middle attack and the requirement for online certification authority (CA). The paper focuses on consortium blockchain and proposes a lightweight CA enabled approach for privacy preservation. The performance evaluations show that the proposed method offers better privacy and security compared to existing approaches.

The different aspects to be considered while implementing blockchain technology using smart city use case are studied and evaluated (Rejeb et al., 2022). The study was conducted and recently published in 148 documents. The paper evaluates the recent trends in the application and focuses on the challenges which need to be addressed for implementing the blockchain in the smart city use case. The discussions will help researchers to gain a thorough understanding of the available research gap and opportunities.

Another recent work in technology tries to incorporate blockchain with the food supply chain to overcome the difficulties faced in supply chain management due to the outbreak of COVID-19. The demonstrations show that the incorporation of blockchain to the food supply chain can greatly reduce the difficulties faced by the supplier, retailer, and the consumer if the peers at all the sends successfully adopt the technology (Yang et al., 2021).

The studies on applications of blockchain in food supply chain were further extended (Friedman & Ormiston, 2022) to understand the role of technology in maintaining the sustainability and related challenges. The studies show that the incorporation can lead to fair supply chains and improve traceability and environmental sustainability.

The impact of artificial intelligence in blockchain is another milestone in the blockchain area. There are different works available on exploring this concept. One of the recent works evaluates the possibilities of incorporating the technology in smart city use case (Singh et al., 2022). The paper shows that many security issues faced by smart city applications can be greatly reduced if blockchain technology along with artificial intelligence can be deployed.

The thorough analysis of different approaches proposed in the literature review shows that blockchain can be used in different scenarios to greatly improve the performance and security of connected applications. One of the key features which decide the working performance is the underlying consensus protocol used. There are numerous protocols designed to fulfill different parameters. Different consensus protocols have different levels of complexity and computational overhead. The consensus protocols and their features have to be deeply explored and evaluated for choosing the appropriate protocol. Section 4 elaborates on common consensus protocols. Before diving deep into consensus protocols, it is mandatory to have an idea about basic blockchain concepts and the common blockchain terminologies. Section 3 discusses on the blockchain concepts.

2.3 BLOCKCHAIN CONCEPTS

Blockchain can be defined as a group of cryptographically verified immutable blocks spanned in a distributed manner across the globe. The blocks are publicly available for anyone who joins the network. The blocks once added to the network can't be modified by anyone else making the technology more reliable and trustworthy. The blockchain network can be classified into two basic categories, such as permissionless and permissioned blockchains.

The blockchain concept was initiated with the introduction of bitcoin. The technology has gained appreciation due to the highly secure and transparent nature of transactions and due to its distributed nature. The bitcoin blockchain is a public blockchain that can be publicly available to anyone who joins the network. Bitcoin technology uses Proof of Work (PoW) consensus protocol where the addition of new blocks to the network purely depends on solving computationally intensive problems. Later on, the public blockchain was modified and permissioned blockchains came into existence to suit the requirements of different industrial projects. The permissioned blockchain will have limited access to the outside world. The access rights and permissions can be set in such a way that only a closed group of people would be able to access the system. This gave wide popularity to the blockchain concept, opening the path for numerous applications.

Permissionless blockchains allow anyone to join the network and commit transactions without any restrictions. Permissioned blockchains on the other hand allow only restricted access to the network. The entry to permissioned network is based on different criteria, and the different participating nodes in the network also will have further restrictions on the actions they can perform on the blocks. Permissioned blockchain includes private and consortium blockchains. Private blockchains are usually controlled by any particular authority like a company. A consortium blockchain is usually controlled by group of people. The main category under permissionless blockchain is public blockchain structure, which doesn't have any central authority to control the working apart from the underlying consensus mechanisms. Hybrid blockchains are formed as a combination of permissioned and permissionless blockchains, which are usually controlled by some authority with support of few permissionless activities. Figure 2.1 depicts the blockchain categories.

Blockchain includes a distributed immutable ledger with multiple blocks connected. The blocks added to the blockchain can be verified publicly and remains intact throughout. The nodes will be added to the blockchain by miners who solve the computational problem in case of bitcoin network and Ethereum. The one who solves the problem first will be the winner and will get the permission to add new blocks to the blockchain. The miner will be awarded with a fee called miner fee as a complement to the work done. The chain generated continues to grow and can be validated anytime by all the peers connected to the network. All the peers in the network will be having a copy of the ledger readily available. This makes the system tamperproof. The attacker will be finding it difficult to induce an attack in this scenario as the copy of the ledger is maintained by everyone and is publicly available.

The functionalities required for the applications can be coded using the smart contracts. Smart contracts add a programming layer to the blockchain network. The

```
                        ┌─────────────────────┐
                        │  BLOCKCHAIN TYPES   │
                        └─────────────────────┘
```

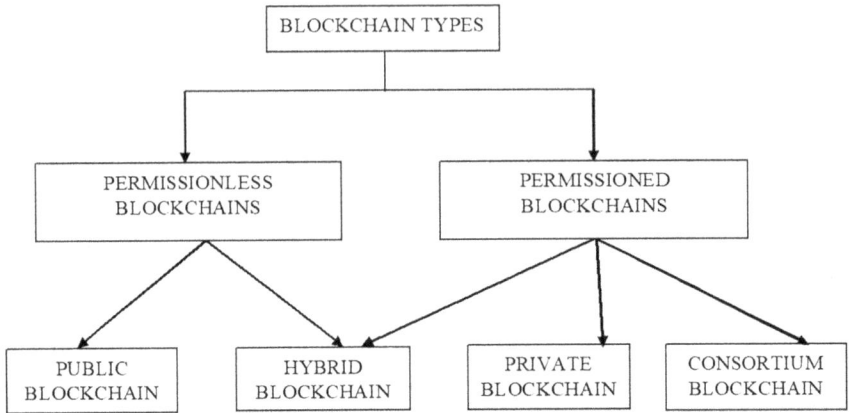

FIGURE 2.1   Blockchain categories.

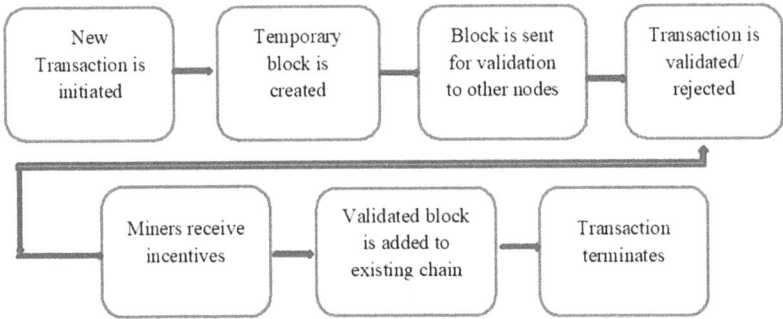

FIGURE 2.2   Working of blockchain.

consensus protocols create trust to the network. The working of the blockchain network can be summarized, as shown in Figure 2.2.

The peers connected to the blockchain network will initiate the transaction, creating a temporary block. The created transaction will be circulated through the network for validation. The validation procedure will be initiated by solving the complex cryptographic puzzle. The miner who solves the puzzle first will receive the miner incentives. The transaction will be verified or rejected by the miner. After verification, set of transactions will be coded as a block. The block will be added to the network, which includes the timestamp, reference to previous nodes, the details of smart contract it has undergone, and the summary of block transactions. This will terminate the transactions.

2.3.1   COMMON TERMINOLOGIES USED IN BLOCKCHAIN TECHNOLOGY

The blockchain uses different terminologies specific to the technology. This section briefs on the different terminologies used.

a. Blockchain

Blockchain is distributed chain of blocks which holds verified transactions. The concept is arranged as a public ledger accessible to anyone but can't be modified. The blockchain follows a hash-linked data structure where hash values are generated cryptographically. The blockchain in short acts like a public ledger where transactions are updated every second. The set of transactions will be validated based on the underlying consensus protocol. After validation, the set of transactions will be recorded as a block in the chain. The block once added to the chain remains permanent, tamperproof, and accessible to the public. Thus, the term blockchain refers to a chain of verified blocks.

b. Block

Blocks are the building units of blockchain networks. A group of transactions after successful validation and verification is recorded as a block. The block includes a creation timestamp, a reference to the previous block in the network, a summary of the transactions coded in it, and the details of smart contract used while creating the block. Each block is securely hashed and designed in such a way that it is not possible to alter any block without altering the previously connected blocks. This feature makes the block immutable and tamperproof.

c. Bitcoin

Bitcoin is abbreviated as BTC. This is the world's first cryptocurrency working using the Proof of Work consensus protocol. Bitcoin was introduced in the year 2009 by Satoshi Nakamoto. Bitcoin is distributed in nature and cryptographically designed to avoid security violations. The transactions will be initiated if both parties are ready to perform fund transfer. There will be no centralized control like normal banks. The normal money transfer depends on the underlying bank policies and regulations. The bitcoin transactions need not wait for the third party making it more popular. There are lots of cryptocurrencies other than bitcoin, such as Litecoin, Ethereum, etc.

d. Mining

Mining is the process of verifying and adding new transactions to the blockchain network. Miners need to solve cryptographically complex problems to win the bitcoin rewards. The miners are required to maintain the validity of bitcoin transactions and to avoid the possibility of double spending. The miner who solves the problem first only will be rewarded. Miners need to invest on expensive and complex hardware and software packages to perform mining process.

e. Double Spending

Double spending is spending the same digital currency twice by making a fake copy of the real one. In blockchain context, double spending refers to sending the same transaction to different entities. The problem of double spending has to be addressed using the underlying consensus protocol. The consensus protocols has to be designed in such a way that the double spending issues has to be identified and prevented to maintain the authenticity of the network.

TABLE 2.1

Common Blockchain Terminologies

Terminology	Brief explanation
Blockchain	Chain of securely connected immutable blocks E.g., Ethereum, Lite chain
Block	Fundamental unit of blockchain which includes set of transactions cryptographically verified
Bitcoin	Cryptocurrency working using Proof of Work consensus protocol
Mining	Computationally intensive procedure to verify the authenticity of transactions
Double Spending	Sending the same transaction multiple times for verification by spending coins
Consensus	Agreement reached by participating entities on how to validate the transactions
Smart Contracts	Self-executing set of rules coded as programs to run automatically when defined conditions are met

f. Consensus

Consensus mechanism can be referred as an agreement reached by participating entities on how to validate any transaction. The consensus protocols help to maintain the legitimacy of transactions. It includes a group of rules used to verify the integrity of communicating entities and contributors. The common consensus protocols used in blockchain network are Proof of Work (PoW), Proof of Stake (PoS), Proof of Capacity (PoC) etc.

g. Smart Contracts

Smart contracts are computer programs which define the steps and actions required for signing a contract between two parties. It is available publicly and needs to be fulfilled to initiate all blockchain transactions. Smart contracts are immutable once signed. It can be created by anyone and must be agreed by participating entities. The contract will be publicly available to read but can't be altered. The smart contract is designed in a self-executing manner where the contract runs automatically when the conditions specified in it are met.

Table 2.1 provides brief notes on the common terminologies used in blockchain technology.

2.4   CONSENSUS PROTOCOLS

The consensus protocols play an important role in deciding the computational cost of blockchain applications. The consensus protocols work using the underlying consensus algorithms. The consensus algorithm can be defined as set of procedures through which different entities in the blockchain network reach to a common

CONSENSUS ALGORITHMS

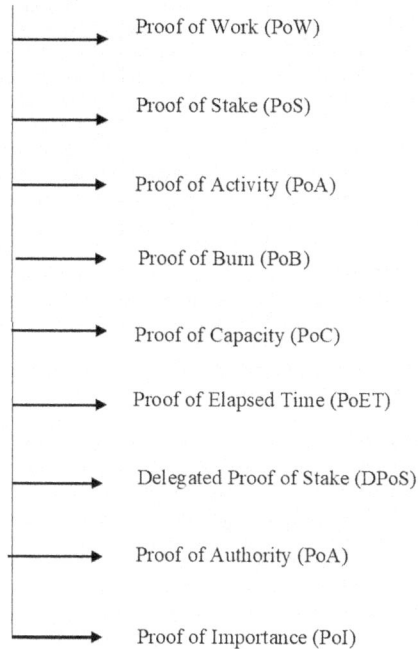

Proof of Work (PoW)

Proof of Stake (PoS)

Proof of Activity (PoA)

Proof of Burn (PoB)

Proof of Capacity (PoC)

Proof of Elapsed Time (PoET)

Delegated Proof of Stake (DPoS)

Proof of Authority (PoA)

Proof of Importance (PoI)

FIGURE 2.3 Common consensus algorithms.

agreement regarding the current state of the distributed public ledger (Bhattacharjya et al., 2022). This helps to build trust among different unknown peers participating in the blockchain network transactions. The blockchain network can be public, private, consortium, or hybrid based on the accessibility and availability of underlying blocks. Different networks use different types of consensus algorithms to verify the transactions as shown in Figure 2.3. Few common consensus algorithms are discussed to understand the working of consensus protocols in depth.

2.4.1 PROOF OF WORK

The most common consensus algorithm is the bitcoins' Proof of Work (PoW) algorithm, which forms the basis of the cryptocurrency systems. The PoW algorithm works on the concept of miners solving computational puzzles. The transactions happening in the network have to be verified and added to the blockchain. Miners are responsible for verifying the transactions happening in the blockchain. This is done by solving complex cryptographic problems. The one who solves the mathematical computation first will be the winner and will receive the miner incentives. The winner will be capable of adding the next block to the chain.

The main criticism faced by the PoW algorithm is that it consumes enormous computational resources. The algorithm is time-consuming as the transaction confirmation roughly consumes 15–60 minutes. The task will be complex as well as resource draining, making the algorithms less attractive for many resources constrained applications. The PoW consensus algorithm is subjected to different attacks like selfish mining, de-anonymization, 51% attack, etc. (M. Shen et al., 2019).

Selfish mining attack is induced by attackers who selfishly mine the blocks to increase the length of blockchain to gain mining rewards. This will result in unfair practices making systems reliable. A 51% attack is invoked when a major share of minted coins will be accumulated by nodes having powerful computational resources. De-anonymization attack is used to extract the transaction details, including the identification information in detail using data analytics techniques in publicly available blockchain. This opened the path to different variations of consensus algorithms. The PoW algorithm is highly feasible for applications which are ready to invest in computational resources. The cryptocurrencies Litecoin, Ethereum, Monero coin, Dogecoin, etc. mainly use PoW.

2.4.2 PROOF OF STAKE

The Proof of Stake (PoS) consensus algorithm was introduced to overcome a few criticisms of the Proof of Work algorithm. The most serious drawback of PoA, like high resource consumption, was rectified in PoS. Here the concept of miners is replaced by the term validators. So instead of spending enormous amounts of resources to perform mining to win coins and incentives, the validators focused more on investing coins. The validators will be paid incentives in the form of transaction fees.

In Proof of Stake, the validators will be chosen for block creation based on the available coin balance in the system. The coin age also plays a significant role in increasing the weightage. A validator with more stock of coins and more weightage is more likely to be chosen for block creation. The strategy behind PoS will remove the requirement of high computational resources for solving complex puzzles.

The Proof of Stake (PoS) has the advantages of being low-cost and lesser energy consuming compared to PoW algorithm. The coins in the network will be fixed and finite from the beginning. The attempt to create fraudulent blocks will result in slashing the entire coins from the validator's account, making further transactions difficult. To conduct attacks like 51% attack, the intruder must gain 51% share of cryptocurrency in the network, making attack more difficult and expensive. One of the main disadvantages of PoS is it promotes coin saving instead of spending. A large number of validators can join together to grab the main share of cryptocurrencies. This will increase the chance of gaining the major share of transaction fees and currencies by specific groups, making the system more centralized. Moreover, coin age can be manipulated by malicious nodes in the network in order to achieve significant network weight for performing a successful double spend, making the system less reliable (Vasin, 2014). Blockchains using PoS includes Ethereum (Casper update), Peercoin, Nxt, etc.

2.4.3 PROOF OF ACTIVITY

Proof of Activity (PoA) was another consensus protocol designed as a combination of PoW and PoS. The algorithm includes the features of both. The bitcoin network usually restricts the number of coins flooding through the system after a point of time to avoid the condition like hyperinflation (Belfer et al., 2020). As a result of this, the miners working using PoW algorithm will fail to get newly minted coins, and the incentives will be transaction fees alone. This may lead to unfair practices, which can be reduced by using the Proof of Activity algorithm. Here the system switches to Proof of Stake after the miner solves the computational problem. A group of validators is chosen to sign the selected block like PoS. The criteria for getting selected as validators depend on the coin balance in their account. After validation the block will be full-fledged and will be capable of getting added to the network. The fee is divided among miners and validators.

The combined protocol has the advantage that it can completely resist 51% attack as it will be expensive and time-consuming for an attacker to gain 51% of computational resources of the network as well as 51% of coin deposit in their account at the same time. The PoA faced the same criticism like PoW as the computational cost and complexity was much higher. The validation process and reaching the consensus consumes too much time, thereby making it less suitable or time-sensitive applications (Salimitari & Chatterjee, 2018). Hence the algorithm is not used in common.

2.4.4 PROOF OF BURN

Proof of Burn (PoB) is another concept which works by burning the coins instead of saving it. Here, the miner or validator will be chosen based on the number of coins burned by him. The more the number of coins burned, the more the chance of getting selected as the next miner.

The PoB is implemented in different ways. Traditionally, the coins owned by anyone will be burned by sending them to an irretrievable address, which remains as a longtime privilege to become a miner. The coins to be burned can be the native currency or the coins from the bitcoin network depending on the implementation. The stake in the system decays over time, which forces for a periodic renewal of the account to get selected as a miner.

Proof of burn (POB) can be thought as an alternative to PoW as it reduces the energy consumption and resource wastage. The algorithm can be customized by including features from PoW and PoS like Slimcoin currency network. The main issue with PoB is the major share of coins can be centralized to those who are willing to burn more coins.

2.4.5 PROOF OF CAPACITY

The Proof of Capacity (PoC) protocol was derived as a completely different concept from the existing consensus algorithms. Here, the miner doesn't want to invest in computational resources or coins. The algorithm works based on the available storage space or hard drive space available with a miner. The more the storage space, the more the chance of getting selected as the next miner.

The PoC works basically in two stages. The first step is the plotting stage where the algorithm generates plots or datasets to be stored in the hard drive. These datasets will be having the possible solutions required for mining the blocks. So if more space is available, more solutions can be stored, increasing the chance of getting selected as the miner. The second stage is the mining stage, where the actual mining happens with the help of stored solutions. The winner will be gaining the rewards.

The Proof of Capacity has the advantage of being low cost as the miner doesn't want to spend on extensive computational resources or coins. The only requirement is to have a large amount of storage space. So investing in terabytes of hard drive space will increase the chance of getting selected as next miner. The Proof of Capacity is currently implemented in the Burstcoin cryptocurrency. Disadvantages of the algorithm include lower adoption rate and the possibility of malware affecting mining activities (Tosh et al., 2017).

2.4.6 PROOF OF ELAPSED TIME (PoET)

Proof of Elapsed Time (PoET) is the consensus algorithm designed by Intel. The algorithm works well in all Intel processors. The protocol uses a trusted execution environment where participants are not responsible for creating new blocks, but the system generates it randomly. The block creation depends on the wait time. Here all the validators will be given a chance to create new blocks (Xiao et al., 2020). The resource consumption is much lower compared to the Proof of Work consensus algorithm.

PoET possesses several advantages over the existing consensus algorithms. The algorithm uses fair policies for choosing the next block and is widely used in permissioned blockchain, which are more of a private nature. The validators in the blockchain will be waiting for a random amount of time, and the proof of the same will be updated by them. The next block chosen to be added to the network will be the one which has the least timer value. The algorithm also ensures that all the validators get equal chance without any dominance and also cross checks whether nodes generate lowest timer value illegally. Thus, the algorithm proves to be efficient and reliable. The only drawback with the consensus protocol is that we must put complete trust on Intel in order to use this environment.

2.4.7 DELEGATED PROOF OF STAKE

Delegated Proof of Stake (DPoS) is derived from the traditional Proof of Stake algorithm. The algorithm was designed in 2014 by Daniel Larimer, founder of BitShares, Steemit, and EOS.

The DPoS consensus algorithm uses the election strategy. Here, the users will either vote directly or transfer their voting power to other entities. The winner of the election will be given a chance for adding new blocks to the blockchain network after successful confirmation of the transactions. On successful verification of transactions, rewards will be given to the winner, which is shared among all those who have given their votes.

DPoS has the advantage of being more democratic in nature and hence promoting decentralization. The power consumption is also less. The validators with fewer stakes can also get selected as the winner in DPoS unlike PoS (Xiao et al., 2020). This happens if they get voted by other entities having larger stakes. But decision-making is very critical in DPoS. If not properly done, this will spoil the entire chain.

So honest and reliable witnesses have to be chosen by the delegators for the smooth functioning of the system.

2.4.8 PROOF OF AUTHORITY

The Proof of Authority (PoA) is a trusted consensus protocol used in permissioned blockchain networks. Here, the block creation rights are given only to the trusted validators. The validators have to prove their trust and maintain their reputation throughout in order to be rewarded as validators. The block creation process is automated, but the validators must continuously monitor their system to make sure it is not compromised, which is a requirement.

The PoA will be able to withstand the common attacks like 51% attack and DoS attacks. The DoS attack can be blocked as the system uses pre-authenticated blocks, and only nodes which can withstand will be given the permission for block creation. The 51% attack can be invoked in PoA only if the attacker can gain control over 51% of nodes in the network, which is practically difficult (Liu et al., 2019). The algorithm is more sustainable than other algorithms due to less computational requirements. The main criticism faced by PoA is that it is more of centralized nature, and all nodes are preselected. This will reduce the decentralization power of the blockchain architecture.

2.4.9 PROOF OF IMPORTANCE

The Proof of Importance (PoI) consensus algorithm gives importance not only to the account holder with maximum stake but also to the entity which transacts more coins to others. Proof of Importance uses the concept called harvesting for performing the mining of blocks. The harvester will be a node with greater importance score. The account should have a minimum 10,000 vested coins for becoming eligible for performing harvesting (Fan & Huo, 2020). Larger and frequent transactions will also impact the importance score.

The algorithm consumes less power and computational resources and is more energy efficient. But it requires high initial investment for becoming eligible for harvesting.

Different consensus algorithms are introduced to generate trust in the blockchain network. The algorithms vary from each other in terms of performance, computational complexity, resource consumption, ability to withstand attacks like 51% attack and DoS, scalability, cost-effectiveness, etc. The algorithms are also classified based on whether they suit for permissioned or public blockchains. Table 2.2 summarizes the features of compared algorithms. Based on the requirements the user can choose the best algorithm which suits the parameters required by the implementation.

2.5 ADVANTAGES AND DISADVANTAGES OF BLOCKCHAIN TECHNOLOGY

Blockchain technology offers numerous benefits to the connected community in terms of security, reliability, transparency, and immutability. The technology being distributed in nature reduces the issues associated with centralized control. The wide

TABLE 2.2

Consensus Protocols

Algorithm / parameters	PoW	PoS	PoA (Proof of Activity)	PoB	PoC	PoET	DPOS	PoA (Proof of Authority)	PoI
Computational complexity	High	Low	High	Less	Less	Less	High	Medium	Less
Time consumption	High	Medium	High	Less	Less	Less	Less	Medium	Less
Accessibility	Public	Public	Public	Public/private	Public/private	Permissioned/private	Public/private	Permissioned/private	Public/private
Methodology	Solving computational puzzle	Coin Staking	Solving computational puzzle + coin staking by validators	Burning coins	Storing solutions in large amount of memory	Using Trusted Execution Environment	Miner selected using election process	Block creation only by trusted validators	Harvesting using importance score
Applications/ Implementations	Bitcoin. Ethereum. Litecoin Monero coin Dogecoin	Ethereum (Casper update). Peercoin. Nxt	Decred, Espers	Slimcoin	Burstcoin	Hyperledger Sawtooth	BitShare. Steemit	Microsoft Azure	NEM

popularity of the technology is mainly due to its distributed and transparent nature. The technology being in earlier phase of development also suffers from a few drawbacks which need to be addressed. The major advantages of the technology and disadvantages are discussed in this section.

2.5.1 ADVANTAGES

- Immutability
  The data recorded in the blockchain network remains immutable throughout. It is not possible for anyone to delete or update the entered data. The blocks are added after complex verification process and once verified and recorded as a block it remains immutable. The identity of data can also be kept anonymous in blockchain, thereby ensuring better privacy for the personnel documents.
- Transparency
  Blockchain transactions are transparent in nature. The data is available publicly for anyone to verify. The centralized database doesn't support transparency and will not allow users to verify or view the entered data without gaining permission (Berdik et al., 2021). The data entered in the network are timestamped and easily accessible. The distributed nature of the network improves the data accessibility and transparency.
- Increased speed and efficiency
  Blockchain technology is completely online in nature, eliminating unnecessary paper exchanges. The data uploading and verification don't require any offline settlements, improving the speed and efficiency. The documentation details and transaction details can be recorded along with the verified blocks, making it easily traceable. Blockchain creates and maintains irreversible audit trails for improved traceability.
- Free from censorship
  Blockchain doesn't require any centralized authority for control. The normal databases are controlled by specific authorities. As blockchain doesn't support it, the problem of censorship doesn't occur. No one will be able to interfere into the operation of the blockchain.
- Automation
  Blockchain technology is fully automated in nature. The smart contracts coded for the transactions are automatically triggered on satisfaction of desired conditions. The entire system works without any third-party intervention, reducing manual works.

2.5.2 DISADVANTAGES

Blockchain technology also suffers from few drawbacks. The common drawbacks of the technology are listed here.

- Higher cost
  Blockchain technology requires more implementation cost when incorporated in different organizations. The business requires careful planning

and execution strategy for incorporating the technology into organization (Shahnaz et al., 2019). The technology finds it difficult to pass through the legal formalities in different counties. The regulations need to be matching with company requirements.

- Higher energy and time consumption
  The technology demands a lot of energy for computation. The verification of transactions in blockchain requires complex and intensive calculations. Millions of people compete each time to verify transactions. This greatly increases the energy consumption. The computation of nonce value for adding new blocks to existing chains takes a lot of time, making the process bit slower.

- Less scalable
  Blockchain technology can incorporate multiple blocks, but the size of each block is usually restricted to 1MB. The size of blocks remains fixed throughout the blockchain. This reduces the transaction count supported by the blocks, making it less scalable.

- Redundancy and storage issues
  Blockchain technology requires every participating node to maintain copy of the existing transactions. This will require more storage requirements. The same data is redundantly stored in multiple locations. The requirements increase with increase in number of transactions.

- Less flexible
  The key feature of blockchain is immutability which adds trust to the network. But this feature can cause issues if the data entered needs to be modified at a later point of time. The editing of any block in blockchain also requires modification of all connected blocks, making the procedure more expensive and complex. The data recording in blockchain is highly critical in nature.

Blockchain technology has numerous advantages, which makes it more popular within few years of its introduction. Like many other technologies, blockchain is also in its infant stage. The technology suffers from few drawbacks, which need to be addressed to make the technology more acceptable. Table 2.3 summarizes the pros and cons of the technology.

TABLE 2.3
Pros and Cons of Blockchain Technology

Advantages	Disadvantages
Immutable in nature	High cost
Highly transparent	Consumes more time and energy
Increased speed and efficiency	Less scalable
Free from censorship	Redundant and requires more storage
Highly automated	Less flexible

2.6 BLOCKCHAIN USE CASES

Blockchain provides immutable, tamperproof, and decentralized power to applications. A large number of applications currently supports blockchain technology. A few examples include healthcare sector, financial institutions, voting applications, the Internet of Things, etc. The introduction of blockchain into these applications had improved the reliability, trustworthiness, and security of applications. The most common blockchain use cases are depicted in Figure 2.4.

2.6.1 HEALTHCARE SECTOR

The healthcare sector is one of the major areas where blockchain is implemented for enhancing the security of data. Patient records can be stored electronically and can be encoded to enable tamperproof and decentralized access to medical records. Lots of people around the world are losing their lives due to delay in getting the past medical records for proceeding further with treatments. This can be greatly reduced by incorporating blockchain features into medical sector. The tampering of electronically stored medical records is another problem to be addressed in handling patient

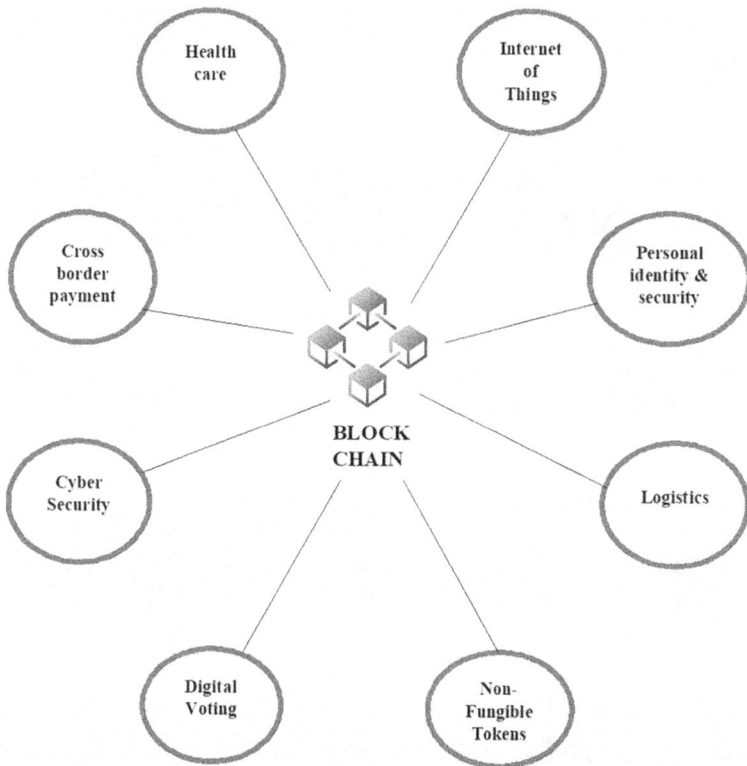

FIGURE 2.4 Blockchain use cases.

data. This can also be rectified by making the blocks encoded using blockchain's underlying cryptographic principles [16].

The blockchain can store the medical records of each patient in a tamperproof block. The permission to access the record will be given only to authorized people using the permissioned blockchain implementation. This prevents unauthorized modification of records and eases the procedures for authorized retrieval of records during emergencies.

2.6.2 INTERNET OF THINGS

Internet of Things (IoT) has gained wide popularity within a short span of time. Technology is highly capable of making our lives easier and comfortable, which has resulted in its adoption in almost all the sectors. Along with the positive aspects of IoT, the technology is highly subjected to numerous security breaches and havoc.

IoT-enabled device manufacturers are in a hurry to grab the market, paying less attention to the security aspects making the system less attractive for secure applications. Incorporation of blockchain to IoT can enhance the security features, making the system more safe and secure. The principles of decentralization can also be used to avoid single point of failure of IoT nodes. The possible implementation models should also be able to address the challenges faced by incorporating blockchain into energy-constrained nodes of IoT. So well designed and coded blockchain models will be a real solution for security issues of IoT applications.

2.6.3 PERSONAL IDENTITY SECURITY USE CASES

Securing personal documents has become an important challenge in the current scenario. Hackers are actively trying to access the stored confidential data from different logins [17]. Numerous cases are being reported regarding the forgery of documents and malpractices related to data hacking.

Blockchain technology can be used as a possible solution for securing the personal records. The technology can be adopted by the government agencies for safeguarding sensitive identity information of citizens including social security numbers, birth certificates, birth dates, and other sensitive information on a decentralized blockchain ledger. This will dramatically reduce the theft and forgery of confidential documents.

The different use cases mentioned previously clearly shows the impact of blockchain technology in securing transactions. The technology is still in its early stage and hence provides a wide range of research opportunities to be explored in the near future. Being a new technology, thorough study has to be made on the challenges of implementing blockchain and has to be addressed for making maximum throughput out of the applications [18].

2.6.4 LOGISTICS USE CASES

The logistics industry is performing a huge number of transactions every single minute. The industry needs to maintain information transparency without soiling the data. This is a critical challenge that needs well-equipped technologies to offer proper solutions. Blockchain can be used for verifying the data source and for maintaining a transparent and tamperproof ledger. The technology offers source integrity and data security in a cost-effective manner. Blockchain-enabled shipping and logistics will help to create a trustworthy industry. The tracking data will be available publicly in the ledger, making it easy to verify and maintain. The coordination between supply chain industries and logistics partners can be made more reliable and easier with the support of blockchain. The shipping services like DHL have already experimented with blockchain in building trust in the logistics chain. This incorporation can be studied and analyzed by researchers for further implementation.

2.6.5 NON-FUNGIBLE TOKENS (NFTs)

Non-fungible tokens (NFTs) are another popular application of blockchain which is used to digitally claim the ownership of digital arts. NFTs are intangible cryptographic tokens available in blockchain to ensure the authenticity of the source of digital assets. It gives complete ownership to respective owners on their digital items available remotely, which includes pictures uploaded, video contents, audio, GIFs, and many more. Blockchain technology maintains the uniqueness of ownership of the digital item as the ledger is maintained publicly and will not support duplicates. The NFTs can be utilized to safeguard the copyright of the item without extra maintenance costs.

2.6.6 DIGITAL VOTING

Blockchain technology has been effectively used in different places to conduct error-free voting. The voting process is highly vulnerable to fraudulent approaches, which greatly influence the voting results and reliability. Blockchain-enabled voting ensures that the cast votes remain safe, and no one will be able to modify it further. Each cast vote will be considered as one unique transaction, and after verification it will be stored as a block in the blockchain, which prevents further manipulation of the record. The blocks in blockchain always remain immutable, and the probability of creating and using fake ID for casting vote is null. The government can decide the rules and regulations for voting, and smart contracts can be designed and embedded accordingly. Thus, the entire voting process can be made more transparent, tamperproof, and trustworthy.

2.6.7 CYBER SECURITY

Cyber security is one of the largest areas where growing requirements never end. Blockchain can be a promising solution in offering cyber security as the technology is highly decentralized and immutable. The data and transactions through the Internet are usually vulnerable to a variety of attacks. Security breaches are very common. The encrypted data generated from the sender is stored as an immutable block in the blockchain, and the attacker won't be able to modify it. As the generated blocks are verified by multiple nodes, attempting to decode and modify the blocks is highly computationally complex and impossible due to permanency nature of the technology.

2.6.8　Cross-Border Payments

Traditional cross-border payments include a lot of extra costs, effort, and are highly time-consuming due to many intermediaries involved in the whole process. This can also result in creeping of errors, unnecessary charges at different levels, and chances of failures at different points. Blockchain-enabled cross-border payments are simple and error-free as it uses encrypted distributed ledgers that are capable of providing on-time verification of transactions without the involvement of intermediaries such as third parties and correspondent banks. As the technology greatly removes the chance of fraudulent attacks, it creates a higher chance of compliance with consumer data privacy regulations. The verified and stored data remains intangible and cannot be modified without permission of multiple verifiers associated in that network, making the approach more secure, durable, and easy to manage.

2.7　FRAMEWORK FOR DATA SECURITY USING BLOCKCHAIN IN IOT HOME AUTOMATION

Blockchain technology has numerous applications, as mentioned previously. This section models a smart home use case built with incorporation of blockchain. The same concept can be applied in other domains as well. The main application of blockchain is to enhance security and provide data transparency. Smart homes are IoT-enabled homes where the different connected sensors and devices communicate with each other, sense different conditions, and notify the connected users in a periodic manner or based on the request. The current smart home use cases have issues with data security as the data available over network is highly subjected to manipulations and forgery, thereby generating false alarms and notifications to users and connected devices. This issue can be resolved if blockchain technology is incorporated in data verification.

2.7.1　System Model

The proposed smart home network works as a three-tier architecture where IoT devices (D1, D2 . . . Dn) lie the in lowermost layer and sense data. The raw data sensed will be transmitted to the gateways (G1,G2 . . . Gn) in the middle layer. The gateways will be responsible for filtering the data and preprocessing data before passing to the remote cloud, which lies in the third layer. The cloud layer performs extended processing and analysis. The three-tier architecture for the model is depicted in Figure 2.5.

The raw data from sensors will be formatted and hashed at the gateway. After preprocessing, the data has to be generated as a block and should be distributed across the blockchain network for verification and validation. The validated data after required quality checks will be permanently added to the blockchain as a valid block. Figure 2.6 depicts this scenario.

The data generated at the gateway will be transmitted to the network after encryption using any cryptographic algorithms such as AES, SHA, etc. The encrypted data can always be subjected to different network attacks and manipulations. The data

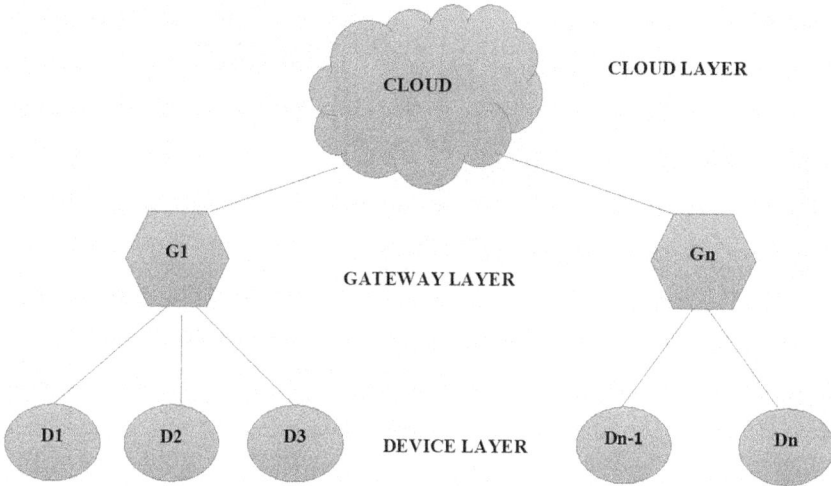

FIGURE 2.5 Smart home architecture.

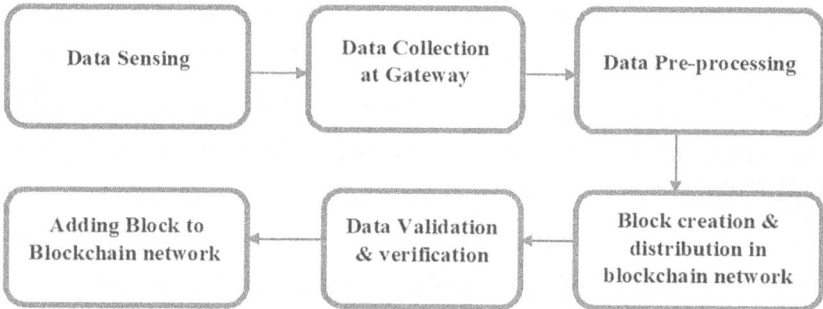

FIGURE 2.6 Blockchain implementation of smart home.

will be received and stored in remote cloud data center, and users can access it anywhere at any time. The data being permanently recorded in blockchain can always be cross-checked for chances of forgery or manipulations. Thus, even if the data has undergone attacks on the way to cloud, it can be easily detected and reversed. Thus blockchain ensures data security and transparency in the smart home network. The proposed method can be applied to different domains to make the data valid and tamperproof.

2.8 CONCLUSION

Blockchain is one of the most promising technologies which have proved to be tamperproof in nature. The technology maintains a distributed ledger that can be verified publicly by anyone. The technology was introduced to overcome the problems faced by centralized control. The distributed system will be useful in many applications which don't trust centralized control and which are subjected to single point of failure. Different applications like wireless applications, IoT, healthcare applications, logistics, banking, cross-border payments, etc. can be improvised to a greater extent by adopting blockchain technologies. The security features of applications can be enhanced using blockchain's consensus protocols.

The consensus protocols are designed in such a way that the attackers will find it difficult to induce threats to the network. All the committed transactions in the blockchain network are visible to everyone through the distributed ledger, making the attacks even tougher. Consensus protocols can be designed separately for public and private blockchains. Different algorithms were designed to meet different application domain requirements. Computationally complex algorithms like Ethereum's PoW focus more on security than resource consumption. The algorithms like PoS, DPoS, etc. focus more to reduce the resource consumption and complexity to make it more user-friendly. The consensus algorithms like Intel's Proof of Elapsed Time reduce the effort of miners by using a system-generated environment like SGX. This is for automatic selection of nodes for block creation and thereby greatly reduces the complexity and saves system resources and power. Different algorithms are designed by keeping different parameters and design requirements. The smart home use case discussed in the chapter provides a clear picture about methodology to be followed in blockchain incorporation. A thorough understanding of blockchain concepts, use cases, and consensus protocols is mandatory to design blockchain applications with high reliability, security, and enhanced performance. This chapter clearly discusses the basics of blockchain, the advantages and disadvantages of the technology, various consensus mechanisms, different application use cases of blockchain, and models the smart home use case for better understanding. Different concepts explained here will help the research community to gain thorough idea about blockchain, which can be extended to design newer innovative applications.

REFERENCES

Afzaal, H., Imran, M., Janjua, M. U., & Gochhayat, S. P. (2022). Formal Modeling and Verification of a Blockchain-Based Crowdsourcing Consensus Protocol. IEEE Access, 10, 8163–8183. https://doi.org/10.1109/ACCESS.2022.3141982.

Belfer, R., Kashtalian, A., Nicheporuk, A., Markowsky, G., & Sachenko, A. (2020). Proof-of-Activity Consensus Protocol Based on a Network's Active Nodes. CEUR Workshop Proceedings, 2623, 239–251.

Berdik, D., Otoum, S., Schmidt, N., Porter, D., & Jararweh, Y. (2021). A Survey on Blockchain for Information Systems Management and Security. Information Processing and Management, 58(1), 102397. https://doi.org/10.1016/j.ipm.2020.102397.

Bhattacharjya, A., Wisniewski, R., & Nidumolu, V. (2022). Holistic Research on Blockchain's Consensus Protocol Mechanisms with Security and Concurrency Analysis Aspects of CPS. Electronics (Switzerland), 11(17). https://doi.org/10.3390/electronics11172760.

Biswas, S., Sharif, K., Li, F., Maharjan, S., Mohanty, S. P., & Wang, Y. (2020). PoBT: A Lightweight Consensus Algorithm for Scalable IoT Business Blockchain. IEEE Internet of Things Journal, 7(3), 2343–2355. https://doi.org/10.1109/JIOT.2019.2958077.

Chang, S. E., & Chen, Y. (2020). When Blockchain Meets Supply Chain: A Systematic Literature Review on Current Development and Potential Applications. IEEE Access, 8, 62478–62494. https://doi.org/10.1109/ACCESS.2020.2983601.

Daraghmi, E. Y., Daraghmi, Y. A., & Yuan, S. M. (2019). MedChain: A Design of Blockchain-Based System for Medical Records Access and Permissions Management. IEEE Access, 7, 164595–164613. https://doi.org/10.1109/ACCESS.2019.2952942.

Fan, X., & Huo, Y. (2020). Blockchain-Based Dynamic Spectrum Access of Non-Real-Time Data in Cyber-Physical-Social Systems. IEEE Access, 8, 64486–64498. https://doi.org/10.1109/ACCESS.2020.2985580.

Friedman, N., & Ormiston, J. (2022). Blockchain as a Sustainability-Oriented Innovation? Opportunities for and Resistance to Blockchain Technology as a Driver of Sustainability in Global Food Supply Chains. Technological Forecasting and Social Change, 175(December 2021), 121403. https://doi.org/10.1016/j.techfore.2021.121403.

Koshy, P., Babu, S., & Manoj, B. S. (2020). Sliding Window Blockchain Architecture for Internet of Things. IEEE Internet of Things Journal, 7(4), 3338–3348. https://doi.org/10.1109/JIOT.2020.2967119.

Liu, Z., Tang, S., Chow, S. S. M., Liu, Z., & Long, Y. (2019). Fork-Free Hybrid Consensus with Flexible Proof-of-Activity. Future Generation Computer Systems, 96, 515–524. https://doi.org/10.1016/j.future.2019.02.059.

Medhane, D. V., Sangaiah, A. K., Hossain, M. S., Muhammad, G., & Wang, J. (2020). Blockchain-Enabled Distributed Security Framework for Next Generation IoT: An Edge-Cloud and Software Defined Network Integrated Approach. IEEE Internet of Things Journal, 4662(c), 1–1. https://doi.org/10.1109/jiot.2020.2977196.

Pyoung, C. K., & Baek, S. J. (2020). Blockchain of Finite-Lifetime Blocks with Applications to Edge-Based IoT. IEEE Internet of Things Journal, 7(3), 2102–2116. https://doi.org/10.1109/JIOT.2019.2959599.

Rejeb, A., Rejeb, K., Simske, S. J., & Keogh, J. G. (2022). Blockchain Technology in the Smart City: A Bibliometric Review. In Quality and Quantity (Vol. 56, Issue 5). Springer Netherlands. https://doi.org/10.1007/s11135-021-01251-2.

Salimitari, M., & Chatterjee, M. (2018). A Survey on Consensus Protocols in Blockchain for IoT Networks. http://arxiv.org/abs/1809.05613.

Shahnaz, A., Qamar, U., & Khalid, A. (2019). Using Blockchain for Electronic Health Records. IEEE Access, 7, 147782–147795. https://doi.org/10.1109/ACCESS.2019.2946373.

Shen, H., Zhou, J., Cao, Z., Dong, X., & Choo, K.-K. R. (2020). Blockchain-Based Lightweight Certificate Authority for Efficient Privacy-Preserving Location-Based Service in Vehicular Social Networks. IEEE Internet of Things Journal, 4662(c), 1–1. https://doi.org/10.1109/jiot.2020.2974874.

Shen, M., Tang, X., Zhu, L., Du, X., & Guizani, M. (2019). Privacy-Preserving Support Vector Machine Training Over Blockchain-Based Encrypted IoT Data in Smart Cities. IEEE Internet of Things Journal, 6(5), 7702–7712. https://doi.org/10.1109/JIOT.2019.2901840.

Singh, J., Sajid, M., Gupta, S. K., & Haidri, R. A. (2022). Artificial Intelligence and Blockchain Technologies for Smart City. Intelligent Green Technologies for Sustainable Smart Cities, August, 317–330. https://doi.org/10.1002/9781119816096.ch15.

Tosh, D. K., Shetty, S., Liang, X., Kamhoua, C., & Njilla, L. (2017). Consensus Protocols for Blockchain-Based Data Provenance: Challenges and Opportunities. 2017 IEEE 8th Annual Ubiquitous Computing, Electronics and Mobile Communication Conference, UEMCON 2017, 2018-Janua, 469–474. https://doi.org/10.1109/UEMCON.2017.8249088.

Vasin, P. (2014). BlackCoin's Proof-of-Stake Protocol v2 Pavel. Self-Published, 2. https://blackcoin.co/blackcoin-pos-protocol-v2-whitepaper.pdf.

Xiao, Y., Zhang, N., Lou, W., & Hou, Y. T. (2020). A Survey of Distributed Consensus Protocols for Blockchain Networks. IEEE Communications Surveys and Tutorials, 22(2), 1432–1465. https://doi.org/10.1109/COMST.2020.2969706.

Xu, L. Da, & Viriyasitavat, W. (2019). Application of Blockchain in Collaborative Internet-of-Things Services. IEEE Transactions on Computational Social Systems, 6(6), 1295–1305. https://doi.org/10.1109/TCSS.2019.2913165.

Yang, L., Zhang, J., & Shi, X. (2021). Can Blockchain Help Food Supply Chains with Platform Operations During the COVID-19 Outbreak? Electronic Commerce Research and Applications, 49(March), 101093. https://doi.org/10.1016/j.elerap.2021.101093.

Zheng, W., Zheng, Z., Chen, X., Dai, K., Li, P., & Chen, R. (2019). NutBaaS: A Blockchain-As-A-Service Platform. IEEE Access, 7, 134422–134433. https://doi.org/10.1109/ACCESS.2019.2941905.

3 Technological Advances in Smart Monitoring Imaging Systems Based on Machine Learning

Abdulla Desmal

3.1 INTRODUCTION

A smart monitoring system is a system that is built to detect failures, anomalies, or specific patterns over input data via modern technologies, for example, intelligent programs, machine learning (ML), or the Internet of Things (IoT). The following summarizes some important applications in smart monitoring imaging systems based on machine learning.

- Vision-based imaging:
  - Detecting and tracking intruders or thieves over CCTVs (Shirsat et al., 2019).
  - Extracting car plate numbers for overspeeding cars (Asaei, Kasaei, & Kasaei, 2010).
  - Monitoring and controlling crowds to detect criminals and reducing crowd congestion (Singh, Determe, Horlin, & De Doncker, 2021).
  - Monitoring fruits and vegetable maturity for classification (Behera, Rath, & Sethy, 2021).
  - Building auto purchase systems that automatically monitor customers' actions and update bills accordingly (Mishra, Mohan, Mandal, Mohanty, & Chowdhury, 2022).
- Thermal and infrared (IR) imaging:
  - Detecting human body temperature through thermal cameras for cancer treatments or flue detection (Chidurala & Li, 2021; Nguyen, Hong, Kim, & Park, 2017).
  - Monitoring solar cell performance or failure detections through IR cameras (Akram et al., 2020; Ali, Khan, Masud, Kallu, & Zafar, 2020).
  - Predicting water-soil content through thermal cameras mounted on an unmanned aerial vehicle (Bertalan et al., 2022).
  - Detection of natural gas methane leakage (Wang, Ji, Ravikumar, Savarese, & Brandt, 2022).

- Microwave imaging:
  - Breast tumor imaging through electromagnetic imaging (Khoshdel, Asefi, Ashraf, & LoVetri, 2020; Ambrosanio, Franceschini, Pascazio, & Baselice, 2022; Desmal, 2020, 2021, 2022a, 2022b).
  - Hydrocarbon reservoir exploration through microwave imaging (Sun et al., 2020).
  - Landmine detection using ground-penetrating radars (GPRs) (Nunez-Nieto, Solla, Gomez-Perez, & Lorenzo, 2014).
- X-ray imaging:
  - Baggage screening (Jain & others, 2019; Wang, Ismail, & Breckon, 2020; Desmal et al., 2019).
  - Breast tumor detection (Sharma & Jindal, 2022; Al-Antari, Han, & Kim, 2020; Alkhaleefah et al., 2022).
  - Classifying dental diseases based on X-ray images (Almalki et al., 2022).
  - Segmentation and visualization of X-ray chest images (visualization, 2020).
- Magnetic resonance imaging (MRI):
  - Brain tumor classification based on MRI images (Zacharaki et al., 2009).
  - Predicting the stage of Alzheimer's from Brain MRI images (Moradi et al., 2015).
  - Detecting prostate cancer through MRI images and machine learning (Cuocolo et al., Machine learning applications in prostate cancer magnetic resonance imaging, 2019).
  - Prostate cancer detection using MRI and machine learning (Li et al., 2022).

Figure 3.1 illustrates the three major stages in any smart monitoring system, sensing, processing, and monitoring. In the sensing stage, sensors detect the required signal like images from vision or thermal cameras, distances through LiDAR sensors, online customer data, or sounds through a microphone. The processing can be categorized into conventional or intelligent processing. Counting the number of cars entering a parking zone and deciding about soil humidity levels are examples of conventional processing. On the other hand, intelligent processing requires artificial intelligent programs like extracting car plate numbers, crowd monitoring, and monitoring online purchases to select items for advertisements. The final stage displays the decision or output from the processing unit through a graphic user interface, short text messages, or any suitable output system. Another task for the final stage could be to send a report to an online server or another program to generate control signals and take further actions.

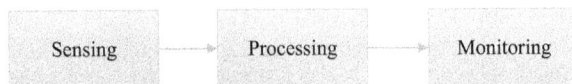

FIGURE 3.1   Elementary stages of smart monitoring systems.

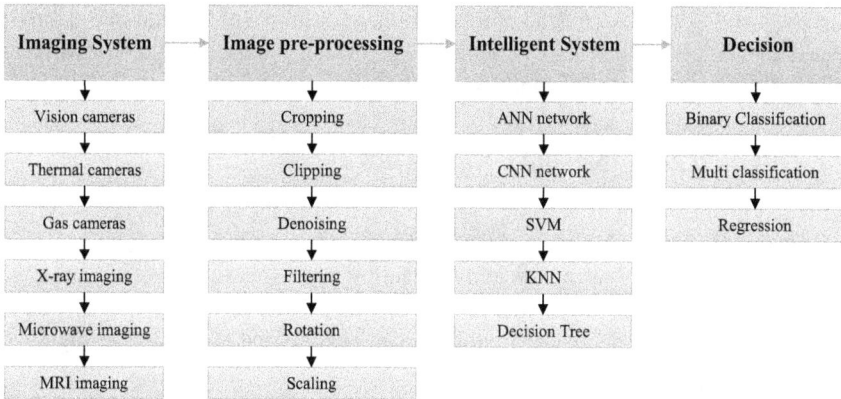

FIGURE 3.2   Dominant approaches followed in smart monitoring imaging systems.

Figure 3.2 shows the basic elements of smart imaging monitoring systems with the possible approaches followed. The first block is the sensing system used, which could be vision cameras, thermal cameras, gas cameras, X-ray systems, microwave systems, or MRI systems. After collecting the image from the sensing system, image preprocessing is usually required. In the preprocessing block, the image properties are adjusted to become suitable for the block that follows; another name suitable for this stage is the data preparation stage. The operations involved in the preprocessing include image cropping, clipping, denoising, filtering, rotation, and scaling. The intelligent system block will further process the image to produce the final prediction. Many approaches have been utilized for this purpose, including multilayered fully connected neural networks, or what is known as artificial neural networks (ANN). Convolutional neural network (CNN) is a more advanced technique and has proven better performance compared to usual ANN networks. Other techniques have been utilized like supported vector machines (SVMs), K-points nearest neighborhood (KNN), and decision trees. The final stage is the output which can be in a different form depending on the problem's nature. The binary classification is used for outputs that are either True/False, Positive/Negative, or 1/0. Multi-classification is used to classify output with more than two levels, like classifying the type of animals or automobile brands.

This chapter focuses on intelligent monitoring imaging systems based on machine learning algorithms using artificial neural networks, which have been applied dominantly. First, a brief explanation of the structure of the biological and artificial single-node neural networks (also known as single perceptron networks) will be introduced, along with their differences and comparisons. The training process of a single-layer perceptron will be discussed to predict the required output. Then, the design of multilayered artificial neural networks (ANNs) and their advantages over single perceptron networks will be explained. Advanced neural networks designated for vision recognition, known as convolutional neural networks (CNNs), will be detailed in subsequent sections. The concept of transfer learning will be introduced, which is an important tool that helps design neural networks with accurate prediction at a low training cost. At the end of the chapter, an example of a vision-based intelligent monitoring system will be explained. The example is used to monitor workers' safety regulations, like wearing masks and helmets.

3.2   BIOLOGICAL NEURAL NETWORKS

Although artificial neural networks are built to mimic the output decisions of biological neural networks, there is still a large gap between the two systems' structure and performance. It is worth indicating the following points about the human brain:

- The human brain performs nonlinear processing and consists of a highly complex structure.
- Unlike the sequential processing in microcontrollers and CPUs, the brain, by nature, processes the information in parallel. The so-called parallel processing in CPUs only runs independent sequential programs in multiple processing units. The latter means that the programs, by nature, are series coding. It is important to point out here that the human brain is essentially different and, simply by nature, has parallel processing.
- The data in the human brain are stored globally throughout the entire brain area. Therefore, while the brain processes the data in parallel (or globally), the data that the human brain processes are either stored globally or due to the human five senses.

Figure 3.3 shows the biological neuron structure, the smallest unit in the human brain. Put in mind that this neuron model is simplified. The human brain consists of biological neurons in the range of ten billion.

The center or body cell of biological neurons is knowns as soma, which is connected to multiple fiber connections known as dendrites. The dendrite that connects two somas of two different neurons is called axon. The synapses produce chemical substances that cause electrochemical signals to propagate throughout the biological neural network.

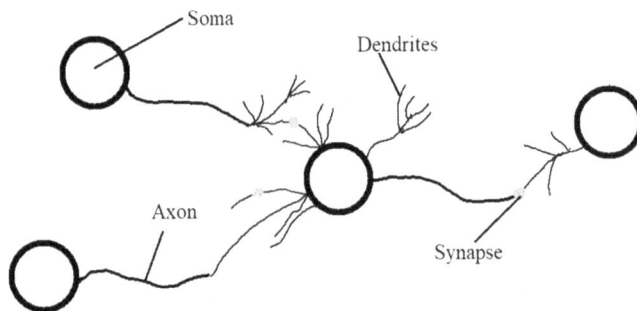

FIGURE 3.3   Structure of a single biological neuron (for more information, see Negnevitsky, A guide to intelligent systems, 2005).

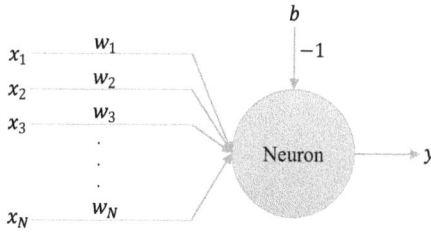

FIGURE 3.4 Structure of a single perceptron network.

3.3 SINGLE PERCEPTRON NETWORK

Figure 3.4 shows the smallest unit in artificial neural network (ANN), which is known as the perceptron. Similar to the soma in biological neurons, the body of perceptrons is called a neuron or simply a node. Inputs $x_1, x_2, \ldots, x_M$ can come from other percep-trons' outputs or the system inputs directly, while N indicates the number of inputs. The inputs are multiplied by their relative weights $w_1, w_2, \ldots, w_M$ respectively, which mimics the task of Synapses. The biasing b is always multiplied by the negative one to give biasing effect. Inside the node or neuron, the weighted inputs are summed and inserted into a nonlinear function known as the activation function. Recall that the human brain processes data nonlinearly; hence a nonlinear process is essential in building an ANN inspired by the biological neural network. The output y of the single perceptron is computed as follows:

$$y = f\left(\bar{x}, \bar{\theta}\right) = \sigma\left(w_1 x_1 + w_2 x_2 + w_3 x_3 + \cdots + w_M x_M - b\right) \qquad (3.1)$$

where $f\left(\bar{x}, \bar{\theta}\right)$ is the perceptron function, $\bar{x} = [x_1, x_2, \ldots, x_M]^t$ is $M \times 1$ column vector carrying the input, and the superscript (t) indicates performing the transposed oper-ation. While $\bar{\theta}$ is a vector of size $(M+1)$ consisting of the perceptron hyperparam-eters. The perceptron hyperparameters are the weights $\bar{w} = [w_1, w_2, \ldots, w_M]$, which is $1 \times M$ row vector, and the biasing (b). The activation function $\sigma(.)$ is implemented to add nonlinearity, as explained next.

Table 3.1 and Figure 3.5 list and plot the most popular activation functions imple-mented in machine learning. Besides nonlinearity, there are specific properties that make an activation function desired: (1) differentiable, (2) squeezed output, and (3) monotonic. As will be seen later, the derivatives of the neural network model con-cerning hyperparameters are needed to optimize the hyperparameters. Selecting non-differentiable activation functions will make the optimization process challeng-ing. Researchers also prefer functions with easy non-complexed derivative forms. The squeeze feature is also desired, in which researchers try to shrink or squeeze the output of each neuron. In large networks that consist of a sequence of neurons, the squeeze feature is important to avoid propagating large numbers that might lead the output to diverge. It is clear in Figure 3.5, the Tanh and Sigmoid activation func-tions, the output is confined in the range of [0,1] or [-1,1]. In addition, monotonicity is the feature that enforces functions to either continuously increase or continuously

TABLE 3.1
Activation Functions

Activation function	Equation form
Relu	$\sigma(x) = \begin{cases} x & for\ x \geq 0 \\ 0 & for\ x \leq 0 \end{cases}$
Leaky Relu	$\sigma(x) = \begin{cases} x & for\ x \geq 0 \\ \alpha x & for\ x \leq 0 \end{cases}$
Sign	$\sigma(x) = \begin{cases} 1 & for\ x \geq 0 \\ -1 & for\ x \leq 0 \end{cases}$
Step	$\sigma(x) = \begin{cases} 1 & for\ x \geq 0 \\ 0 & for\ x \leq 0 \end{cases}$
Tanh	$\sigma(x) = \tanh(x) = \dfrac{\exp(x) - \exp(-x)}{\exp(x) + \exp(-x)}$
Sigmoid	$\sigma(x) = \dfrac{1}{1 + \exp(-x)}$

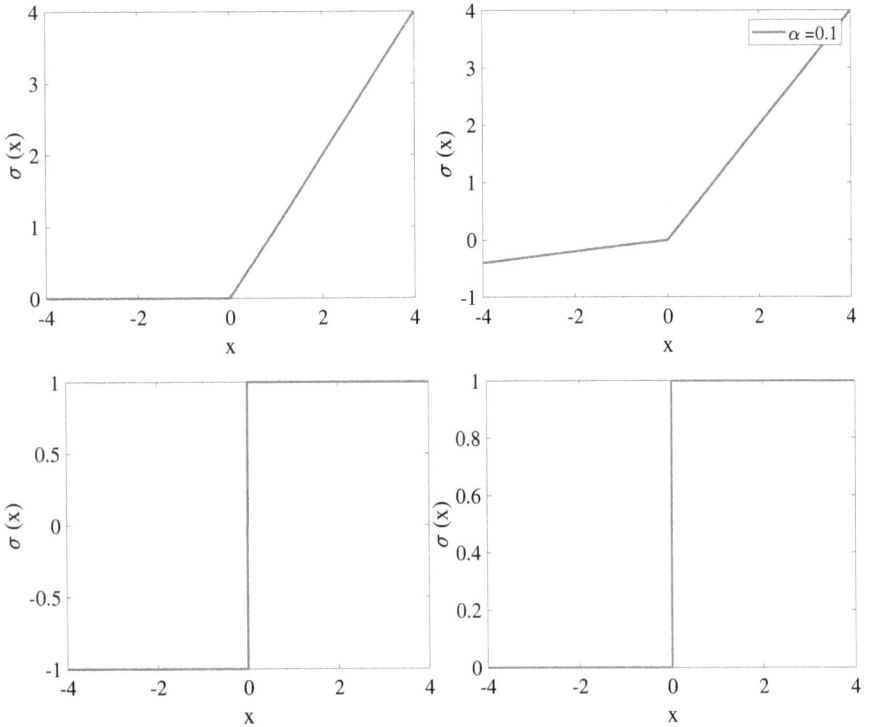

FIGURE 3.5   Plots of different activation functions.

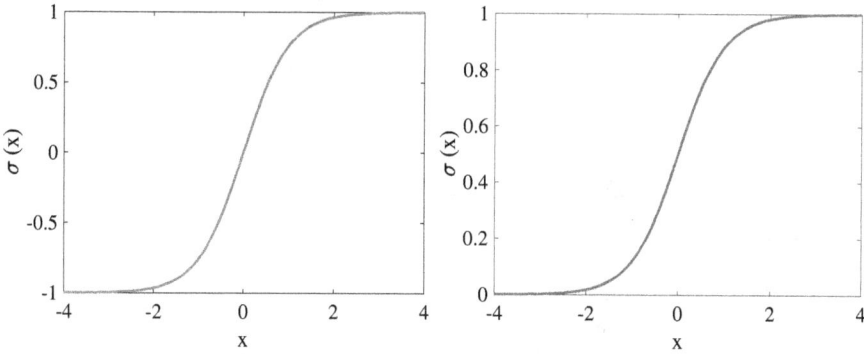

FIGURE 3.5   (Continued)

decrease over the entire function range, for example, leaky Relu, Tanh, and Sigmoid functions are monotonic.

Equation (3.1) can be rewritten in vector form, as shown here.

$$y = f\left(\overline{x},\overline{\theta}\right) = \sigma\left(\overline{w}'\overline{x} - b\right) \tag{3.2}$$

The process of fitting the hyperparameters in $\overline{\theta} = (\overline{w}, b)$ to achieve the desired output is called the training process, and it is explained in the next section.

3.4   TRAINING NEURAL NETWORKS

In this section, the training process is explained for single perceptron networks. Nevertheless, the same concepts are applied in multilayered and convolutional neural networks. Assume N training examples $\left(\overline{x}_1, y_1\right), \left(\overline{x}_2, y_2\right), \ldots, \left(\overline{x}_N, y_N\right)$, let

$$\overline{\overline{X}} = \begin{bmatrix} \overline{x}_1 & \cdots & \overline{x}_N \end{bmatrix} \tag{3.3}$$

$$\overline{Y} = \begin{bmatrix} y_1 & \cdots & y_N \end{bmatrix} \tag{3.4}$$

where $\overline{\overline{X}}$ is $M \times N$ matrix consisting of inputs of all the examples while $\overline{Y}$ is $1 \times N$ consisting of the outputs of all the examples. Equation (3.2) can be rewritten in terms of $\overline{\overline{X}}$ and $\overline{Y}$ as follows:

$$\overline{Y} = F\left(\overline{\overline{X}}, \overline{\theta}\right) = \sigma\left(\overline{w}'\overline{\overline{X}} - b\right) \tag{3.5}$$

where the function $F\left(\overline{x}, \overline{\theta}\right)$ computes the outputs of all examples, which is $1 \times N$ vector. The training process aims to find the hyperparameters $\overline{\theta}$ that will fit the outputs $\overline{Y}$ given the inputs $\overline{\overline{X}}$. But before this step, one important question shall be raised,

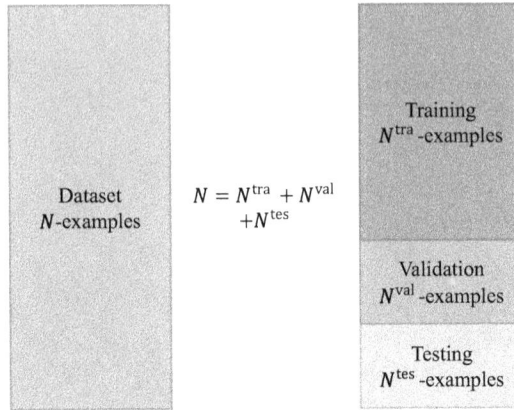

FIGURE 3.6 Distributing the main dataset with N examples into N^{tra} training examples, N^{val} validation examples, and N^{tes} testing examples.

how shall we justify that the optimized $\bar{\theta}$ has a good or excellent prediction of the output, or in other words, has a good fitting? To answer this question, examples in the dataset will be divided into three groups, (1) training, (2) validation, and (3) testing, see Figure 3.6.

Training examples are used in the fitting process to find the hyperparameters that will fit the outputs given the inputs, let $\left(\bar{\bar{X}}^{\text{tra}}, \bar{Y}^{\text{tra}}\right)$, $\left(\bar{\bar{X}}^{\text{val}}, \bar{Y}^{\text{val}}\right)$, and $\left(\bar{\bar{X}}^{\text{tes}}, \bar{Y}^{\text{tes}}\right)$ be the subsets of training, validation, and testing examples. The training process minimizes the least square differences between the model and the outputs in terms of the hyperparameters considering the training subset only $\left(\bar{\bar{X}}^{\text{tra}}, \bar{Y}^{\text{tra}}\right)$.

$$\min_{\theta} \left\| \bar{Y}^{\text{tra}} - F\left(\bar{\bar{X}}^{\text{tra}}, \bar{\theta}\right) \right\|_2^2 \tag{3.6}$$

The validation and testing sets are only used to verify the model accuracy using examples that have not been processed in training. During the optimization process in (3.6), the error or accuracy of the validation subset is computed to assure that the training process is on track. As further confirmation, the error or accuracy of the testing subset is computed at the end of the training process. In the training process, if the validation error is minimized along with the training error, the model is said to achieve globalization. Globalization means that the model is learning the problem features in general, not only those in training examples. Thus the model can produce small errors in the validation examples. On the other hand, if the model achieves a small error in the training examples while the error for the validation examples is worst, then the model is said to achieve overfitting or memorization. This means the model memorizes or fits the training examples and their outputs well but fails to make good predictions for examples outside the training subset.

To solve (3.6), different approaches can be applied, like conjugate gradients, Newton-Raphson, or the most dominantly applied approach, the steepest descent and its variants (Bottou, 2012). The steepest descent starts with an initial value $\bar{\theta}_0$, usually random, and updates the hyperparameters using the descent path defined by the negative gradient of the objective function. The steepest descent is written as (Bottou, 2012; Sandhu, Desmal, & Bagci, 2021; Desmal et al., 2019)

$$\bar{\theta}_{i+1} = \bar{\theta}_i - \alpha \partial_{\bar{\theta}} C\left(\bar{\bar{X}}^{\text{tra}}, \bar{Y}^{\text{tra}}, \bar{\theta}_i\right) \tag{3.7}$$

where $C\left(\bar{\bar{X}}^{\text{tra}}, \bar{Y}^{\text{tra}}, \bar{\theta}\right) = \left\|\bar{Y}^{\text{tra}} - F\left(\bar{\bar{X}}^{\text{tra}}, \bar{\theta}\right)\right\|_2^2$ is the cost or objective function. The parameter α is the learning parameter, and it is a user-specified parameter. In machine learning problems, the cost function is nonlinear and consists of multiple minima. Consequently, specifying α is crucial in terms of achieving global minimization. If the learning parameter α is too large, the steepest descent path might skip local and/or global minima. On the other hand, if the learning parameter α is too small, there is a good chance that the search path will be confined to a local minimum. Besides selecting the learning parameter empirically, three issues in (6) need to be addressed in detail:

- How to compute the gradient of the cost term?
- Is it practical to assume all the training examples in each steepest descent step?
- How shall we seek global minimization?

These points are addressed in the subsequent sections.

3.5 BACKPROPAGATION

Regarding the first point, the gradient of the cost term assuming a single perceptron network can be derived as shown in the next equation:

$$\partial_{\bar{\theta}} C\left(\bar{\bar{X}}^{\text{tra}}, \bar{Y}^{\text{tra}}, \bar{\theta}_i\right) = -C\left(\bar{\bar{X}}^{\text{tra}}, \bar{Y}^{\text{tra}}, \bar{\theta}\right) \partial_{\bar{\theta}} F\left(\bar{\bar{X}}^{\text{tra}}, \bar{\theta}\right) \tag{3.8}$$

$$\partial_{\bar{\theta}} F\left(\bar{\bar{X}}^{\text{tra}}, \bar{\theta}\right) = \begin{cases} \partial_{\bar{w}} F\left(\bar{\bar{X}}^{\text{tra}}, \bar{\theta}\right) = \sigma'\left(\bar{w}^t \bar{\bar{X}} - b\right) \bar{\bar{X}}^t \\ \partial_b F\left(\bar{\bar{X}}^{\text{tra}}, \bar{\theta}\right) = -\sigma'\left(\bar{w}^t \bar{\bar{X}} - b\right) \bar{1}_{N\times1} \end{cases} \tag{3.9}$$

where $\partial_{\bar{\theta}} F\left(\bar{\bar{X}}^{\text{tra}}, \bar{\theta}\right)$ is the gradient of the neural network concerning the hyperparameters $\bar{\theta}$, and $\bar{1}_{N\times1}$ is a row vector of ones and size N. Equations (3.8, 3.9) can be derived using definitions of matrix derivatives. However, this is not the usual way; as the neural network model gets complex, computing the gradient directly becomes impractical. The backpropagation is implemented in machine learning programs like

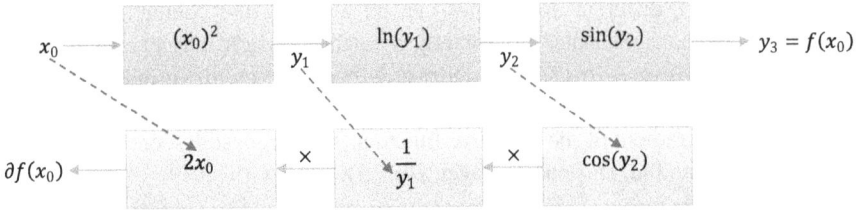

FIGURE 3.7 The upper path shows the forward model computation, while the lower path shows the backpropagation computation.

Pytorch and Tensorflow that simplify the gradient computations and establish the gradient path once the neural network model, the forward path, is coded. The following shows a simplified example of the backpropagation process.

As an example, let $f(x) = \sin\left(\ln\left(x^2\right)\right)$, the following show how to compute the gradient $\partial f(x_0)$ defined at point x_c using backpropagation.

Figure 3.7 shows the computation of the forward and backward paths. Notice, some values in the backpropagation path are taken from the forward path. The latter suggests that computing the forward path first is necessary for computing the backpropagation. To compute the forward path of $f(x) = \sin\left(\ln\left(x_0^2\right)\right)$, start from the inner function to the outer one. On the first stage compute $y_1 = x_0^2$, then $y_2 = \ln(y_1)$, then $y_3 = \sin(y_2)$. The backpropagation uses the chain rule to compute the function derivatives starting from the output to the input. At the output, we have $y_3 = \sin(y_2)$, according to the chain rule $\partial y_3 / \partial x = \cos(y_2)\partial y_2 / \partial x$, while $\partial y_2 / \partial x = (1/y_1)\partial y_1 / \partial x$, and at the input $\partial y_1 / \partial x = 2x_0$. Thus the boxes here are derivative of the functions for the boxes in the previous section, and to compute the total gradient $\partial f(x_0)$, multiply the boxes here, which results in $\partial f(x_0) = \cos(y_2)(1/y_1)2x_0$. The same process is applied to more complex functions.

3.6 BATCH STEEPEST DESCENT

In most machine learning applications, computing the gradient $\partial_\theta C\left(\overline{\overline{X}}^{\,tra}, \overline{Y}^{\,tra}, \overline{\theta}_i\right)$ in equation (7) using all the training data $\overline{\overline{X}}^{\,tra}$ in each steepest descent step (i) is expensive in terms of memory and computation. The idea of batch steepest descent is to divide the training data into small batches $\left(\overline{\overline{X}}_b^{\,tra}, \overline{Y}_b^{\,tra}\right)$ such that $b = 1,.., N^{bat}$, where N^{bat} is the number of batches. Equation (7) is rewritten as

$$\overline{\theta}_{i+1} = \overline{\theta}_i - \alpha\partial_\theta C\left(\overline{\overline{X}}_b^{\,tra}, \overline{Y}_b^{\,tra}, \overline{\theta}_i\right) \tag{3.10}$$

First, equation (3.10) is executed for all the batches, and the index (i) will be updated on each batch. If all the batches in the training data are executed, the batch index will cycle again through the training data. In machine learning, the term epoch refers to updating the hyperparameters using all the training data, mainly when the batch index cycles around, and the epoch index is incremented. The batch steepest descent has been approved to significantly reduce the computation and memory costs while still achieving good fits to the training data (Raschka, 2015; Aggarwal, 2018). The batch steepest descent has linear convergence, which is slow. Other versions of the steepest descent have been designed to overcome the slow rate of convergence like Adam and RMSprop that are not covered in this chapter, and the reader is encouraged to read them (Aggarwal, 2018; Raschka, 2015).

3.7 GLOBALIZATION VERSUS LOCALIZATION

Once the training process is finished, multiple questions have to be addressed to assure efficient modeling, and one of the most important questions to answer is "Does the model work well on general data outside the examples used in the training process?". If the answer is yes, we say the model is generalized, which means the model learns patterns from the training examples that will identify the answer generally regardless of which example is used. The generalization feature is essential to determine if the model can be applied in real life as a successful machine learning program.

As indicated earlier, the dataset is divided into three subsets, as shown in Figure 3.6, training, testing, and validation. The training examples are the ones used throughout the training process. The validation dataset is optional and usually involved if the training process is computationally costly. The error from the validation subset is computed after a bunch of epochs as a short test to make sure that the training is processing well. Once the training process finishes, the curve of the error per epoch is compared for both the training and validation subsets. Figure 3.8 and Figure 3.9 show a comparison between the two curves of training and validation. In

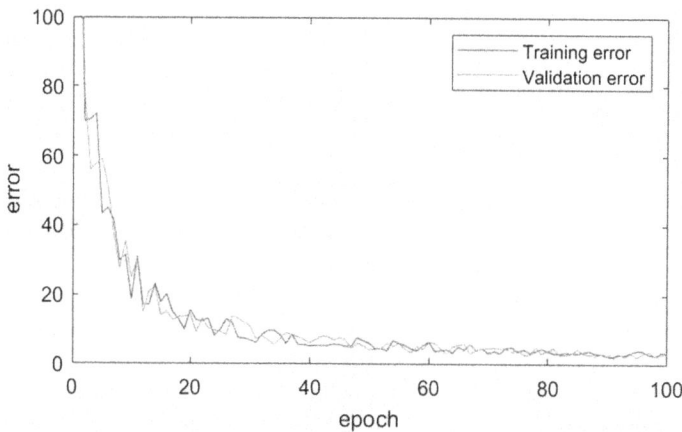

FIGURE 3.8 Generalization case, the validation error curve tracks the reduction in the training error curve.

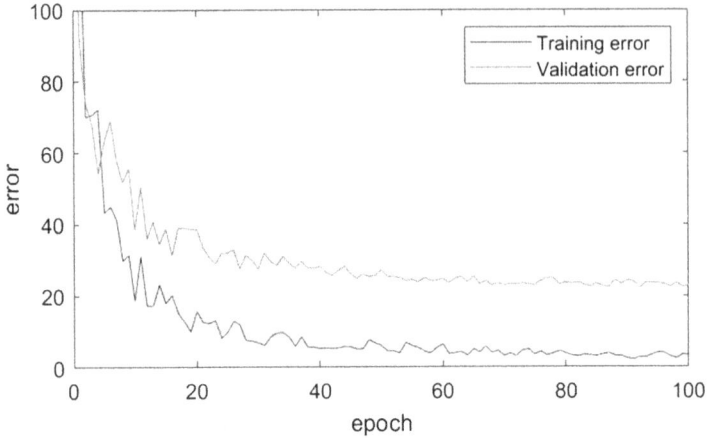

FIGURE 3.9 Memorization case, the validation error curve has a higher error than the training error curve.

Figure 3.8, the validation error follows the track of the training error. This means that the model performance on unseen examples is improving as the training process evolves. This case is known as a generalization, as the model seems to generalize a conclusion about the features and patterns of the required problem. The second case shows that while the training error improves, the validation error saturates at a certain level. This is the case of memorization, which means the model memorizes the answers in the training examples while failing to make a correct conclusion in unseen examples.

In terms of optimization, the problem in (3.6) is nonlinear and consist of different local minima and global minima. Figure 3.10 shows some problems that might be faced while using the steepest descent approach. Usually, optimization problems arising from machine learning consist of multiple local minima. The goal is to achieve globalization, which means approaching the global minimum. This task is not easy as different scenarios might rise, as clarified in the points here:

1) Falling in local minima: once the steepest descent iteration step reaches a local minimum, the gradient on the local minimum valley is zero, and the iterative solver will not be updated according to Equation (3.7).
2) Small gradient: some parts of the training curve might be flat with small slop; if the iterative solver falls into these areas, it will face extremely small updates and will be frozen. See Figure 3.10.
3) Skipping a global minimum: this problem occurs if the global minimum is located in a narrow region, while the learning rate is large enough to escape the whole region.
4) Oscillations: oscillation occurs if the steepest descent falls into a region where minima exist. And the learning rate is not small to approach the minima and not large enough to escape the region which causes oscillation.

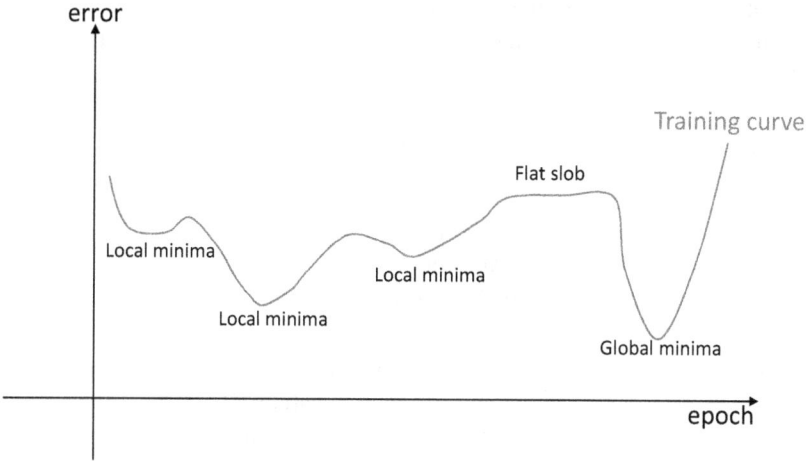

FIGURE 3.10   Local versus global minimization.

Points (1)–(3) cause the training error to converge to a bad level. Regarding point (IV), the oscillation causes instability in the validation and training curves.

Different approaches have been followed to avoid memorization or falling into one of the problems listed previously. The regularization approach explained in the next section has been adopted in multiple research, and it illustrates an effective approach to overcome the localization or memorization problem. Another effective approach for generalization is by inserting dropout layers. The dropout layer affects the training process only, and it works by neglecting some random nodes while updating the steepest descent. Since a random process is involved, a user must specify a probability of whether nodes will be counted or not. This probability is known as the dropping rate.

3.8   REGULARIZATION

Different approaches have been studied to achieve model globalization and avoid memorization or overfitting. These techniques are known as regularization. One common practice in regularization is adding a regularization term, also known as a penalty term, to the minimization problem in (3.6):

$$\min_{\bar{\theta}} \left\| \bar{Y}^{\text{tra}} - F\left(\bar{\bar{X}}^{\text{tra}}, \bar{\theta}\right) \right\|_2^2 + \lambda \mathcal{R}\left(\bar{\theta}\right) \tag{3.11}$$

where λ is the regularization parameter selected by the user to achieve a balance between the fitting data term and the regularization term $\lambda \mathcal{R}\left(\bar{\theta}\right)$. The regularization function has been implemented in different forms, most commonly as a second or first norm $\mathcal{R}\left(\bar{\theta}\right) = \left\| \bar{\theta} \right\|_p$, such that $p = 1, 2$ (Acot, Gabriel, & Hongler, 2018; Wei, Lee, Liu, & Ma, 2019). Another approach is to perform regularization via structured neural networks. In structured neural networks, the neural network structure is designed to benefit from the nature of the application. In vision recognition, the convolutional neural network (CNN) is an example of a structured neural network, and it will be detailed later in this chapter.

3.9 MULTILAYERED NEURAL NETWORK

In Figure 3.11, the general structure of the multilayered neural network is shown. Besides input and output layers, the artificial neural network consists of L hidden layers. Each hidden layer has a different number of perceptrons called nodes. The first layer has A nodes, the second has B nodes, and the last L^{th} layer has Z nodes. The input layer consists of M inputs, while the output layer consists of O outputs.

Notice that the output of each node is written as $y_i^{(k)}$ to indicate the output of the i-th node in k-th layer, and it is computed using equations (1–2). The final output is written as y_i without superscripts to indicate the ith output. The next equations show how to compute the outputs of the first hidden layer in Figure 3.11 using (1):

$$y_a^{(1)} = \sigma\left(w_{a1}^{(1)}x_1 + w_{a2}^{(1)}x_2 + w_{a3}^{(1)}x_3 + \cdots + w_{aM}^{(1)}x_M - b_a^{(1)}\right) \tag{3.12}$$

where $(a = 1, 2.., A)$ is the node index in the first layer. The previous equation can be written in the matrix form as

$$\overline{y}^{(1)} = \sigma\left(\overline{\overline{W}}^{(1)}\overline{x} - \overline{b}^{(1)}\right) \tag{3.13}$$

where the ith row and j^{th} column of the $A \times M$ matrix $\overline{\overline{W}}^{(1)}$ is evaluated as $\{\overline{\overline{W}}^{(1)}\}_{ij} = w_{ij}^{(1)}$, while the ith row in the $A \times 1$ vectors' $\overline{y}^{(1)}$ and $\overline{b}^{(1)}$ are evaluated as $\{\overline{y}^{(1)}\}_i = y_i^{(1)}$ and $\{\overline{b}^{(1)}\}_i = b_i^{(1)}$, respectively. Notice, the activation function is applied element-wise over $A \times 1$ vector that results from $\overline{\overline{W}}^{(1)}\overline{x} - \overline{b}^{(1)}$. Moreover, the activation function doesn't

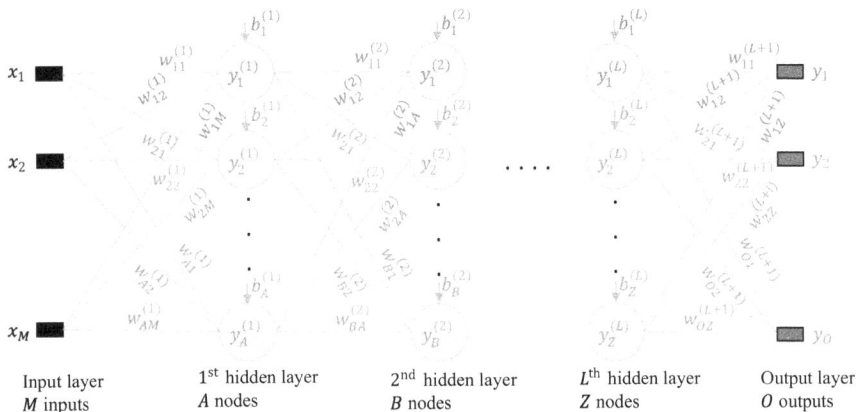

| Input layer
M inputs | 1st hidden layer
A nodes | 2nd hidden layer
B nodes | L^{th} hidden layer
Z nodes | Output layer
O outputs |

FIGURE 3.11 Multilayered artificial neural network.

need to be the same over each node in a layer or from one layer to another. The equation in (3.13) is repeated for the other hidden layers to compute the final output.

$$\overline{y}^{(1)} = \sigma\left(\overline{\overline{W}}^{(1)}\overline{x} - \overline{b}^{(1)}\right) \tag{3.14}$$

$$\overline{y}^{(1)} = \sigma\left(\overline{\overline{W}}^{(1)}\overline{y}^{(l-1)} - \overline{b}^{(2)}\right) \tag{3.15}$$

$$\overline{y} = \overline{\overline{W}}^{(L+1)}\overline{y}^{(L)} \tag{3.16}$$

where (14) computes the output of the first layer, (3.15) computes the outputs for the lth layer, $l = (2,..,L)$, and (3.16) compute the output of the output layer. The previous recursive equations can be written in one line.

$$\overline{y} = f\left(\overline{x},\overline{\theta}\right) = \overline{\overline{W}}^{(L+1)}\sigma\left(\overline{\overline{W}}^{(L)}.....\sigma\left(\overline{\overline{W}}^{(2)}\sigma\left(\overline{\overline{W}}^{(1)}\overline{x} - \overline{b}^{(1)}\right) - \overline{b}^{(2)}\right) - \overline{b}^{(L)}\right) \tag{3.17}$$

The previous model computes one example only, to compute multiple examples in a single shot, the input vector $\overline{x}$ is replaced by $\overline{\overline{X}}$, $\overline{\overline{X}}^{tra}$, $\overline{\overline{X}}^{Val}$, or $\overline{\overline{X}}^{tes}$ to compute $\overline{Y}$, $\overline{Y}^{tra}$, $\overline{Y}^{val}$, or $\overline{Y}^{tes}$, respectively. Equations (11) or (6) are implemented then to train the hyperparameters. It has been shown that multilayered neural networks can extract nonlinear features more than single perceptron networks. In Negnevitsky (A guide to intelligent systems, 2005), it was shown that single perceptron networks could be trained to accurately predict the outputs of all binary digital logic gates except for XOR and XNOR, which requires multilayered neural networks.

It is worth indicating that neural networks are categorized into continuous output (regression) or discrete outputs (classification). Predicting whether it will rain tomorrow or not or determining if an image contains a cat or dog are discrete classification machine learning problems. While predicting land prices or the atmosphere temperature is a continuous regression problem. The minimization problem in (3.6) and (3.11) uses the least square minimization, which is a loss function usually applied for regression problems. On the other hand, for classification problems, there are also multiple loss functions like cross-entropy, binary cross-entropy, and hinge loss (Wang, Ma, Zhao, & Tian, 2022).

3.10 CONVOLUTIONAL NEURAL NETWORKS

Convolutional neural networks (CNNs) are machine learning models that consist of convolution operations. As will be shown in this part, the convolutional layers do not perform the same convolution process studied in signals and systems or digital signal processing. It applies "trained" filters that extract specific features from an image. CNN networks have been dominantly used for image applications like classification, image restoration, captioning, object detection, and segmentation, demonstrating super performance (Albawi, Mohammed, & Al-Zawi, 2017; Sandhu, Shaukat,

Desmal, & Bagci, 2021; Gdeisat et al., 2020). As indicated earlier, the CNNs are considered structured neural networks, meaning the network structure is inspired by the nature of the problem, which forces the regularization of the problem.

Figure 3.12 shows the convolution process that assumes a small image of 7-by-7 pixels convolved with a 3-by-3 filter that results in a 5-by-5 output image. The notations I_{ij}, W_{ij}, and Y_{ij} refers to the ith row and j^{th} column of an input image pixel, filter weight, and output image pixel. In Figure 3.12(a), the highlighted part in the top right of the input image is multiplied by the filter weights elementwise and summed with biasing added. The biasing is then added, which is optional, followed by an activation function to produce the output Y_{11}. The computations of Y_{12} and Y_{13} are similar to Y_{11} except that the filtering window is shifted one and two pixels to the right.

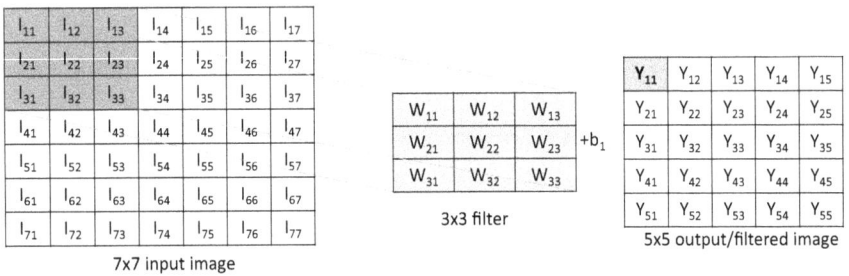

(a) The convolution process of output Y_{11}.

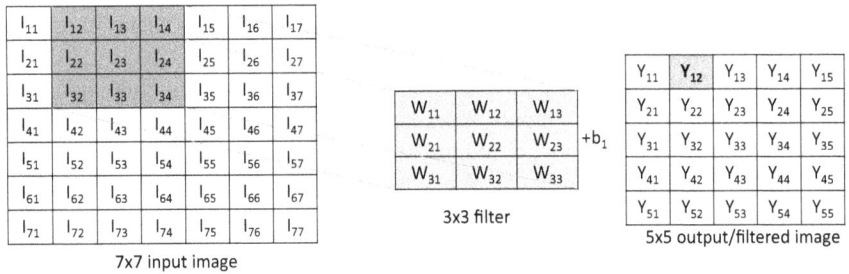

(b) The convolution process of output Y_{12}.

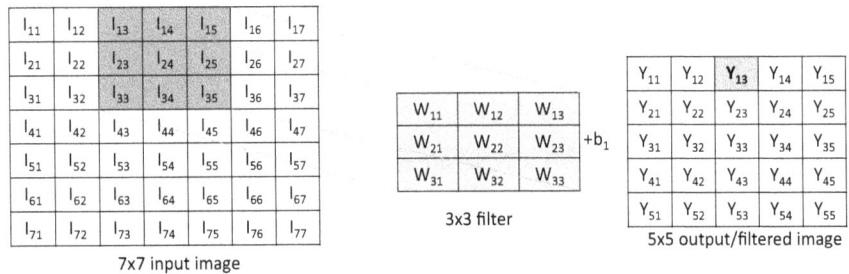

(c) The convolution process of output Y_{13}.

FIGURE 3.12   Convolution process at different output locations.

$$Y_{11} = \sigma \left(\begin{array}{l} I_{11}W_{11} + I_{12}W_{12} + I_{13}W_{13} \\ +I_{21}W_{21} + I_{22}W_{22} + I_{23}W_{23} + b_1 \\ +I_{31}W_{31} + I_{32}W_{32} + I_{33}W_{33} \end{array} \right) \qquad (3.18)$$

$$Y_{12} = \sigma \left(\begin{array}{l} I_{12}W_{11} + I_{13}W_{12} + I_{14}W_{13} \\ +I_{22}W_{21} + I_{23}W_{22} + I_{24}W_{23} + b_1 \\ +I_{32}W_{31} + I_{33}W_{32} + I_{34}W_{33} \end{array} \right) \qquad (3.19)$$

$$Y_{13} = \sigma \left(\begin{array}{l} I_{13}W_{11} + I_{14}W_{12} + I_{15}W_{13} \\ +I_{23}W_{21} + I_{24}W_{22} + I_{25}W_{23} + b_1 \\ +I_{33}W_{31} + I_{34}W_{32} + I_{35}W_{33} \end{array} \right) \qquad (3.20)$$

The scanning continues vertically and horizontally, resulting in a 5-by-5 output image.

The previous scanning assumes a jump of one pixel either vertically or horizontally, which is said to have stride one. If the stride is two or three, the jumping will be for every two or three pixels, respectively. In addition, sometimes, it is preferred to have the training filters centered on the corner or image frame pixels to extract more features about the image. To achieve this, zero padding is inserted, with an additional frame of pixels around the image. The padding is one if one frame depth of zero pixels is inserted, and so on. The following formula calculates the output image size:

$$O = \frac{N^{\text{ima}} - F + 2P}{S} + 1 \qquad (3.21)$$

where O, F, N^{IMA}, P, and S are the output size, filter size, input size, number of zero padding, and number of strides. In applications like classifications, it is usually desired to reduce the output image size. Thus, after performing convolution, image downsampling is applied, which is known as pooling. Figure 3.13 illustrates the two mostly applied downsampling schemes, the max-pooling, and the average pooling. The figure shows two-by-two pooling that divides the image into two-by-two windows and replaces each with either the element's maximum or average's value. Therefore, the resulting image size is the input size divided by two. If three-by-three window pooling is applied, the resulting image size is the input size divided by three, and so on.

So far, the images are assumed to be grayscale or consist of single channels. The filters are three-dimensional for multichannel or colored images, as shown in Figure 3.14. Each time the convolution product is applied, only one value is obtained. Thus, the three-dimensional convolutional scanning produces a single output sheet of one channel, see Figure 3.14. Multiple three-dimensional filters are applied to extract multiple output sheets, as illustrated in Figure 3.15.

In image classification, the output ranges from one to thousands. For example, in handwritten digit classification, ten outputs are considered, one for each digit. More outputs are needed for English letter classification (Cohen, Afshar, Tapson, & Van Schaik, 2017). The output from the convolution process is an image with height, width, and channels. This image is first reshaped into a vector in a process known as

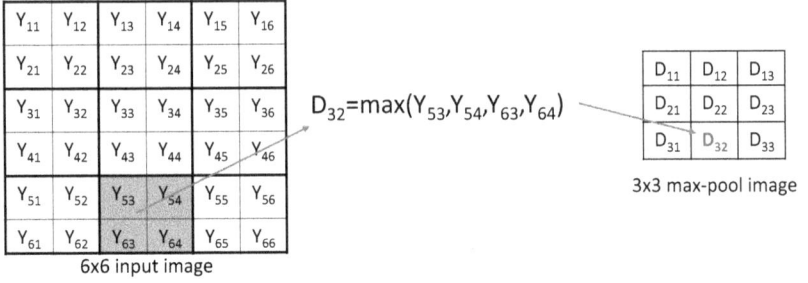

(a) The max-pooling process of output D_{31}.

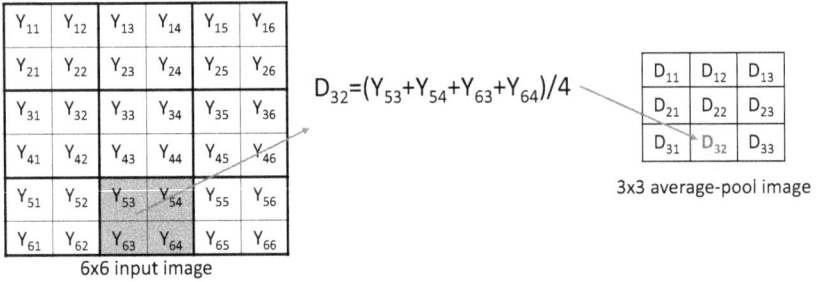

(b) The average-pooling process of output D_{31}.

FIGURE 3.13   An example of computing (a) max pooling and (b) average pooling.

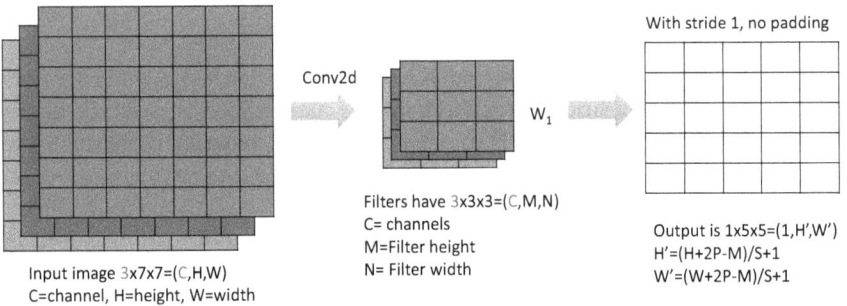

FIGURE 3.14   Multichannel convolution.

flattening. They are then fed into a multilayered neural network with an output size equal to the required number of classifications. For example, in handwritten digit classification, it will become ten. These outputs are called scores. The output with the highest score will refer to a relative class. For an instant, the first output refers to digit zero, the second to digit one, and so on. In the training process, examples are given in the neural network, and the optimization is carried over a loss function, which can be considered an error function to minimize. As mentioned, in regression (continuous) problems, the least square minimization is applied, see equations (3.6) and (3.11). For classification (discrete) problems, cross-entropy has been mostly used (Wang, Ma, Zhao, & Tian, 2022).

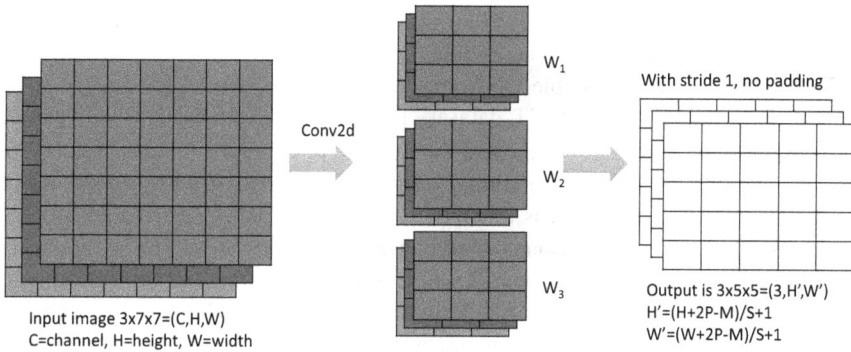

FIGURE 3.15 Multichannel multi-filter convolution.

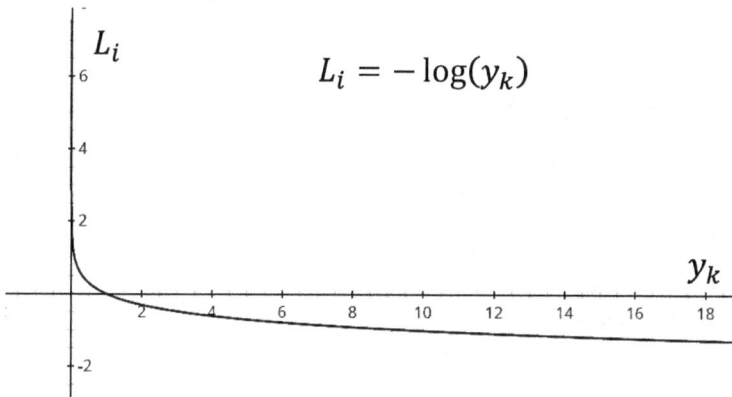

FIGURE 3.16 Plot of single example loss in cross-entropy.

To compute the cross-entropy, assume one training example only (i^{th} example), and let L_i be the measure of loss of that example that is computed as shown here:

$$L_i = -\sum_m \hat{y}_m \log(y_m) = -\hat{y}_k \log(y_k) = -\log(y_k) \tag{3.22}$$

where y_m is the mth output from the multilayered neural network. $\hat{y}_m$ is a binary output that is one if y_m is the highest score and zero otherwise. Assume that the kth element has the highest output, the sum in equation (3.22) will collapse into the kth element only $-\log(y_k)$. Notice that the loss L_i has to be minimized, which means $-\log(y_k)$ has to be minimized as we hit the highest scores in y_k. Figure 3.16 shows the plot of $-\log(y_k)$ as a function of y_k that shows as y_k grows, the loss reduces.

Equation (3.22) calculates the cross-entropy over one example. To compute the cross-entropy over all of the training examples, the average of (3.22) overall examples is taken:

$$L = \frac{1}{N} \sum_{i=1}^{i=N^{tra}} L_i \tag{3.23}$$

where L is the total loss. When L is minimized, L will be minimized, and the correct score over each output will be maximized.

Figure 3.17 shows a possible CNN structure that can be used to classify hand-written digits like in the MNIST dataset. The notation $\otimes$ is used to indicate the convolution process and N^{bch} indicates the number of examples in each batch. The network consists of three convolutional layers of 32 filters with 3-by-3 sizes and a stride of one. Notice the input is assumed to be a grayscale of one channel. Hence the filters in the first convolutional should have one channel to match the input. The second and third convolutional layers have 32 channels, equal to the number of filters in the previous convolutional layer. A Relu activation function follows all three convolutional layers. To downsample the feature maps, the second and third activation functions are followed by max-pooling layers. After the second max-pooling layer, the feature maps are converted into vectors using a flattening layer and then given to a multilayered neural network, also known as a fully connected neural network. In the final stage, cross-entropy is involved as a loss function.

3.11   TRANSFER LEARNING

Transfer learning is an approach that allows fast training (Torrey & Shavlik, 2010; Pan & Yang, 2009). It consumes low memory allocation and low computations while achieving excellent results. It has been illustrated in many researches and books that the few first convolutional layers in any trained CNN learn the basic elements of figures like dots with different positions and edges with different skewness and positions. The latter is concluded in any trained CNN regardless of which trained database is involved. Thus, it is concluded that it is unnecessary to train the whole hyperparameters in CNN if it will be used for different applications. For instant, assume that the CNN in Figure 3.17 was trained to recognize English letters, and we need to train it again to recognize numbers. Since the first CNN layers will always recognize the basic elements of figures like dots and lines no matter what training

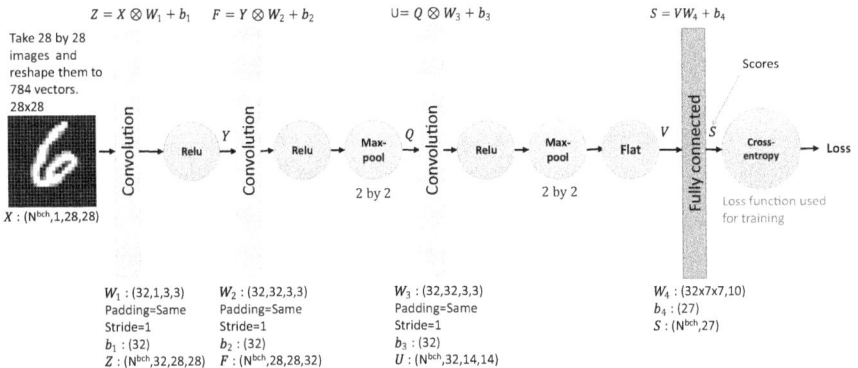

FIGURE 3.17   An example of a convolutional neural network that can be implemented for digit regression.

database is used, one can freeze the first three convolutional layers that were trained to recognize English letters and only change the last fully connected layer to another one with ten output to be trained for digit recognition. This will reduce the number of hyperparameters needed in the training process, which reduces memory and computation consumption.

The transfer learning approach in the previous section freezes the first layers while training the last fully connected layer for the new application. There is another approach in which we allow fine tweaking to update the weights in the first layers instead of completely freezing the initial layers. This approach enables better fitting for the new application but further computations and memory if compared to the freezing transfer learning approach.

One of the most used databases in transfer learning is the ImageNet made by Google, which consists of millions of images. Every year, Google runs a competition based on its database for the best neural network design that produces better accuracy. A breakthrough was achieved by a network known as AlexNet in 2012, showing that neural networks can surpass average human performance (Alom et al., 2018; Lu, Lu, & Zhang, 2019). Later, different neural networks were suggested with less memory and computation requirement while achieving better accuracies than AlexNet, like VGGNet in 2014, ResNet in 2015 (Muhammad, Wang, Chattha, & Ali, 2018; Yu et al., 2016; Targ, Almeida, & Lyman, 2016).

3.12　MASK AND HELMET DETECTION

This section introduces an application for smart monitoring systems based on transfer learning. The system monitors safety regulations in labor working environments by detecting workers' masks and helmets (Alshehhi, Almansoori, Alnaqbi, Aljewari, & Desmal, 2022). Figure 3.18 shows the components used to build the system: (1) a PC with Matlab software, (2) a Wi-Fi router connection, (3) Raspberry Pi, and (4) a five-megapixel camera. The Raspberry Pi is connected to the PC through a Wi-Fi link and controlled by Matlab.

Figure 3.19 shows the flow chart of the system. First, the Matlab program requests an image from the Raspberry Pi through a Wi-Fi link. A face detection algorithm based on KLT (Kanade-Lucas-Tomasi) algorithm is applied to the image that specifies

FIGURE 3.18　Hardware used in helmet and mask detection system (Alshehhi, Almansoori, Alnaqbi, Aljewari, & Desmal, 2022).

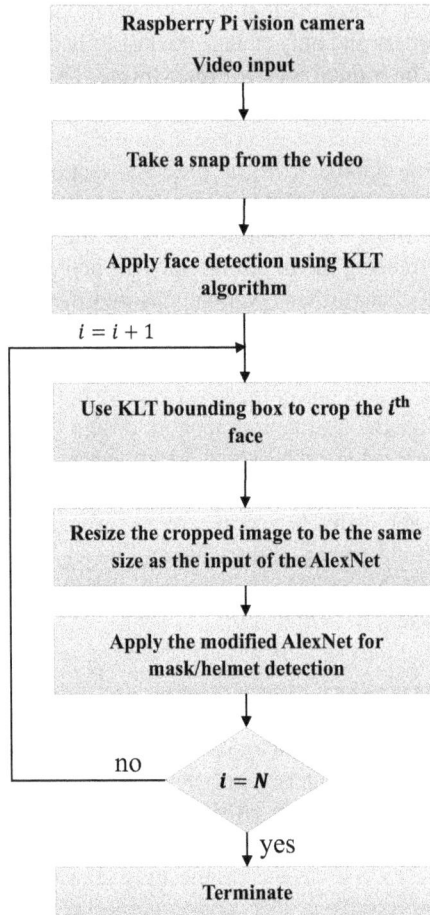

FIGURE 3.19   Flow chart of the mask and helmet detection system. N is the number of faces inside the captured image.

the bounding boxes of all faces in the image. It was noticed that the KLT sometimes crops some part of the face that might be valuable information for the final output decision. Thus, after extracting the bounding boxes from the KLT algorithm, the height and width of the bounding boxes are widened by a small percentage. Let N be the number of faces involved in the captured image. As the chart shows, the algorithm will loop over each face. First, the cropped faces are scaled into images of 277-by-277 pixels, which is the input size required by AlexNet in the next step. The image is then given to two modified AlexNet networks. The first determines if a face mask exists, while the second determines if a helmet exists. The two networks were trained following the transfer learning approach, in which the last fully connected layer was replaced with a new layer of two outputs (to classify with or without), and the training was executed only over the last layer. Since the procedure followed for both tasks is similar, this section details only the mask detection network results. The

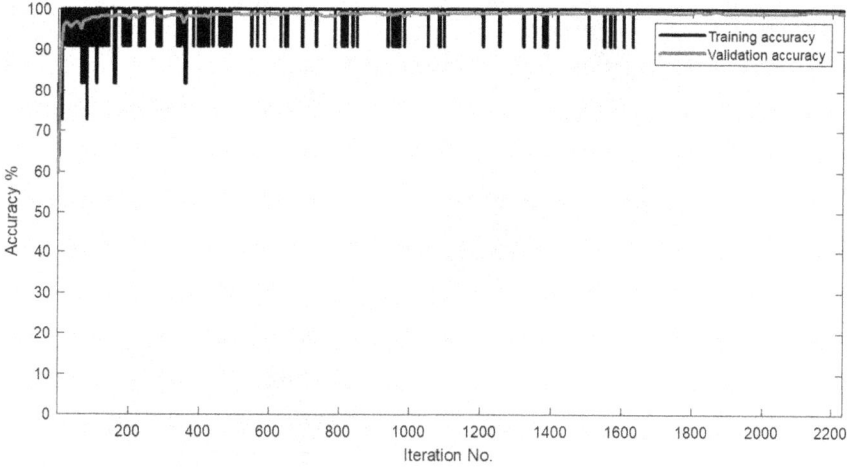

FIGURE 3.20 Training and validation accuracy for AlexNet transfer learning using mask dataset.

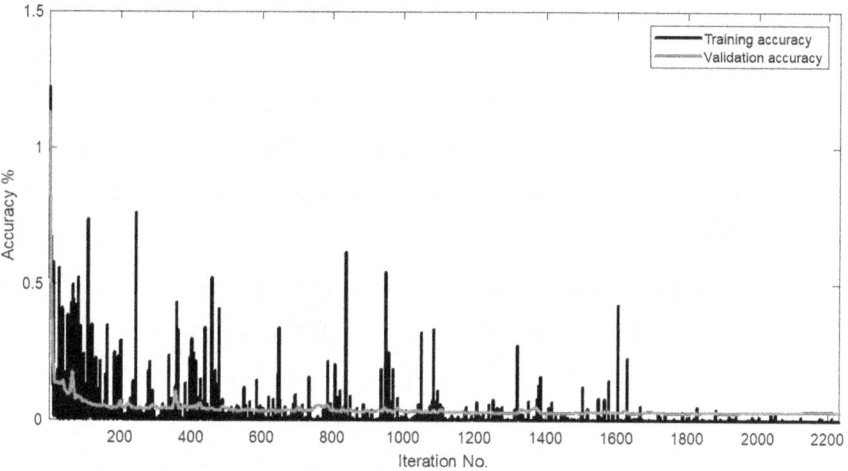

FIGURE 3.21 Training and validation loss for AlexNet transfer learning using mask dataset.

dataset was collected from a GitHub repository (Deb, 2021). The dataset is composed of 4,095 cropped face images, 2,165 images with masks, and 1,930 images without masks, 252 images from the total images with equal on-mask and off-mask number of images. The rest of the images, 20% of them were used for validation, selected randomly, and the rest 80% for training.

Figure 3.20 and Figure 3.21 show the training and validation accuracies and loss function, respectively, of the AlexNet for mask detection, where up to eight epochs were used. The learning rate was 0.0001. At the end of the iteration, the maximum validation accuracy reaches 99.09%.

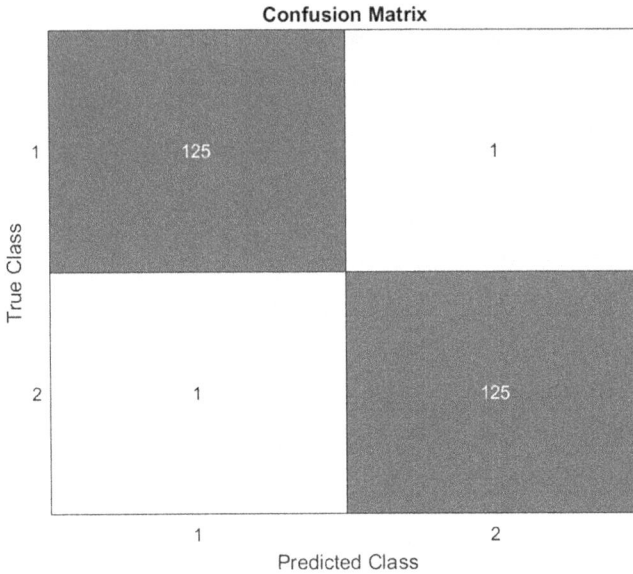

FIGURE 3.22 Confusion matrix for mask detection testing dataset where 252 images are involved.

The confusion matrix of the testing data is shown in Figure 3.22 for mask detection. The confusion matrix shows four cases for two classes, (1) with a mask and (2) without a mask:

- The top-left number shows the True-Positive (TP) images, images predicted by AlexNet to contain masks, and in the actual images, they have masks.
- The below-right number shows the True-Negative (TN), the number of images predicted not to contain masks, and in the actual images, they are without masks.
- The top-right number shows the False-Negative (FN), the number of images predicted not to contain masks, and in the actual images, they are with masks.
- The below-left number shows the False-Positive (FP), the number of images predicted to contain masks, and in the actual images, they are without masks.

The numbers TP and TN show the correctly classified images, while FN and FP show the wrongly classified images. The accuracy is computed as the correctly classified images over all the images.

$$Accuracy = \frac{TP + TN}{TP + TN + FP + FN} \times 100\% \tag{3.24}$$

The error is the total of wrongly classified images over the total number of images.

TABLE 3.2

Evaluation Metrics for Mask Detection

Evaluation metric	Values for mask detection
Accuracy	99.21%
Error	0.79%
Precision	99.21%
Recall	99.21%

$$Error = \frac{FP + FN}{TP + TN + FP + FN} \times 100\% \tag{3.25}$$

Two other criteria used are precision and recall. These two criteria are helpful when the dataset is unbalanced, which means that the number of images wearing masks is considerably different from the number of images without masks. The precision is defined as the true-positive images over the total number of images predicted as positive.

$$Precision = \frac{TP}{TP + FP} \times 100\% \tag{3.26}$$

The recall is defined as the true-positive images over the total number of images that were in the actual images positive.

$$Recall = \frac{TP}{TP + FN} \times 100\% \tag{3.27}$$

The previous evaluation metrics are calculated for mask detection and shown in Table 3.2. As shown, the accuracy, precision, and recall are almost the same. This is because the used dataset is balanced.

The results in this section show a high performance of the transfer learning approach followed. Nevertheless, in machine learning, the problem of detecting masks from cropped faces is not considered challenging if compared to other problems, like in the ImageNet competition, where the task is to detect more than 20,000 classes (Google, n.d.). As the number of classes grows, the nonlinearity of the optimization problem is expected to worsen, and the performance metrics will start to degrade. In the latest years, substantial research has been carried out to advance machine learning models. The transformer models are models that were developed initially for language processing and have proven there to achieve unrealizable performance if compared to traditional algorithms (Vaswani et al., 2017). Recently, these models have been applied in image processing and show better performance than CNN-based algorithms under huge datasets (Lin et al., 2022).

Another limitation factor that shall be considered here is the fact that this approach depends on KLT face detection. The performance of the KLT affects the final output.

The KLT crops the faces only which makes the output easier for classification, but on the other hand, failing to detect the faces will cause no output to generate. This can be avoided by applying the ML algorithms directly to the entire image without applying face detection and cropping. Although the latter approach solves the problem, it increases the complexity of the input images and makes the classification problem more nonlinear, which will worsen the accuracy.

3.13 CONCLUSION

Overall, smart imaging monitoring systems have gained substantially increasing interest over the last two decades. Integrating these systems with machine learning has noticeably improved their efficiency. This chapter introduces transfer learning, which is an innovative approach that provides engineers and researchers to use advanced machine learning under limited resources in memory and processing. In addition, the chapter details the training process of convolutional and fully connected neural networks that are considered the backbones of recent machine learning systems. At the end of the chapter, an application for intelligent monitoring systems is introduced. The system's task is to detect irregularities in labor working environments. The system was built based on transfer learning applied to the well-known AlexNet. Two modified AlexNet were designed, one to detect wearing masks while the other to detect wearing helmets. The numerical results and performance metric measures show achieving high accuracies for the suggested tasks. The tradeoff and limitations of applying the face detection algorithm in the preprocessing stage that considerably enhances the detection performance is discussed. Further, future works that could advance this field were introduced at the end of the chapter.

REFERENCES

Acot, A., Gabriel, F., & Hongler, C. (2018). Neural tangent kernel: Convergence and generalization in neural networks. *Advances in Neural Information Processing Systems, 31.*

Aggarwal, C. C. (2018). *Neural networks and deep learning.* Springer.

Akram, M. W., Li, G., Jin, Y., Chen, X., Zhu, C., & Ahmad, A. (2020). Automatic detection of photovoltaic module defects in infrared images with isolated and develop-model transfer deep learning. *Solar Energy,* 175–186.

Al-Antari, M. A., Han, S.-M., & Kim, T.-S. (2020). Evaluation of deep learning detection and classification towards computer-aided diagnosis of breast lesions in digital X-ray mammograms. *Computer Methods and Programs in Biomedicine,* 105584.

Albawi, S., Mohammed, T. A., & Al-Zawi, S. (2017). Understanding of a convolutional neural network. *2017 International conference on engineering and technology (ICET).*

Ali, M. U., Khan, H. F., Masud, M., Kallu, K. D., & Zafar, A. (2020). A machine learning framework to identify the hotspot in photovoltaic module using infrared thermography. *Solar Energy,* 643–651.

Alkhaleefah, M., Tan, T.-H., Chang, C.-H., Wang, T.-C., Ma, S.-C., Chang, L., & Chang, Y.-L. (2022). Connected-SegNets: A deep learning model for breast tumor segmentation from X-ray images. *Cancers,* 4030.

Almalki, Y. E., Din, A. I., Ramzan, M., Irfan, M., Aamir, K. M., Almalki, A., . . . Rahman, S. (2022). Deep learning models for classification of dental diseases using orthopantomography X-ray OPG images. *Sensors,* 7370.

Alom, M. Z., Taha, T. M., Yakopcic, C., Westberg, S., Sidike, P., Nasrin, M. S., Esesn, B. C., Awwal, A. A., & Asari, V. K. (2018). *The history began from alexnet: A comprehensive survey on deep learning approaches.*

Alshehhi, K. A., Almansoori, M. Y., Alnaqbi, M. K., Aljewari, Y. H., & Desmal, A. (2022). Mask and helmet detection using transfer learning. *2022 Advances in science and engineering technology international conferences (ASET).* Dubai.

Ambrosanio, M., Franceschini, S., Pascazio, V., & Baselice, F. (2022). An end-to-end deep learning approach for quantitative microwave breast imaging in real-time applications. *Bioengineering*, 651.

Asaei, S. H., Kasaei, S. M., & Kasaei, S. A. (2010). New morphology-based method for robustiranian car plate detection and recognition. *International Journal of Computer Theory and Engineering*, 2(2), 264.

Behera, S. K., Rath, A. K., & Sethy, P. K. (2021). Maturity status classification of papaya fruits based on machine learning and transfer learning approach. *Information Processing in Agriculture*, 8(2), 244–250.

Bertalan, L., Holb, I., Pataki, A., Szabo, G., Szaloki, A. K., & Szabo, S. (2022). UAV-based multispectral and thermal cameras to predict soil water content—A machine learning approach. *Computers and Electronics in Agriculture*, 107262.

Bottou, L. (2012). Stochastic gradient descent tricks. In *Neural networks: Tricks of the trade.* Springer.

Chidurala, V., & Li, X. (2021). Occupancy estimation using thermal imaging sensors and machine learning algorithms. *IEEE Sensors Journal*, 8627–8638.

Cohen, G., Afshar, S., Tapson, J., & Van Schaik, A. (2017). EMNIST: Extending MNIST to handwritten letters. *2017 International joint conference on neural networks (IJCNN).*

Cuocolo, R., Cipullo, M. B., Stanzione, A., Ugga, L., Romeo, V., Radice, L., . . . Imbriaco, M. (2019). Machine learning applications in prostate cancer magnetic resonance imaging. *European Radiology Experimental*, 1–8.

Deb, C. (2021). *Chandrikadeb7.* Retrieved from GitHub: https://github.com/chandrikadeb7/Face-Mask-Detection.

Desmal, A. (2020). Deep learning enhanced electromagnetic imaging scheme. *2020 International conference on innovation and intelligence for informatics, computing and technologies (3ICT).*

Desmal, A. (2021). High-quality self-contained electromagnetic imaging scheme based on projected nonlinear landweber and machine learning. *IEEE Transactions on Antennas and Propagation*, 70(2), 1380–1388.

Desmal, A. (2022a). A contrast-source electromagnetic imaging scheme based on projected nonlinear Landweber-Kaczmarz. *IEEE Transactions on Antennas and Propagation*, 8666–8670.

Desmal, A. (2022b). A trained iterative shrinkage approach based on born iterative method for electromagnetic imaging. *IEEE Transactions on Microwave Theory and Techniques*, 70(11), 4991–4999.

Desmal, A., Schubert, J. R., Denker, J., Kisner, S. J., Rezaee, H., Couture, A., . . . Tracey, B. H. (2019). Limited-view X-Ray tomography combining attenuation and compton scatter data: Approach and experimental results. *IEEE Access*, 7(IEEE), 165734–165747.

Gdeisat, M., Desmal, A., Moumouni, Y., Al-Aubaidy, Z. S., Al Khodary, A., Hindash, A., & Wavegedara, C. (2020). Kernel symmetry for convolution neural networks. *2020 Advances in science and engineering technology international conferences (ASET).* Dubai.

Google. (n.d.). *ImageNet.* Retrieved from www.image-net.org/.

Jain, D. K., & others. (2019). An evaluation of deep learning based object detection strategies for threat object detection in baggage security imagery. *Pattern Recognition Letters*, 112–119.

Khoshdel, V., Asefi, M., Ashraf, A., & LoVetri, J. (2020). Full 3D microwave breast imaging using a deep-learning technique. *Journal of Imaging*, 80.

Li, H., Lee, C. H., Chia, D., Lin, Z., Huang, W., & Tan, C. H. (2022). Machine learning in prostate MRI for prostate cancer: Current status and future opportunities. *Diagnostics*, 289.

Lin, A., Chen, B., Xu, J., Zhang, Z., Lu, G., & Zhang, D. (2022). Ds-transunet: Dual swin transformer u-net for medical image segmentation. *IEEE Transactions on Instrumentation and Measurement*, 1–15.

Lu, S., Lu, Z., & Zhang, Y.-D. (2019). Pathological brain detection based on AlexNet and transfer learning. *Journal of Computational Science*, *30*, 41–47.

Mishra, A., Mohan, A., Mandal, A., Mohanty, A., & Chowdhury, A. (2022). Utomation in retail 'follow-me-auto shopping cart': A self-propelled computer vision-based shopper following cart with auto-billing feature using IIoT. In *Recent advances in manufacturing, automation, design and energy technologies* (pp. 475–483). Springer.

Moradi, E., Pepe, A., Gaser, C., Huttunen, H., Tohka, J., Initiative, A. D., & others. (2015). Machine learning framework for early MRI-based Alzheimer's conversion prediction in MCI subjects. *Neuroimage*, 398–412.

Muhammad, U., Wang, W., Chattha, S. P., & Ali, S. (2018). Pre-trained VGGNet architecture for remote-sensing image scene classification. *2018 24th International conference on pattern recognition (ICPR)*.

Negnevitsky, M. (2005). *A guide to intelligent systems*. Pearson Education.

Nguyen, D. T., Hong, H. G., Kim, K. W., & Park, K. R. (2017). Person recognition system based on a combination of body images from visible light and thermal cameras. *Sensors*, *17*.

Nunez-Nieto, X., Solla, M., Gomez-Perez, P., & Lorenzo, H. (2014). GPR signal characterization for automated landmine and UXO detection based on machine learning techniques. *Remote Sensing*, 9729–9748.

Pan, S. J., & Yang, Q. (2009). A survey on transfer learning. *IEEE Transactions on Knowledge and Data Engineering*, *22*(10), 1345–1359.

Raschka, S. (2015). *Python machine learning*. Packt Publishing Ltd.

Sandhu, A. I., Desmal, A., & Bagci, H. (2021). An accelerated nonlinear contrast source inversion scheme for sparse electromagnetic imaging. *IEEE Access*, *9*, 54811–54819.

Sandhu, A. I., Shaukat, S. A., Desmal, A., & Bagci, H. (2021). ANN-assisted CoSaMP algorithm for linear electromagnetic imaging of spatially sparse domains. *IEEE Transactions on Antennas and Propagation*, *69*(9), 6093–6098.

Sharma, G., & Jindal, N. (2022). Breast tumour detection using machine learning: Review of selected methods from 2015 to 2021. *Multimedia Tools and Applications*, 32161–32189.

Shirsat, S., Naik, A., Tamse, D., Yadav, J., Shetgaonkar, P., & Aswale, S. (2019). Proposed system for criminal detection and recognition on CCTV data using cloud and machine learning. *2019 International conference on vision towards emerging trends in communication and networking (ViTECoN)*. Vellore, India.

Singh, U., Determe, J.-F., Horlin, F., & De Doncker, P. (2021). Crowd monitoring: State-of-the-art and future directions. *IETE Technical Review*, *38*(6), 578–594.

Sun, Y., Denel, B., Daril, N., Evano, L., Williamson, P., & Araya-Polo, M. (2020). Deep learning joint inversion of seismic and electromagnetic data for salt reconstruction. *SEG Technical Program Expanded Abstracts 2020*, 550–554.

Targ, S., Almeida, D., & Lyman, K. (2016). Resnet in resnet: Generalizing residual architectures. *arXiv preprint arXiv:1603.08029*.

Torrey, L., & Shavlik, J. (2010). *Handbook of research on machine learning applications and trends: Algorithms, methods, and techniques*. IGI Global.

Vaswani, A., Shazeer, N., Parmar, N. A., Jones, L., Gomez, A. N., Kaiser, L., & Polosukhin, I. (2017). Attention is all you need. In *Advances in neural information processing systems*. MIT Press.

Visualization, R. t.-r. (2020). Rahman, Tawsifur; Khandakar, Amith; Kadir, Muhammad Abdul; Islam, Khandaker Rejaul; Islam, Khandakar F; Mazhar, Rashid; Hamid, Tahir; Islam, Mohammad Tariqul; Kashem, Saad; Mahbub, Zaid Bin; others. *IEEE Access*, 191586–191601.

Wang, J., Ji, J., Ravikumar, A. P., Savarese, S., & Brandt, A. R. (2022). VideoGasNet: Deep learning for natural gas methane leak classification using an infrared camera. *Energy*, 238.

Wang, Q., Ismail, K. N., & Breckon, T. P. (2020). An approach for adaptive automatic threat recognition within 3D computed tomography images for baggage security screening. *Journal of X-ray Science and Technology*, 35–58.

Wang, Q., Ma, Y., Zhao, K., & Tian, Y. (2022). A comprehensive survey of loss functions in machine learning. *Annals of Data Science*, 9(2), 187–212.

Wei, C., Lee, J. D., Liu, Q., & Ma, T. (2019). Regularization matters: Generalization and optimization of neural nets vs their induced kernel. *Advances in Neural Information Processing Systems*, 32.

Yu, W., Yang, K., Bai, Y., Xiao, T., Yao, H., & Rui, Y. (2016). Visualizing and comparing AlexNet and VGG using deconvolutional layers. *Proceedings of the 33rd international conference on machine learning*.

Zacharaki, E. I., Wang, S., Chawla, S., Soo Yoo, D., Wolf, R., Melhem, E. R., & Davatzikos, C. (2009). Classification of brain tumor type and grade using MRI texture and shape in a machine learning scheme. *Magnetic Resonance in Medicine: An Official Journal of the International Society for Magnetic Resonance in Medicine*, 1609–1618.

4 IoT-Based Seaweed Cultivation for a Sustainable Economy

Jeevanantham Lenin and Charles Savarimuthu

4.1 INTRODUCTION

Both the government and private sectors have recently begun paying more attention to the global aquaculture industry. Aquaculture has been highlighted as a major pillar of global economic policy in an effort to better diversify the economy beyond the available potential resources. In many nations, it was advised to involve small businesses and local women in the cultivation and use of seaweed. In addition, it was proposed to grow seaweed and fish together. Algae will use fish waste to increase their growth and clean the environment in such farms.

The world's second-largest freshwater farming sector is thought to be that of seaweed. Seaweeds are consumed both uncooked and through the industrial extraction of some of their constituent parts for use in other dishes. Carrageenan is an extraction of around 25% of dried Eucheuma, also referred to as seaweed flour. Seaweed is used in its semi-refined or refined form to produce food and drink products for both human and animal consumption, in addition to being used in the mass customization of ice cream, chocolates, custards, toppings, and fillings for cakes, milkshakes, yoghurts, dessert gel, canned foods, fish gel, sauces, and many other products. Manufacturing water paints, toothpastes, lotions, shampoos, and a number of pharmaceutical goods all include some refined extraction. Within six to eight weeks, they had become ten times heavier. Because seaweed farming doesn't require fertilizers or clean water, it has no influence on the environment.

Farming seaweed is a new job opportunity. It will provide us with the opportunity to raise our standard of living. Seaweed is easy to grow, sustainable, and nutritious. Seaweed farming is not harmful to the environment. The use of IoT to cultivate seaweed at smaller scales in smarter ways. The water level, pH value, light sensor value, and temperature which suit the planted crop can be easily monitored with the help of sensors, and a course of action can be taken accordingly. The key objectives of the research effort are as follows: (1) improve living conditions in coastal communities, (2) encourage the sustainable use of marine and coastal resources, (3) introduction of Internet of Things–based seaweed production as a green type of revenue, particularly for women and young company owners.

DOI: 10.1201/9781003343332-4

4.2 MATERIALS AND METHODS

The second-largest freshwater farming sector worldwide is seaweed farming. In addition to being consumed raw, seaweed also has parts that can be harvested industrially and added to other dishes. The four species under consideration were primarily present throughout the year, despite varying seaweed population records in the study area [Kaviarasan 2018; Nakhate 2021]. Seaweed farming is a significant and expanding industry. Although aquaculture has been practiced for over a thousand years [Nitaigour Premchand Mahalik 2014; Robert,Huu and Chien 2022]. Oceans are home to a wide variety of edible seaweeds that are acceptable for human eating [Mahadevan 2022]. Seaweed aquaculture accounts for 51.3% of the world's mariculture and is growing at a 6.2% annual rate from 2000 to 2018. It offers various organizations a stock of food and regular things notwithstanding an extensive variety of biological system administrations. Additionally, it offers an adaptable, all-natural solution to the problems of eutrophication, the loss of biodiversity, and coping with and adapting to climate change [Torres, Kraab and Dominguez 2020; Simona, Kim and Charles2021]. Products made from seaweed should do well on the market because people are becoming healthier by using seaweed. It has been determined that the use of seaweed in functional foods has the potential to improve future international security [Rajaa,Kadirvel and Subramaniyana 2022]. There are historical records of the cultivation and harvesting of seaweed, but as knowledge of the crucial compounds in these plants has increased, so has aquaculture. The farming practices used onshore and offshore are different. Additionally, seaweed may be employed in IMTA*, a farming method that faces both considerable obstacles and bright futures. [Sarkar, Rekha and Biswas 2022]. An IoT-based technique for monitoring the environment to increase the yield of seaweed farming [Muthumanickam, Poongodi and Kumaraperumal 2022]. There are numerous remote mon-itoring options for studying seaweed growth, but none that track multiple biotic and abiotic parameters. While reducing the system's cost and deployment time, a sensor system would provide pertinent data on the dynamic forces affecting the plants andharvest biomass [Jeroen Gerlo, Kooijman and Wieling 2022; Nakhate 2021].

Computing systems with tightly integrated hardware and software integration that are created to carry out a specific function are known as embedded systems. Embedded systems can occasionally perform as independent systems. One category of embedded CPUs prioritizes size, power use, and cost. Because of this, certain embedded processors are functionally constrained; that is, a processor may be competent for the class of applications for which it was built but may not be for other application classes.

A plan for an inserted framework frequently utilizes a small microcontroller, for example, the PIC microcontroller (MCU) or dsPIC computerized signal regulator (DSC) from CPU. These microcontrollers make a smaller control module that requires not many other outside gadgets by consolidating a microchip unit (like the central processor in a PC) with a couple of different circuits known as peripherals and extra circuits on a similar chip. For low-cost digital control, this one component can then be incorporated into various mechanical and electronic equipment [Barnett 2003].

4.2.1   EMBEDDED CONTROLLER VS. PERSONAL COMPUTER

A personal computer and an embedded controller differ primarily in that the latter is focused on a single task or collection of tasks, while the former is not. PC is made to run a variety of applications and link to a wide range of external devices. Due to the fact that an embedded controller only runs one program, it is possible to create them affordably and just provide them with the hardware and processing resources needed to do their specific duty. A personal computer's central processing unit (CPU), which is quite expensive, is surrounded by a variety of external components, including memory, disc drives, video controllers, network interface circuits, etc. A low-cost microcontroller unit (MCU) powers the intelligence of an embedded system, which also contains a limited number of external devices and many peripheral circuits on a single chip. A low-cost microcontroller unit (MCU) is used as the brains of an embedded system, together with a large number of peripheral circuits that are all located on a single chip. A cordless drill, refrigerator, garage door opener, or an embedded system can be an invisible sub-module or component in any object. These products' controllers perform a very small percentage of the device's overall functions. Some of the crucial components of these devices' subsystems receive low-cost intelligence from the controller. A smoke detector is one type of embedded system. When signals from a sensor indicate the presence of smoke, it analyzes the signals and sounds an alarm. The smoke detector contains a small program that either sleeps in a low power "Sleep" mode until it receives a signal from the smoke sensor or runs indefinitely while sampling the signal from the smoke sensor. Next, the warning is sounded by the program. User test and charge low alert could probably be included in the program's list of additional features. In these products, the controller only manages a portion of the total product's functions. Embedded controller provides intelligence at a low cost to a few of the most important subsystems. Its responsibility is to examine sensor signal data and sound an alarm if smoke is found.

4.2.2   CHARACTERISTICS

1) Unlike the general-purpose computer for many purposes, embedded systems are designed to perform one task at a time. Only a basic level of performance which would simplify the hardware and reduce costs could be requested by others. For elements such as safety and usability, some of them also have real-time performance requirements.

2) In embedded systems, separate devices are not always used. They're often embedded in the machinery that they operate.

3) The firmware developed for embedded systems—often referred to as software—is stored in read-only memory or Flash memory chips as opposed to being retained on a disc drive. It frequently operates with scan computer gear, including a tiny keyboard, screen, and little memory.

4.2.3   EMBEDDED REAL-TIME OPERATING SYSTEM

Figure 4.1 shows the development of real-time embedded systems for a range of uses, including data processing and data control, a common practice. Meeting deadlines at

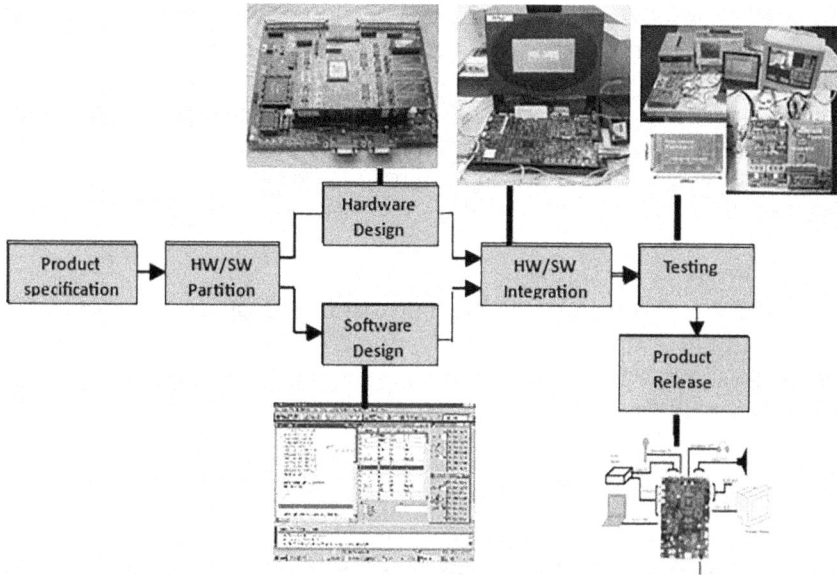

FIGURE 4.1 Embedded system development life cycle.

the proper time is one of the traits of a real-time system. RTOS is frequently employed to accomplish this goal and has a number of APIs that allow users to create apps. The use of an RTOS does not guarantee that the system will be consistently able to meet its deadlines because it is dependent on the overall scheme's design and organization. Although the most commonly used RTOS for microcontroller embedded systems are high-end microprocessors or CPUs with a 32-bit central processing unit, there is an increasing trend towards incorporating similar functions in the mid-range of 16 bit and 8-bit processor systems. A piece of software called an operating system (OS) manages how resources are distributed across computers. RTOS is frequently distinguished from ordinary OS because it was built expressly for scheduling to achieve real-time responses. There are several RTOS varieties available today, ranging from open source to proprietary, commercial, and other sorts. It is a common misconception that small embedded systems using microcontrollers with average ROM sizes of 128 KB and RAM sizes of 4 KB do not require an RTOS.

4.2.3.1 RTOS

Code optimization in software development: The key challenge, owing to time constraints for the market and a shortened development cycle, is also increasing software efficiency when developing systems using small microcontrollers. An RTOS can be used to manage code and create jobs for developers as the complexity of codes increases.

4.2.3.2 Synchronization

Global variables are frequently used for the synchronization between modules and functions when creating small embedded systems without using a real-time operating system. However, employing global variables results in defects and issues with software safety, particularly in systems that are heavily interrupt-driven. The likelihood of these global variables becoming corrupted at any point during the execution of the program is very high because they are frequently shared and accessed by the functions. The presence of an RTOS will ensure that synchronization is safe and tasks are able to communicate messages or synchronize with one another without any sort of problems.

4.2.3.3   Resource Management

In most RTOSs, developers' APIs are offered to manage system resources. These techniques cover communication, synchronization, tasking, and management of memory pools, time controls, and interruptions. These features allow programmers the freedom to organize their software as they choose, enabling them to develop clean code.

4.2.3.4   RTOS and Time Management

Software developers can implement task deferral, timer handling, or time-triggered processing using time management features without having to be aware of the underlying hardware techniques. In contrast to a tiny system without an RTOS, it might be challenging to implement timing-related features since the software developer must be familiar with the underlying peripherals (such as timers), know how to use them, and know how to connect them to the top-level application code.

4.3   WORKING PRINCIPLE

Current aquaculture seaweed cultivation does not use any technology other than the post-processing regions. Our goal is to introduce sensor-based control in seaweed farming to maximize the production.

4.3.1   System Design

In order to achieve the stated objectives, the following methodology is proposed:

A control system that comprises sensors, motors, pumps, electronics, computers, and software should be prepared to monitor the farming area, which is shown in Figure 4.1.

To grow seaweed on a small scale, the following are needed:

- At least one larger tank.
- Access to seawater or freshwater: A water circulation pump can be used to circulate the water. The pH level can be monitored by a salinity sensor, and the level of the water can be monitored by a level sensor.
- Appropriate lighting, PAR (photosynthetically active radiation) lamps like LEDs because they are durable, do not emit heat and use little energy. This can be monitored with light sensors.
- Aeration: This can be a simple air pump connected to a tube with holes at one side (bottom) of the tank. This also creates currents.

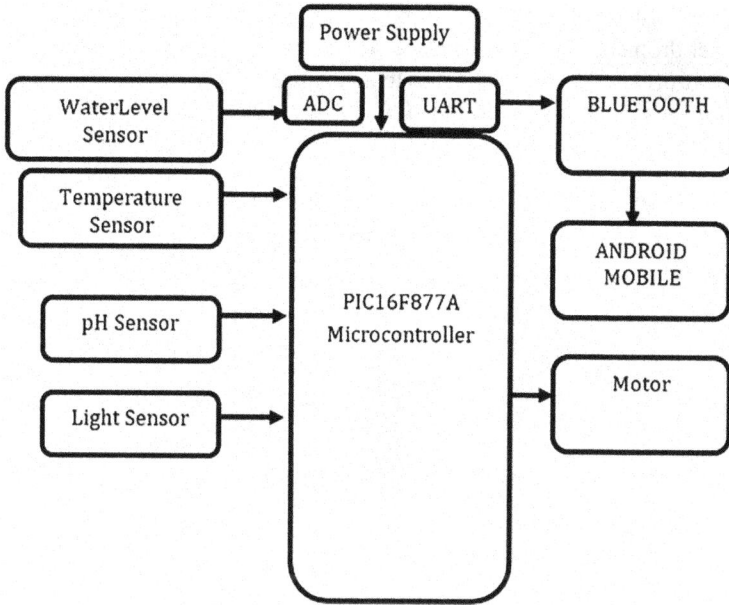

FIGURE 4.2 Architecture (4.2).

- Temperature stability (at least not too much fluctuation): (A few degrees is fine.) Look at the natural temperature optimum for the best choice; otherwise, a mini water heater has to be used. This can be monitored by a water temperature sensor.
- Seaweed species: The recorded sensor data will be sent in a user-understandable form to the laptop or mobile app to notify the farmer for further action on the farm.

4.3.1.1 Power Supply

No matter how the voltage changes, the output of a power supply remains constant. When the load is less than about 7 amperes, to provide fixed voltages of plus or minus 3 V, plus or minus 5 V, 9 V, 12 V, or 15 V, three terminal linear regulators are frequently available. The "79xx" series of 7080, 7812, etc. regulate a positive voltage whereas the "77xx" series of 7905, 7912, etc. regulates a negatively charged voltage. The output voltage is frequently indicated by the device's final two digits; for instance, a regulator with an output voltage of +5 V would have a 7805 code, while one with a -15 V code would have a 7915 code. Depending on the model, the 78x series ICs can supply up to 1.5 amps which is shown in Figure 4.2.

4.3.1.2 Water Level Sensor

- The water sensor includes a high-cost, high-performance water identification detection sensor and is straightforward and simple to use.
- The sensor type is simulation, the working voltage is DC 3–5V, and the working current is 20 mA.

- A series of parallel wire line marks are employed to measure the size of the water droplets and reveal the water level.
- The process of converting water to an analogue signal and achieving water level sensing is simple, and the PIC controller board can read the output of the simulation data directly which is shown in Figure 4.2.

4.3.1.3 pH Sensor

The pH sensor depicted in Figure 4.2 has the following features

- pH range: 2 to 12 (or, for brief durations, 0 to 14)
- The temperature ranges from 14 to 135°F (−10 to 135)
- 58 psig (600 Pa abs, 4 bar) is the highest pressure possible.
- Wetted materials include EPDM and glass.

4.3.1.4 LDR

The LDR shown in Figure 4.2 has the following features

- 3.3V–5V operating voltage
- Analogue voltage output type -A0
- Digital switching output (D0, 1)

4.3.1.5 Temperature Sensor

The Temperature sensor shown in Figure 4.2 has the following features

- Directly calibrated in Celsius
- 0.5°C ensured accuracy (at 25°C)
- Linear + 10-mV/°C Scale Factor
- Rated for the entire 55°C to 150°C temperature range
- Suitable for remote applications
- Low-cost due to wafer-level trimming
- Operates from 4 V to 30 V
- Less than 60-A current drain
- Low self-heating, 0.08°C in still air
- Low-impedance output for a 1-mA load

4.4 RESULTS AND DISCUSSION

4.4.1 INTRODUCTION ABOUT MPLAB X IDE

A software development environment for Microchip digital signal controllers and micro-controllers is called MPLAB® X IDE. In view of offering a single integrated development environment that allows the creation of code for embedded microcontrollers.

4.4.2 IMPLEMENTATION OF AN EMBEDDED SYSTEM DESIGN

The brains of embedded systems applications—the code—are written, edited, debugged, and programmed into a microcontroller using a suite of computer programs referred to as an embedded controller development system. One such solution is the MPLAB X IDE, which provides all the components necessary for the development and delivery of embedded system applications.

4.4.3 CREATE THE HIGH-LEVEL DESIGN

Select the PIC MCU or dsPIC DSC product which is best suited for your application according to features and efficiency requirements, develop relevant hardware circuitry. After determining which peripheral and pin is controlling hardware, write the firmware that will be used to manage hardware parts of an embedded application.

Compilers and assemblers help to simplify the code by providing a list of functions for identifying codes routines with variables that have names which correspond to their intended use as well as features, such as those enabling them to be stored in a stable structure.

Use the assembler, compiler, and/or linker to compile, assemble, and link the program to turn the code into machine code for the PIC MCUs. The firmware (the code loaded into the microcontroller) will eventually be created from this machine code.

4.4.4 TESTING THE CODE

Generally speaking, a complex program does not operate as expected, and a "flaw" in the design must be addressed to obtain desired results. A "1 and 0" correspond to the program you wrote. Execute using the debugger, along with codes and procedure names from the program. Burn code into a microcontroller and make sure that the final program runs correctly.

4.4.5 THE DEVELOPMENT CYCLE

Figure 4.3 depicts the process of generating applications as a development cycle, because it is uncommon for all steps from concept to implementation to be completed perfectly in the first place. In order to make a correctly functioning program, the code is frequently developed, tested, and updated. An embedded system design engineer can progress through this cycle without being distracted by moving between multiple instruments by using the integrated development environment cycle without being sidetracked by switching between different instruments. The engineer is free to focus on finishing the application without distractions from more than one tool or mode of operation, as all functionalities merge in the MPLAB X IDE. A "wrapper" that usually automatically combines all the tools into one Graphic User Interface is an MPLAB X IDE. For example, when you write code, it can be translated into an executed instruction

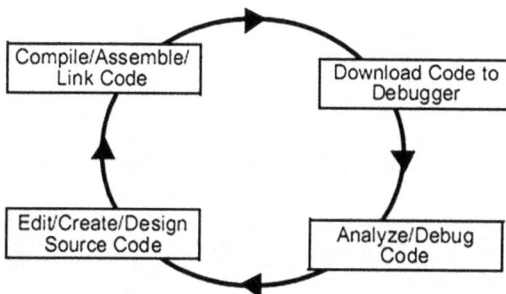

FIGURE 4.3 Development cycle.

and used to test its functionality on a microcontroller. This method will necessitate the use of a code editor, a project manager to modify files and settings, a compiler or assembler to decode source code into machine code, and some form of hardware or software that connects to or simulates the behavior of your target microcontroller.

4.4.6 APPLICATION MODULE

Figure 4.4[A-G] illustrates the different steps involved in creating and setting up MPLAB X IDE

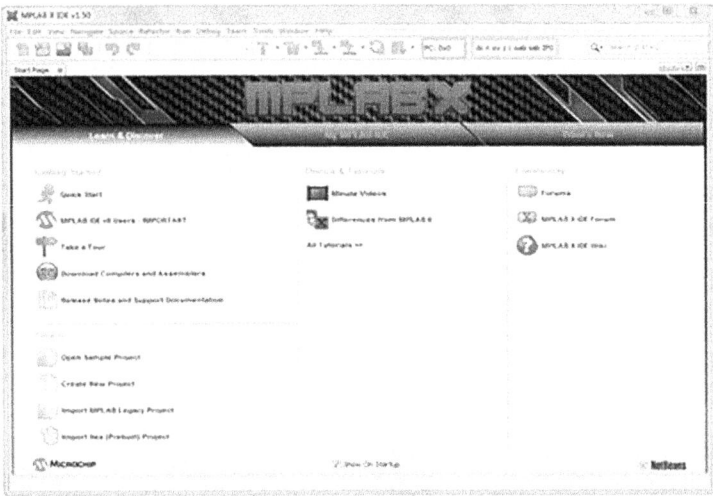

FIGURE 4.4A   Creating and setting up MPLAB X.

Step 1: Setting up the environment.

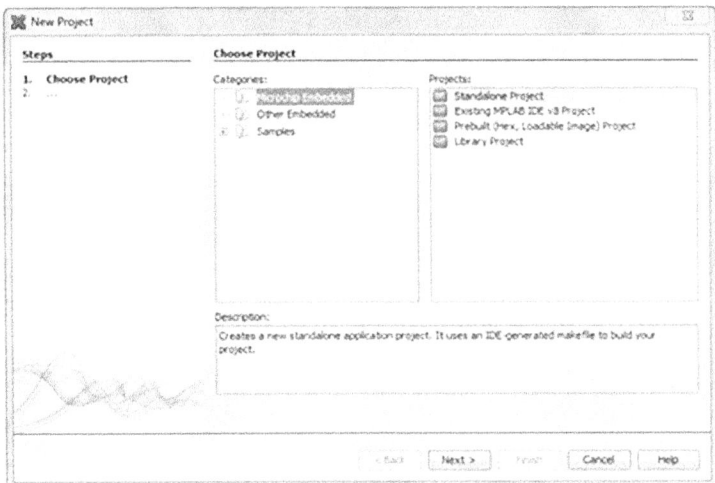

FIGURE 4.4B   Creating and setting up MPLAB X.

Step 2: Launches an MPLAB X IDE dialogue that differs from the one in Step 1. Select your device, in this case the PIC32MX360F512L, and then click Next.

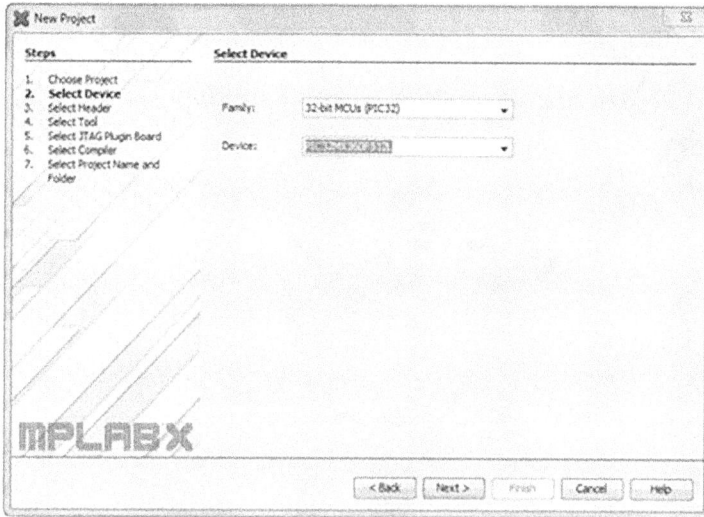

FIGURE 4.4C Creating and setting up MPLAB X.

Step 3: It only appears if a header for your selected device is available. MPLAB X IDE understands to skip this step because there is no header for the PIC32MX360F512L device

Step 4: Choose the tool. Colored circles (lights) in front of each tool name indicate support for the specified device. Green color denotes full support (implemented and fully tested), yellow color denotes beta support (implemented but not fully tested), and red color denotes no support.

Light	Color	Support
○	Green	Full (Implemented and fully tested)
○	Yellow	Beta (Implemented but not fully tested)
◎	Red	None

Step 5: Only by using an MPLAB ICE REAL circuit emulator as the tool will this happen. The plug-in board is a circuit board (MPLAB REAL ICE) placed in an emulator's driver board slot. You can't empty the "supported Plugin Boards" box because Explorer 16 may work with either Standard or HI Speed Communications driver boards. After selecting your tool, click next.

Step 6: In this step, either a C compiler or an assembler is chosen. Once again, the level of device support is indicated by the color of the circle (Refer to Step 4 for Color), which appears before the compiler's name. To view the

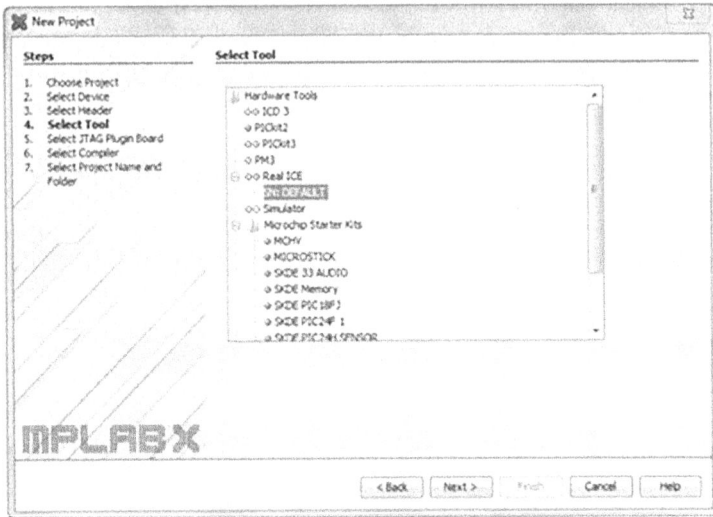

FIGURE 4.4D   Creating and setting up MPLAB X.

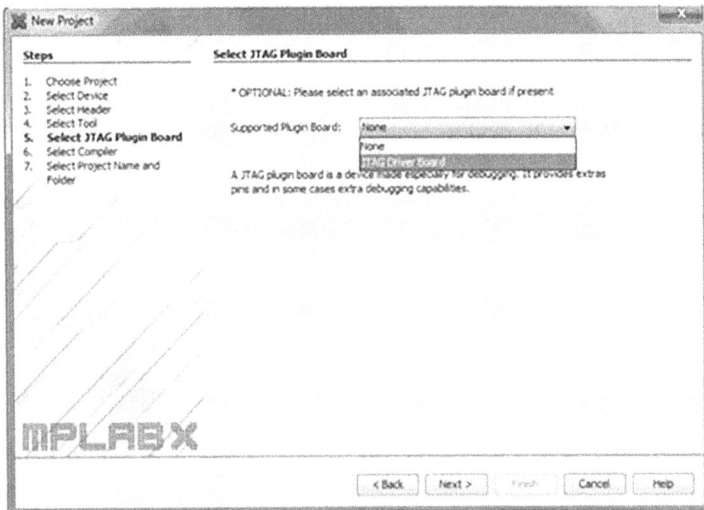

FIGURE 4.4E   Creating and setting up MPLAB X.

text, hover your mouse cursor over the image. Version and installation information for a language is also provided.

Step 7. This step will choose the project name, location, and other options. Type My Project as the project name.

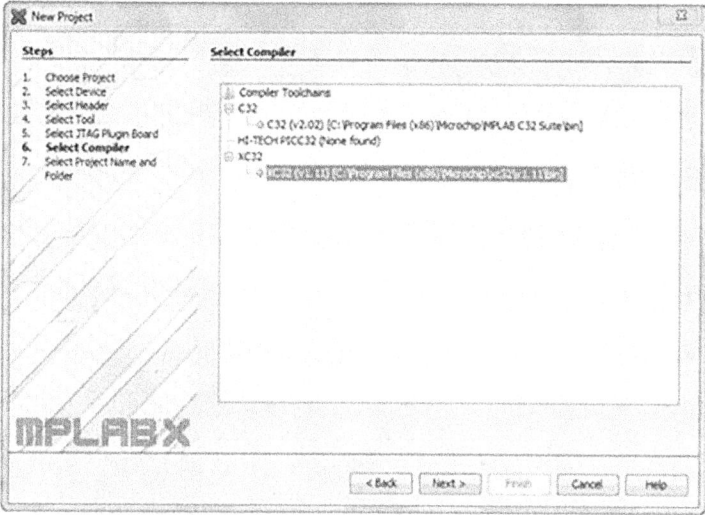

FIGURE 4.4F Creating and setting up MPLAB X.

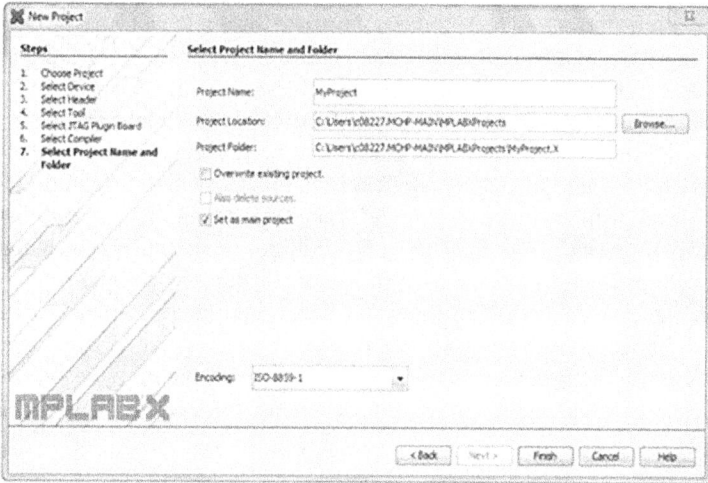

FIGURE 4.4G Creating and setting up MPLAB X.

4.4.7 DATA ANALYSIS AND OBSERVATION

4.4.7.1 Cultivation Method

In this research, we set up the seaweed plants to be spaced 5 cms apart, resulting in a water tank. Environmental factors like pH, temperature, light and water level were measured on a daily basis using the floating method of growing seaweed.

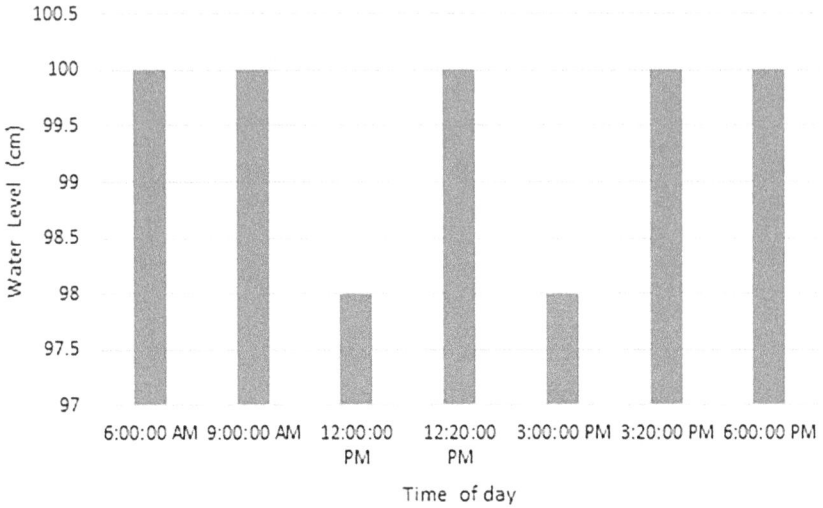

FIGURE 4.5 Analysis of water level per day.

4.4.7.2 Experimental Analysis

A study was conducted to identify the appropriate threshold value based on the lowest and highest values of each important parameter for water level, light, pH, and temperature. When water parameter levels beyond the range are present, seaweed may undergo disruptions that hinder good growth and are uncomfortable.

Figure 4.5 depicted that the water level has been observed in the water tank's regular daily routine. The temperature causes a noticeable decrease in frequency of the water level, with the range of water levels between 98 and 100 cm from 6:00 a.m. to 6:00 p.m. The water level range probably indicates the ideal storage level at which the tank will be filled with water at the range of 100 cm.

Figure 4.6 reveals that pH of the water is an important parameter, and it has a significant impact on seaweed cultivation. The pH ranges from 7.2 to 8.4 is ideal for seaweed cultivation. Research on the absorption of heavy metals by seaweed showed that the clearance rate and pH were inversely related from 6:00 a.m. to 6:00 p.m.

4.5 CONCLUSION

As stated earlier, the only technology systems used in the current aquaculture-based seaweed cultivation are in the post-processing regions. In this research, sensor-based control system for seaweed farming, which comprises water level sensor, light sensor, pH sensor, and temperature sensor, which are controlled by microcontroller. The previously said sensors monitored the actual values, and the same was communicated to owners' mobile. When the set threshold values of water level sensor and light sensor exceed, the motor which is connected will automatically pump the water from the upper tank to the main tank to balance the water level, and LED lights will be on to maintain a consistent level of light. Seaweed farming with this sensor-controlled

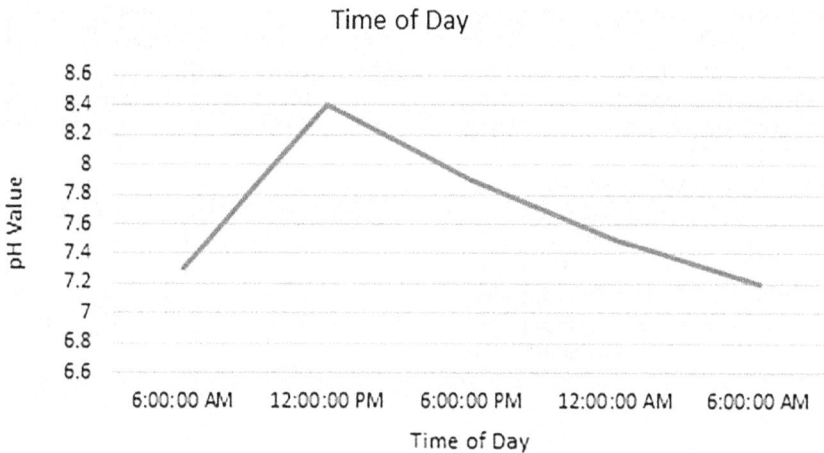

FIGURE 4.6   Analysis of pH level per day.

environment will maximize the production, which increases the export market. It will improve coastal community living conditions and encourage the sustainable use of marine and coastal resources. It is an eco-friendly livelihood activity, especially for women and young entrepreneurs. Since it can be done in a small scale at home, women have become economically active to support for sustainable economy and employment generation.

REFERENCES

Barnett, Richard H., 2003, *Embedded C Programming and the Microchip PIC*, Delmar Cengage Learning.

Jeroen Gerlo, Dennis G. Kooijman, Ivo W. Wieling, Ritchie Heirmans, and Steve Vanlanduit "Seaweed Growth Monitoring with a Low-Cost Vision-Based System", *Sensors*, 23(22), p. 9197, November 15, 2023.

Kaviarasan, T., SanthoshGokul, M., Henciya, S., Muthukumar, K., Hans-Uwe, Dahms and James, R. A., 2018, "Trace metal inference on seaweeds in Wandoor area, Southern Andaman Island", *Bulletin of Environmental Contamination and Toxicology,* 100(5), p. 614619.

Mahadevan, Kritika, 2022, "Seaweeds: A sustainable food source", *Seaweed Sustainability*, pp. 347–364.

Muthumanickam Dhanaraju, Poongodi Chenniappan, Kumaraperumal Ramalingam, Sellaperumal Pazhanivelan, and Ragunath Kaliaperumal," Smart Farming: Internet of Things (IoT)-Based Sustainable Agriculture", *Agriculture*, 12(10), October 21, 2022.

Nakhate, P. and Van der Meer Y., 2021, "A systematic review on seaweed functionality: A sustainable bio-based material", *Sustainability*, 13, p. 6174.

Nitaigour Premchand Mahalik, Kiseon Kim, 2014, "Aquaculture monitoring and control systems for seaweed and fish farming", *World Journal of Agricultural Research*, 2(4), pp. 176–182.

Rajaa, K., Kadirvel, V. and Subramaniyana, T. 2022, "Seaweeds, an aquatic plant-based protein for sustainable nutrition—A review", *Elsevier, Future Foods*, 5.

Robert Maxwell Tullberg, Huu Phu Nguyen, and Chien Ming Wang," Review of the Status and Developments in Seaweed Farming Infrastructure", *Journal of Marine Science and Engine and Engineering (JMSE)*, 10(10), p. 1447, October 7, 2022.

Simona Augyte, Jang K. Kim, and Charles Yarish "Seaweed aquaculture—From historic trends to current innovation", *Journal of the World Aquaculture Society*, 52(5), pp. 1004–1008, October 2021.

Soumyabrata Sarkar, P. Nila Rekha, G. Biswas, R. Nishan Raja, Integrated Multi-Trophic Aquaculture (IMTA): A Potential Farming System to Enhance Production of the Red Seawee Agarophyton tenuistipitatum in Brackishwater, Transforming Coastal Zone for Sustainable Food and Income Security, Springer, pp. 537–552, 2021, DOI:10.1007/978-3-030-95618-9_40.

Torres, M. D., Kraan, S. and Dominguez, H., 2020, *Sustainable Seaweed Technologies, Radarweg 29, PO Box 211, 1000 AE*, Elsevier.

5 Security Analysis and Planning for Enterprise Networks
Incorporating Modern Security Design Principles

Osama Hosam, Rasha Abousamra,
Mohammed Hassouna, and Rula Azzawi

5.1 INTRODUCTION

Cyberspace is characterized as a globally diverse and dynamic area. Cyberspace exists to generate, store, exchange, share, edit, extract, use, and destroy information. In a nutshell, cyberspace is an enhanced medium for technical communication. Informational resources, communication, social networking, and opportunities are some effects of cyberspace. Whereas Cybersecurity [1] is a method of ensuring the security of cyberspace. Cybersecurity aims to achieve and maintain the security features of an organization's and its users' assets over security-related dangers in the cyber environment. It provides protection to the organization and its customers, increases productivity, and gains the customer's confidence [2].

On average, a data breach nowadays costs almost $4 million globally and $8.2 million specifically in the United States. The number of breaches that take place each year is on an exponential rise. This is pushing businesses to constantly update their measure for data security. Appropriate security control planning and analysis is the key to determining the enterprise IT infrastructure's security posture, risk mitigation, and effectiveness of security policies and processes, without which there is a massive threat to the organization's data and service continuity.

The various threats and attack vectors pose a significant risk to the company's network; therefore, researching various security controls and employing them to combat security challenges will massively reduce the company's risk [3]. To secure the enterprise network, security controls such as firewalls, intrusion detection systems, intrusion prevention systems, event logging, incident management systems, access control, physical controls, system hardening, and relevant security frameworks should be implemented.

Security controls can be generally divided into the following categories: Administrative controls, technical controls, and physical controls. Administrative controls reflect the organization's policies, processes, and procedures in place that

DOI: 10.1201/9781003343332-5

direct the actions of the company employees, stakeholders, and top management. Having good administrative controls ensures good cyber hygiene throughout the organization, reducing the probability of risks caused due to human error. Technical or logical controls are where the actual system hardening takes place; these controls include firewalls, IPS, IDS, cryptographic controls, VPNs, DMZ, etc. The physical controls include the measures in place to physically protect the organization and its assets. These controls include CCTVs, anti-theft devices, physical entry controls, etc.

The CIA Triad is a foundational cybersecurity paradigm that serves as a basis for the creation of security regulations intended to safeguard data. The sole purpose of implementing security controls is to safeguard and protect the data. To understand the different types of security controls, one must first understand the fundamental principles of data protection [4]. The fundamental principles of data protection are confidentiality, integrity, and availability.

Confidentiality is most associated with privacy and secrecy. The most common technique to ensure the privacy and secrecy of data is encryption. In this context, secrecy indicates that only parties with authorization have access to the data. When data is kept confidential, it indicates that it cannot be accessed by other parties unless specified. It ensures that data cannot be violated by other parties, sensitive data that is not made accessible to anyone who does not need it or shouldn't have access to it. The objective of ensuring the confidentiality of data is that even if it gets into the hands of the wrong person, they are unable to make sense of it.

Interception, man-in-the-middle attacks, packet sniffing, password stuffing, key logger, and social engineering attacks are some of the most common ways to violate confidentiality. Updated ACLs, stringent password standards, encryption, two-factor authentication, configuration management, the principle of least privilege, and security monitoring are examples of data security controls that protect against compromise of confidentiality. Establishing standards for sufficient authorization and preventing illegal access is a critical component of confidentiality.

Data integrity is the assurance that the data hasn't been manipulated or altered before, during, or after submission. It is the understanding that there hasn't been any unauthorized modification or change of the data, either intentionally or accidentally. The integrity of the data is at risk whenever the data is at rest or in transit. In transit, there are two specific points where the integrity could be risked: during the upload or transfer of information or when data is backed up. Data at risk can be modified and altered by an unauthorized entity if it is not sufficiently protected.

Interception, man-in-the-middle, packet sniffing, and session hijacking are the most common type of security attacks disrupting the integrity of the data. The most effective way to prevent and protect the data against these attacks is by using digital signatures and hashing methods.

Availability is the presence of uninterrupted access to the network [5]. It ensures that authorized users have timely, uninterrupted access to information and information systems. Organizations need to provide connectivity to their authorized users to ensure the services are up and running.

The most common type of attacks that contribute toward the unavailability of the system or network is denial of service, distributed denial of service, SYN flood attack, ICMP attack, and power outage.

Next is the discussion about cybersecurity controls [6] including what these are, their types, how they are implemented, and why they are important to implement.

The mechanisms that are implemented for detecting, preventing, and mitigating cyberattacks and cyber threats from happening in organizations are called cybersecurity controls, which have types including physical and technical controls. Physical controls include having security guards for the physical protection of organizations, also camera surveillance to keep an eye on the physical safety of the organizations. Technical controls include technical things and mechanisms like MFA for multi-factor authentication to prevent unauthorized access and firewalls to block unintended access and traffic.

These mechanisms act as countermeasures and are implemented by businesses to detect, prevent, lessen, or combat security hazards. They are the mechanisms that a company uses to manage threats to networks and computer systems. Controls are always evolving to respond to an ever-changing cyber environment. Controls are put in place to guarantee the integrity, confidentiality, and availability of an organization's information and technological resources. Additionally, controls are based on the four pillars of strategies, procedures, technologies, and personnel.

It is discussed next how these controls are implemented.

Most businesses use frameworks, strict procedures, and controls to minimize cybersecurity risks for preventing or mitigating the associated cyberattacks. The organizations should follow the following steps to implement the controls in the organization: choose a suitable standard; controls should be integrated by doing the risk assessment, data classification, prioritization, considering design constraints, training customers about the procedures of controls implementation, and by at last assembling the monitoring function.

Cybersecurity controls are implemented to lessen the cybersecurity risks to an organization due to evolving technologies and cybersecurity attacks. It helps mitigate the effects of cybersecurity incidents that can collapse the organization if these cybersecurity controls are not in place and configured properly, that's why implementing and configuring these cybersecurity controls properly is important. These security controls, if in place, can help in preventing, detecting, correcting, compensating, and deterrent the cause of cybersecurity breaches. The detail about these terms is discussed in the next sections.

Nowadays, every firm should be prepared for a threat at any time because of evolving technologies and cybersecurity threats. As a result, the controls build systems for identifying, responding to, and recuperating from cybersecurity incidents and breaches [7]. The main objective of putting security controls in place is to stop or lessen the effects of a security incident. Also, it is a requirement under various standards like NIST, ISO, HITRUST, and HIPAA that various security controls should be implemented by organizations to be compliant with these standards, otherwise a heavy fine should be paid by the organizations for not being able to comply with the standards and not implementing the security controls properly. These are all the reasons why it is important to have security controls in place for organizations. One of the main reasons for breaches and incidents today is not being aware of cybersecurity and the mechanisms that can help prevent or mitigate the incidents and breaches. So organizations should conduct awareness activities and drills to equip staff with the importance of cybersecurity controls.

5.2 THE PROPOSED APPROACH

This paper presents a case study of a renowned bank in Saudi Arabia that has recently hired an information security analyst to improve its security posture, identify vulnerabilities, and take actions to detect and prevent their exploitation. The bank is also looking to comply with regulatory requirements from the central bank covering four core areas: cybersecurity leadership and governance, cyber security risk management and compliance, cyber security operations and technology, and third-party cyber security.

The bank has 54 branches across different cities and more than 75 ATMs, with an online digital banking service. The core banking application is executed in the data centers managed by the bank's IT department. However, the disaster recovery site is located in another city and is provided by a third-party service provider as a service. The bank has also utilized the services of an outsourced SOC service provider. IBM Qradar is being used to forward all logs to SOC. Connection with SOC and all remote sites is through VPN.

Figure 5.1 shows the general structure of the proposed approach. The main idea is to use IBM Qradar to send a passive copy of the bank data in the datacenters to third party SOC center for further doing GAP analysis.

The paper focuses on the actions taken by the security analyst, which are explained in terms of core phases. The first step taken by the analyst was to conduct a gap assessment to determine the current state of cyber security of the bank and compare it with the regulatory controls. The gap assessment helped to identify the administrative

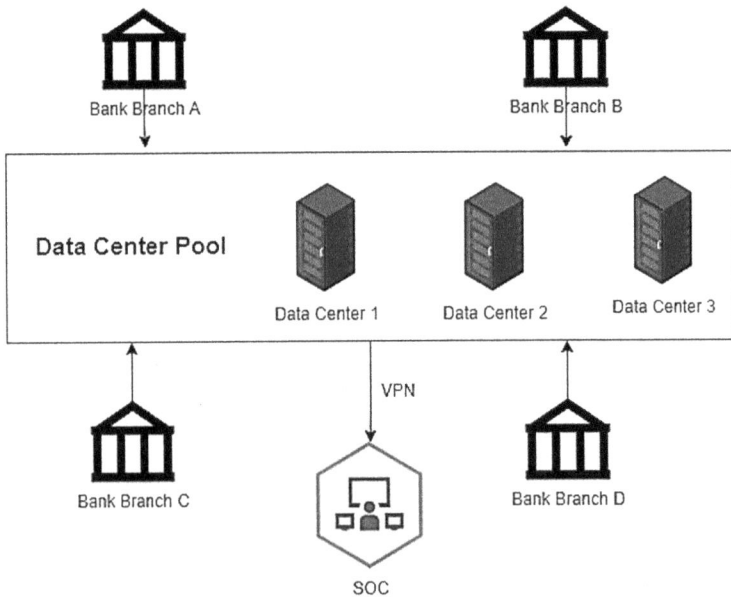

FIGURE 5.1 The proposed approach showing how SOC is going to be adopted to monitor the performance of bank branches across cities.

controls in place addressing the requirements of the standard/framework such as policies, procedures, processes, guidelines, and governance. Along with other controls, the gap report also contained a list of technical as well as physical controls that were already implemented in the bank.

To conduct a comprehensive gap assessment, the security analyst interviewed key stakeholders like the compliance department, IT department, and risk department. He also met business owners to understand core business processes. The gap sheet gave an idea about missing or inefficient controls. The gap dashboard was used for the analysis of leadership and compliance in addition to risk assessment compliance.

The paper also discusses the steps taken by the analyst to address the identified gaps and improve the bank's security posture. These steps include improving security policies, implementing new technical controls, conducting security awareness training, and enhancing the incident response plan.

5.3 TYPES OF CYBERSECURITY CONTROLS

Cybersecurity controls can be grouped into the following categories:

5.3.1 INTERNAL CONTROLS

Internal controls are a category of cybersecurity controls that are designed to ensure the reliable and effective working of the organization's tasks and workings. It can be further categorized into the following types:

5.3.1.1 Administrative Controls

Administrative security controls are procedures, policies, and standards that govern people or business operations in compliance with the security organization's objectives. These controls are called soft controls because their main goal is prevention and detection against cyber threats, cybersecurity incidents, and breaches. Those controls can be applied while auditing, in the incident response plan, or the data classification, also for monitoring and supervising or at the time of job rotation, also at the separation of duties, or during the personnel procedure.

Policies and guidelines are detailed declarations that are intended to affect decisions and actions and add to the security organization. They establish boundaries and define the repercussions of noncompliance with the policy.

These are some policies that are essential for an organization to establish under administrative controls: policy regarding credentials, policy regarding data security, comprehensive awareness security training, policy about social media use and privacy, procedure of respond to cyberattacks incidents, policy regarding mobile device usage, and policy towards crisis management. [8]

5.3.1.2 Technical Controls

Another category of internal controls is technical controls [9]. It requires either software or hardware defenses for assets. These are also called logical controls. Technical controls execute a variety of key tasks, such as restricting unauthorized users from accessing a system and recognizing security breaches. Technical controls are the whole of cybersecurity. Technical control applies technology to mitigate risks. An administrator installs and configures a technological control, and then technical control therefore automatically provides protection.

Some examples of technical controls that can be implemented in organizations include antivirus, access controls lists ~ ACLs, routers, encryption, server images, audit logs, IDS/IPS, data backup, smart cards, and dial-up call back system.

Technical controls should be structured and configured in such a manner that they safeguard all the data at rest, for example, data stored on a storage device and also data transmitted over the Internet. A dedicated skilled staff is usually required in organizations to configure and install these types of controls perfectly.

5.3.1.3 Physical Controls

Physical controls are preventative measures put into place in a specific structure to discourage or restrict unwanted physical access to sensitive data. It refers to any kind of physical device or even humans like guards that are used to restrict or detect unauthorized access to physical spaces, systems, or assets; therefore, it mitigates the risks to physical assets.

Some of the physical controls are fences, locks, badge systems, security guards, biometric systems, mantrap doors, motion detectors, closed circuit TVs, offsite facilities, CCTV cameras, and GPS systems.

"The concept of choice for physical protection is 'defense in depth,'" says Erik van der Heijden.

Physical protective measures must be designed using a process for risk management. The risk evaluation heat map can be relocated to the proposed site to demonstrate where physical safeguards are most required. [10]

5.3.2 INCIDENT-FOCUSED CONTROLS

These measures are designed to prevent, respond to, and detect security breaches. An illustration is shown in Figure 5.2

Following are the types of these controls:

5.3.2.1 Preventive Controls

It describes any security mechanism intended to prevent undesired or unauthorized activities or access. Physical controls like fences, locks, and alarm systems are examples; technical controls like antivirus, firewalls, and intrusion detection systems are examples; and administrative controls such as separation of roles, data classification, and audits are examples. Preventive controls are frequently process-oriented, increasing the time required to carry out specific system tasks but helping in limiting unauthorized activities.

Preventive controls may prevent the breach of the system: the system is secured by the framework and may not have to be checked extensively for security considerations as with other control types. This could be evaluated during an asset risk assessment and/or a control and risks assessment. [11]

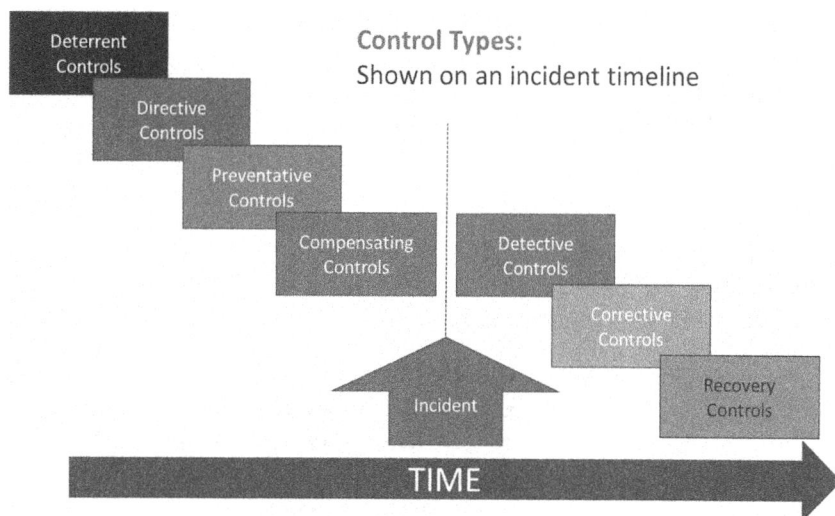

Control Types:
Shown on an incident timeline

Deterrent Controls
Directive Controls
Preventative Controls
Compensating Controls
Detective Controls
Corrective Controls
Recovery Controls
Incident
TIME

FIGURE 5.2   An illustration of the incident-focused controls.

5.3.2.2 Detective Controls

The detective function specifies the operations required to detect the possibility and severity of a cybersecurity incident. The Detective Function allows timely detection of cybersecurity occurrences either threats, incidents, or breaches. They include continuous network access and usage monitoring, intrusion detection technologies, SIEM, motion detector, log monitoring, and analysis to detect odd behavior.

It ensures the detection of anomalies and events and the comprehension of their potential effects, keeping detection processes up to date to alert people to unusual happenings. Examples: physical controls include alarms or notifications from physical sensors (such as door or fire alarms) that notify security personnel, law enforcement, or system administrators. Technical detective controls include honeypots and IDSs.

5.3.2.3 Corrective Controls

Any actions performed to fix harm or return capabilities and resources to their prior state after an unwanted or undesired activity are considered corrective controls. Technical corrective measures include restarting the system, quarantining a virus, ending a process, and applying patches to systems. An illustration of executive corrective control is the implementation of an incident response plan after the happening of an incident. It fixes damaged components or systems for effective use after an incident occurred. For example, IPS (detecting malfunctions in the data flow to immediately prevent unwanted activities) and backup and recovery procedures. [12]

5.3.2.4 Deterrent Controls

These are designed to warn a potential attacker to refrain from attacking and are considered deterrent controls. Deterrent measures, which typically take the shape of a physical thing or person, work to lessen the possibility of a planned attack. Anything that can prevent or deter an attack, such as locks on doors, cable locks, video surveillance, guards, barricades, illumination, or a public warning, could be used in this situation.

5.3.2.5   Recovery Controls

Recovery control is intended to restore the environment to normal functioning. Since data is the most significant asset for any company, having a reliable backup plan is essential for disaster recovery planning. When some problem occurs, you need to ensure that you always have access to at least one reliable backup of your data, and you need to be able to swiftly restore that copy.

So it comprises of the following two main aspects:

Backup strategy. For the various types of data covered by backups, specify both recovery point objective (RPO) as well as recovery time objectives (RTO). RPO denotes how current your backups should have been (what recent information can you afford to lose), whereas RTO specifies how effectively data should have been recovered from backups (how much you can delay for recovery). Select the type of data backups for various data types based on these goals, aligning the performance of the system and data protection with your financial and organizational constraints.

Restore method testing. Even though you regularly back up your data, it won't help you much if something happens to it unless you can successfully restore it within your RPO and RTO targets. The employees must be properly trained, and staff backup restoration processes must be routinely checked and updated. Additionally, keep in mind to routinely verify that your backup is undamaged and always reachable within the RTO duration. [13]

5.3.2.6   Compensating Controls

These are substitute controls [14] used to make up for the main control. Unless the smart cards are created and distributed to every employee, a business that wants to use smart cards for authenticating or dual authentication can utilize a hardware token or OTP. An alternate technique for meeting the security criteria is compensating control. Additionally, some security measures are currently impracticable or costly to deploy. A one-time password is a password that must only be used once. Its algorithm generates a code and incorporates that time of day as a part of its authentication elements.

5.4   STRATEGY FOR CYBERSECURITY CONTROLS

A strategy for protecting a company from external and internal threats through the selection and application of standard practices is known as a cyber security strategy. The cyber security plan should consider establishing defense in depth to efficiently manage current evolving threats and dangers. The layers of security protections are the aim of executing this method, and the zero trust model indicates that do not trust the policies and structures before verifying. Figure 5.3 shows the zero-trust model.

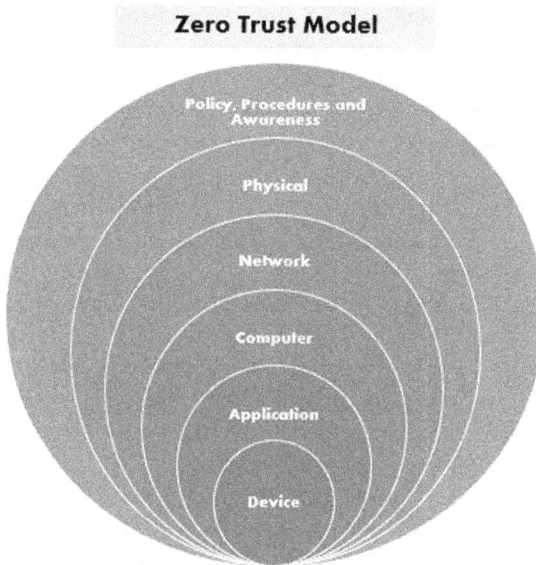

FIGURE 5.3 The zero-trust model shows the layered model and the defense-in-depth strategy.

Things to consider before developing a cybersecurity strategy. What are the potential threats that are most likely to occur? What should employees do if they believe there has been a cyberattack or breach? What procedures should we set up to direct our response to a cyberattack? How often can we go without our information technology systems? In case our systems fall, how will we continue to function? Have we ranked our systems according to how crucial they are to everyday operations? How will we assess the damage and identify which data was affected? What will we accomplish if we lose our data forever? How will we get back up and running after restoring our systems? How will we use communication during damage control to preserve trust and uphold our reputation?

The data should be classified into the following three categories, public only, internal only, and restricted only. [15]

Public only. Public information is accessible to the public both locally and over the Internet. Public data does not necessitate a high level of security as its revelation would not violate compliance or confidentiality. It can be posted on a public website or openly discussed with anyone. It can be openly used and redistributed without ramifications. Publicly available information could be of an individual, a business, or an organization. For example, a data sheet containing information about the company's services and products, names of employees, company names and information about the founders or executives, incorporation dates and date of birth, email addresses, phone numbers, and addresses, job descriptions and open positions, press releases, organizational diagrams, etc.

Internal only. This type of data is only available to the organization's internal staff or employees who have been granted access. This information is frequently related to a company, organization, or business. For example, internal memorandums or email accounts, business strategies and policies, archived documents, platforms for company intranets, spreadsheet budgets and revenue projections, URLs, and IP addresses.

When it comes to internal data, employees can have different levels of security and access. For example, an entry-level accountant may not have the same level of access to historical files or forecasting reports as a financial executive. [16]

Restricted only. The most sensitive information in a company is regarded to be restricted data, which also carries the greatest danger of revelation. It frequently has strict security controls in place to restrict the number of people who have access to the data, as well as backup systems, such as data encryption, to prevent malicious users from accessing or reading content on restricted platforms. Restricted data, if breached or compromised, may endanger public health and wellness or a company's or organization's proprietary information. For example, data that is kept by confidential agreements, information regarding federal taxes, and health information that is private (PHI).

5.5   VULNERABILITIES EXPLOITED BY CYBER THREATS

The following section shows different vulnerabilities found in an enterprise. Not all vulnerabilities are listed here, instead, the most common vulnerabilities that arise challenging the security team of the enterprise.

5.5.1   PHISHING

Phishing is the act of malicious software sending a false email pretending to be from a reliable, trustworthy source. The purpose of the message is to deceive the receiver into downloading malware or disclosing sensitive information. For example, a person gets a malicious link from an email when he/she tries to access it, then it directs to a fake site asking for private information. Spear phishing is one type of phishing assault that is extremely targeted. While both phishing uses emails to contact their targets, a spear phishing email is tailored to a specific individual. Before sending the email, the attacker investigates the target's interests. Vulnerabilities that could be exploited include minimum spam email filtering, improper credential and identity management, and employees unable to detect phishing. And if there are no business policies in place to verify financial transactions. [17]

5.5.2   SOCIAL ENGINEERING

Social engineering is one type of access attack that aims to persuade people into completing actions or disclosing sensitive information. Social engineers frequently rely on people's enthusiasm to help, but they also exploit people's flaws. Figure 5.4 shows different stages in the social engineering attack. Social engineering can be accomplished with:

Social Engineering Attack

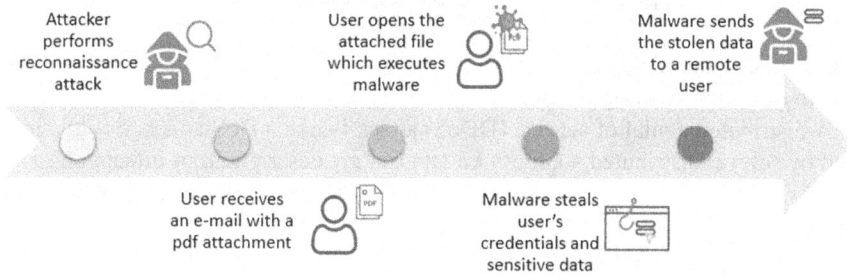

FIGURE 5.4 Social engineering attack shows the sequence that a victim follows to fall into the social engineering trap.

Pretexting: When an intruder calls a person and pretends to them to get access to sensitive data. An example would be an attacker pretending to require personal or financial information to validate the recipient's identity.

Tailgating: When an attacker rapidly follows a legitimate individual into a secure environment.

Quid Pro Quo: When an attacker demands personal information from an individual in exchange for something, such as a gift, this is referred to as social engineering. The vulnerabilities that could be exploited include minimum spam email filtering, improper credential and identity management, and employees unable to detect phishing. And if there are no business policies in place to verify financial transactions. [18]

5.5.3 INSIDER THREATS AND PHYSICAL SECURITY

Attacks can not only originate from the outside but also from inside an organization. Insiders will have immediate access to its infrastructure components and facility. Insider attacks are more dangerous because they have access and knowledge of the company network system and its private information. Vulnerabilities to be exploited: Unauthorized access to physical devices, Unauthorized access to availability servers, Unauthorized access to servers and desktops, Someone with legitimate user credentials misusing their access.

5.5.4 DENIAL-OF-SERVICE DOS

A DoS attack makes its resources/services unavailable to its intended users by disrupting network service for users, devices, or applications, Types of DOS attack:

Overwhelming Amount of Traffic: When the amount of traffic sent to a server, network, or application is greater than the resources it has. This will result in transmission or response lag or device/service crash.

Maliciously Prepared Packets: When a maliciously encoded packet is transmitted to a host, then the receiving device will either crash or engage its resources in processing the result and will become slow.

DoS attacks can quickly stop communication and can make the resources unavailable to the users resulting in considerable time and financial losses.

A distributed denial of service (DDoS) attack is like a DoS attack, but it is initiated by several distributed sources. An attacker creates a group of infected hosts or botnets. These infected hosts can also refer to as zombies and are controlled by the attacker.

These zombie systems are constantly scanning and infecting new hosts. The hacker then launches a DDoS attack using the zombie botnet from distributed machines. It is difficult to block this attack because different machines are sending attack packets. [19]

5.5.5   SEO Poisoning (Search Engine Poisoning)

SEO poisoning is an attack in which cybercriminals create malicious websites using several SEO tactics to make them rank high in search engine results. SEO poisoning can be used by genuine websites to artificially increase their rank position and by harmful websites to target visitors. The attacker aims to install malicious programs like a trojan horse, attack the user's machine, or gain the user's sensitive data.

One of the most common black hat SEO tactics is to detect that a website is accessed by a web search crawler or a regular user. If a crawler visits the page, then high-ranking material is provided, and if a user visits the page, then malicious content is displayed instead.

It aims to bring traffic to malicious websites that contain malware or engage in some social engineering activity. Attackers mostly use popular search queries to rank a malicious website in popular search results. [20]

The most common SEO Poisoning vulnerability is cross-site scripting (XSS).

5.5.6   Types of Malware

Malware is a short form of malicious software and is any code that has the potential to steal sensitive and confidential information, disrupt access controls, or harm or compromise a system. Malware types are shown in Figure 5.5.

A few types of malware are explained here:

Spyware: Spyware tracks and eavesdrops on the user. It runs secretly on a computer and reports back to a remote user. It gathers users' activity using activity trackers, keystroke capture, and data collecting. Spyware alters security settings to evade security measures. It is often combined with other legitimate applications or Trojan horses.

Adware: Also known as advertisement-supported software and is intended to display advertisements automatically. These are highly manipulative and can create an open door for malicious programs. Adware is mostly included with software updates. Adware is also combined with spyware.

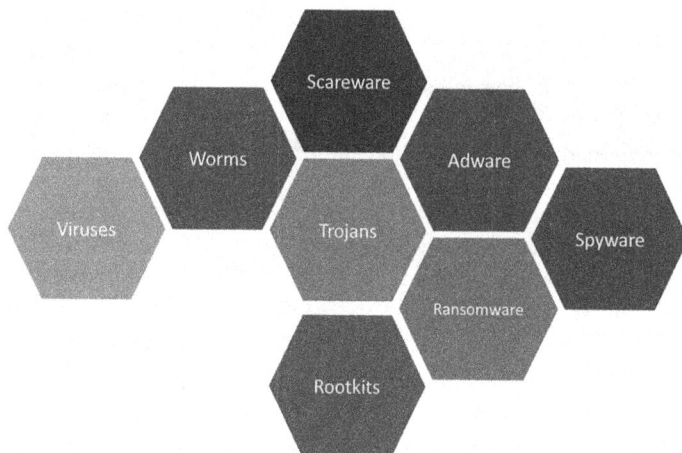

FIGURE 5.5 The common types of malware.

Bot: A bot is designed to perform actions automatically, usually online. Although most bots are not harmful, they are attacked and used in botnets for harmful purposes. Several computers have already been infected by the bots programmed by attackers.

Virus: is an executable malicious code and is linked to another executable file, usually from a trustworthy program. Some viruses are set to activate at a prespecified time or date and can also require end-user activation. Viruses can alter or erase data, but in some cases, they are harmless and only display some graphic content. Viruses can be made to change to prevent detection. Most viruses are spread via network shares, email, USB drives, and optical discs.

Trojan: A Trojan appears to be a legitimate program while it performs harmful actions. Malware exploits user privileges. Trojans are usually found in music files, games, and image files. Trojans are not designed to self-replicate. [21]

Worm: Worms are malicious programs that can independently exploit a network flaw and can replicate themselves. They typically cause network lag. They don't need a host program to function like viruses. They just need user involvement just for the initial infection. The worms can transmit over a network very quickly after host infection. They have a mechanism for spreading themselves and exploiting a vulnerability. Many severe attacks are caused by worms on the Internet.

Ransomware: This kind of malware imprisons a computer, a whole network, or the information it has until a ransom amount is paid. Ransomware encrypts the computer's data with a private key that the user does not know, and that key is given to the user once the ransom is paid. Ransomware use system weaknesses to lock down the system. Ransomware spreads through a software vulnerability or an executable.

Scareware: This kind of malware scares users into taking certain actions. Scareware creates dialogue boxes that look like the operating system's dialogue boxes. It displays fake messages portraying that the system is in danger or has to execute a certain program to prevent it from damage. The user's machine is then infected with malware if they allow the indicated process to run.

Rootkit: This malware alters the operating system to add a back door. The back door is then used to get remote access to the computer by attackers. Rootkits usually manipulate system files and escalate privileges by using the software's flaws. Rootkits frequently alter system forensics and monitoring tools to escape from detection. A rootkit-infected machine has to be cleaned and reinstalled. [22]

5.6 LIST OF SECURITY CONTROLS

Security controls are the parameters that are implemented to protect various forms of important data and infrastructure in an organization. Security controls that are most applicable to the cybersecurity programs of many small businesses are listed here:

5.6.1 BASIC TECHNICAL CONTROLS

In this section, an overview of basic technical controls will be listed. The focus will be on the controls that need to be considered in an enterprise.

Anti-virus: Malware and computer are pervasive. Computers must be protected from malicious malware, viruses, and unauthorized code. Viruses can slow down your computer or delete important data and remain unnoticeable. Antivirus software is essential for your machine. It will identify these real-time threats and protect your data. Antiviruses also offer automatic updates to protect your system against new threats. Antivirus applications must be installed in all the machines of an organization. To keep your system virus-free, regular virus scans must be done.

Anti-malware: Software that not only just scans files for threats, anti-malware programs can also perform these tasks: they can prevent users from accessing websites containing malware, they can prevent malware from spreading to other devices in a computer system, can give information about the extent of infections and the time required to remove them, and can give an overview of the changes a malware has made in a network or device. [23]

To defend a system from malware, anti-malware software employs these strategies:

- Signature-based detection: Software manufacturers create signatures to identify harmful software patterns. These signatures are used to find out similar types of malware that have already been found harmful and to classify new software as malicious.
- Behavior-based detection: Cybersecurity professionals also discover, prevent, and remove malware using behavior-based malware detection. It helps in finding the zero-day vulnerability.

- Sandboxing: Sandboxing is a technique for testing potentially harmful files in a safe isolated environment before they have a chance to cause damage to a real system.

Firewalls: Firewalls isolate your data from the outside world and are already included with Windows and macOS. Firewalls protect your company's network against unauthorized access and notify you in case of any incursion. A firewall can be installed on one computer to safeguard that particular machine (host-based firewalls), or it can be a standalone network device that safeguards an entire group of computers as well as every host device on that network (network-based firewall). [24]

Some of the firewall types:

- Firewall for network layer: Filter using source and destination IP addresses.
- Firewall for transport layer: Filter using connection statuses, destination, and source data ports.
- Firewall for application layer: Perform application, software, or service-based filtering.
- Firewall for context-aware application: Filter person, devices, position, product category, and threat profile.
- Proxy server: Filtering web-based URLs, domains, media, etc.
- NAT firewall: Conceals or masks the private network host addresses.
- Host-based firewall: A single computer system that can filter endpoints and system service requests.

Patching: Hackers are constantly looking for different ways to exploit weaknesses in operating systems and web browsers. Browser's and computer's security settings must be set to medium or higher security to safeguard your device and your data. Operating systems and web browsers on your computer must be updated, and the newest vendor security updates must be frequently installed and downloaded.

The operating systems must be updated, and software on servers, clients, and network devices must be patched as soon as possible. Unpatched software is the reason for almost 60% of the breaches that businesses experienced. Organizations should verify these things about their software:

- It has a license.
- It is removed when no longer offered from devices.
- When a patch is provided to address a vulnerability, whose severity is "critical" or "high risk," the update must be updated within 14 days of its publication.

SIEM: It is software that gathers and analyzes security events, alarms, logs, and other real-time and historic data from the security network's devices.

Security information management gathers information from event logs for assessment and reports on cyber threats and events, and security event management (SEM), which monitors systems in real-time, alerts network administrators to critical

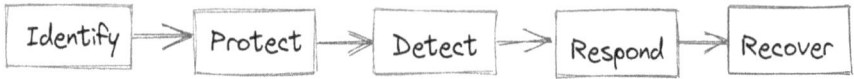

FIGURE 5.6   Threat detection and response cycle.

problems, and creates correlations between security events, are the two technologies that make up SIEM. [25]

The majority of SIEM systems include security dashboards and other direct notification options.

Threat Detection and Response: Threat detection, shown in Figure 5.6, is to identify risks quickly and accurately to a network, its applications, or other assets. When a threat is recognized, a reaction is initiated. To enable quick action, risk countermeasures should be prepared beforehand.

To effectively identify threats, cybersecurity systems with the following factors are considered: fully visible vectors attack, malware detection with a full spectrum, detection of high accuracy, advanced and innovative data analytics, and integration of threat intelligence.

5.6.2 ENDPOINT SECURITY PROTECTION, DETECTION, AND RESPONSE

The pull style of communication is the one used in agent-based systems, where the user is the central main server that requests information from the agents as needed. After an automated process, agents normally need to be deployed on each system. After setting up, the central server can send queries to them for status updates and the outcomes of security-related activities.

Some of the benefits of implementing this control in the system:

- Enable thorough host scanning and monitoring: Agents can carry out more sophisticated host component and service scanning.
- Serve as a firewall since it can restrict network connections according to filtering criteria.
- Provide application or host-specific runtime security.
- Offer security safeguards, like the potential to resist assaults and to update live systems.
- It will carry out tasks without the requirement for a central host: Once it is deployed, the agent will carry out its predetermined tasks whenever needed without the need to first connect to a server, even while it is not connected to the corporate network.

Mobile Device Management: The process of controlling mobile devices, in terms of utilization and security, is known as mobile device management (MDM). A plan should be designed for managing mobile devices and keeping track of each device's

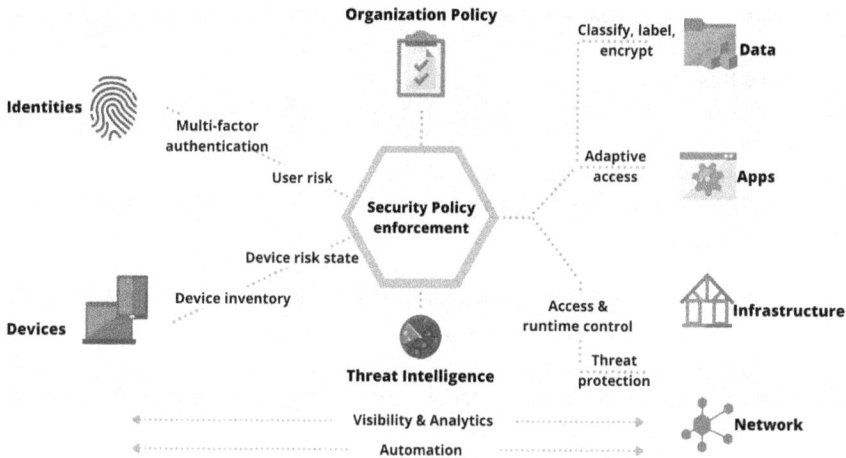

FIGURE 5.7 Security policy enforcement in an enterprise.

crucial data and deciding which programs could be installed. I also remotely protects mobile devices if they are stolen or lost.

Users of any mobile device linked to the MDM server are called clients. Each linked device receives configurations, programs, and policies remotely from the MDM server. IT administrators manage all the endpoints using the MDM server. Smartphones, tablets, computers, and iPods are examples of endpoints. [26]

Security Policies: An organization's network is protected by a set of standardized processes and procedures called cybersecurity policy [27]. Management of cybersecurity policy's rules, methods, and guidelines is referred to as cybersecurity policy identification, implementation, and management. These rules can be kept current by evaluating new IT resources and assets and can help in protecting newly discovered threats to your company. These are some of the policies [28] which should be implemented by the organization to stay up to date with cyber security controls. For example, policy for acceptable use, policy for handling data breaches, implementation of a disaster recovery plan, plan for business continuity, policy for remote access, and policy for access control. Figure 5.7 shows how to enforce the security policy in an enterprise. As shown in the figure, the security policy is the barrier between the access management modules and the system resources.

5.6.3 ACCESS CONTROL

Access control can be done in three stages, authentication, authorization, and accounting (AAA). A separate server can handle the triple-A functions. Cisco has a dedicated AAA server that centrally handles the three functions. In this section, we concentrate only on the authentication part.

Password Management: Passwords are a great source of authentication. Password cracking can occur if they are not managed properly. These are the possible ways of occurrence this threat:

- Dictionary attacks
- reusing passwords on numerous websites
- Cracking security logins questions
- social engineering

Some principles for password management: create a strong, long passphrase; implementation of encrypted passwords, two-factor authentication; use of dictionary words should be avoided; choose a unique password for every account; change or block the credentials if an employee leaves; accounts access can be ensured with privileges; and storing of passwords should be avoided. [29]

Multi-Factor Authentication: MFA is a strategy to protect your Internet accounts and the information they contain by using multiple ways of authentication together. When you implement MFA in your Internet services, you should present a combination of two or more authenticators before it will grant you access. So that even if one of the methods is compromised, unauthorized individuals will not be able to fulfill the second authentication criterion, preventing them from accessing your accounts.

The second authentication is received via the following:

- SMS or Email
- Authenticator app
- FIDO Authentication

If credentials are acquired by phishing attacks or any other way, enabling MFA will make it more challenging for an attacker to get access to systems. To gain illegal access, adversaries are highly proficient in phishing or stealing credentials. They make use of passwords that you have already used on other platforms. MFA provides additional security against data theft by significantly enhancing the level of complexity for adversaries.

Biometrics: Organizations use these biometric characteristics to create a robust authentication program. The aim is to have a strong authentication process and to minimize identity fraud and theft. These characteristics are distinct and differentiate everyone. They are constant over time and in diverse environmental conditions, and they can be quantified. For example, fingerprints, iris, facial patterns, etc. [30]

5.6.4 SECURING CLOUD APPLICATIONS

In this section, we provide an overview of the techniques used to protect the cloud.

Virtual Private Network (VPN): The virtual private network is an encrypted Internet connection that connects a device to a network. The encrypted connection helps in the safe transmission of sensitive data. It keeps unauthorized parties from listening in on data and enables the user to work remotely. VPN technology is commonly used in a corporate environment.

FIGURE 5.8 The encryption procedure. A plaintext is converted to ciphertext using an encryption function.

There are two types of VPNs:

- Remotes Access: It links a device securely located outside the office. VPN technology advancements have enabled security checks on endpoints that fulfill a specific posture before connecting.
- Site to site: It connects the corporate office to subbranch offices through the Internet. Site-to-site VPNs are deployed when the distance between these offices makes direct network connections impossible.

Encryption: Encryption is the method of transforming data so that an unauthorized person cannot read it. Only a trusted, authorized individual who possesses the secret password or key can decode the information and access it in its initial form. The encryption does not prevent the data from being intercepted. Encryption could only prevent unauthorized individuals from reading or retrieving the content. Encryption software is applied to encrypt data, folders, and even whole drives.

Encryption, shown in Figure 5.8, works by taking readable data and altering it so that it appears random. It requires cryptographic keys: a mathematical expression that both the sender and the recipient of the encrypted message have agreed on.

The encrypted data appears random, but encryption is a logical process that is predictable and reversible. Any party having the right encryption key will be able to decrypt the data. The complexity of the key determines the strength of the encryption algorithm.

A cryptographic key can be described as a string of characters used within an encryption algorithm for altering data so that it appears random (ciphertext). Like a physical key, an encryption key locks (encrypts) data, and only someone with the right key can unlock (decrypt) it. [31]

The encryption algorithm adds security to the data that has been exchanged between two parties. Here are some of the important encryption algorithms:

- AES: It stands for Advanced Encryption Standard and is the most common mode of encryption. It is considered the best encryption algorithm to protect systems against all attacks except brute force attacks.
- RSA: It is an asymmetric algorithm, using a set of public and private keys to encrypt and decrypt the data.
- Triple DES: It is a symmetric algorithm and depends on a single key for encryption and decryption.

Data Backup: Data backup is the act of copying critical organizational data to a secondary storage location. It also helps to protect businesses against data loss by providing redundancy to the primary storage location in case that location is compromised or fails. In addition to interrupting business operations, data loss can result in serious repercussions for businesses, ranging from financial penalties to loss of customer or partner trust.

5.6.5 CYBER FORENSIC TOOLS

In this section, we cover cyber security forensics tools. The tools used extensively in an enterprise will be explored and detailed.

5.6.5.1 Logging and Security Information and Event Management (SIEM)

SIEM, shown in Figure 5.9, is a collective term for security software packages ranging from log management systems to security log/event management, security information management, and security event correlation. In most cases, these features are combined to achieve a 360-degree view.

SIEM system is not a panacea, but it is one of the key indicators that an organization has a well-defined cybersecurity policy. Nine out of ten cyberattacks have no obvious clues on the surface. Using log files is a more effective way to detect threats. A SIEM's superior log management capabilities make it a central hub for network visibility. Most security programs operate at a micro-scale, addressing small-scale threats but overlooking the full picture of cyber threats. An intrusion detection system (IDS) alone can do little more than monitor packets and IP addresses. Similarly, service logs only show user sessions and configuration changes. A SIEM brings

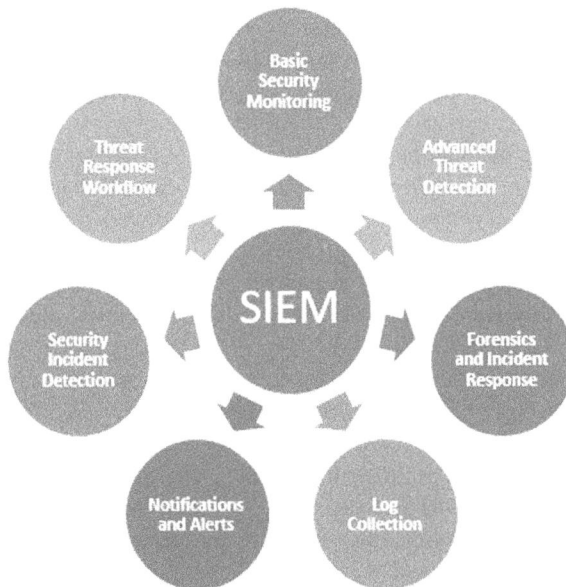

FIGURE 5.9 The SIEM functions.

together these and similar systems to provide a complete view of security incidents through real-time monitoring and event log analysis. [32]

SIEM's basic capabilities are as follows:

- Log Collection
- Normalization—Collecting logs and normalizing them into a standard format
- Notifications and Alerts—Notifying the user when security threats are identified
- Security Incident Detection
- Threat Response Workflow—Workflow for handling past security events

A SIEM records data from across your internal tool network to identify potential problems and attacks. The system analyzes log entries using statistical models. SIEMs deploy collection agents to retrieve data from networks, devices, servers, and firewalls.

All this information is passed to the management console, where it can be analyzed and responded to new threats. Advanced SIEM systems often use automated response, entity behavior analysis, and security orchestration. This allows SIEM technology to monitor and remediate vulnerabilities among cybersecurity tools.

5.6.5.2 Compliance Reporting

From a convenience and regulatory requirements standpoint, a SIEM with extensive compliance reporting capabilities is very important. Generally, most SIEM systems have some form of built-in reporting system to help meet your compliance needs.

The source of standard requirements that must be met has a great impact on the SIEM system you install. When security standards are dictated by customer contracts, there isn't much leeway in choosing which SIEM system. If it doesn't support the required standards, it's not familiar. You may need to demonstrate compliance with PCI DSS, FISMA, FERPA, HIPAA, SOX, ISO, NCUA, GLBA, NERC CIP, GPG13, DISA STIG, or one of many other industry standards.

5.6.5.3 Backup Schedules

The facts healing alternatives for a commercial enterprise significantly rely on the backup schedules which assist to decide how a great deal of time passes among healing points.

Full-backup—as the call implies, it is a complete backup of all facts on all systems, that is the maximum thorough but is the maximum impactful in phrases of time and sources so ought to be performed regularly, however now no longer daily.

Differential—this backup agenda copies documents which have been modified because of the ultimate complete backup; if the overall backup is taken each Sunday, Tuesday's differential backup will include adjustments from Monday and Tuesday, and Wednesday's differential backup will include Monday, Tuesday, and Wednesday's adjustments. This approach has far simpler healing as all this is required to repair is the ultimate complete backup and the ultimate differential.

(a)

(b)

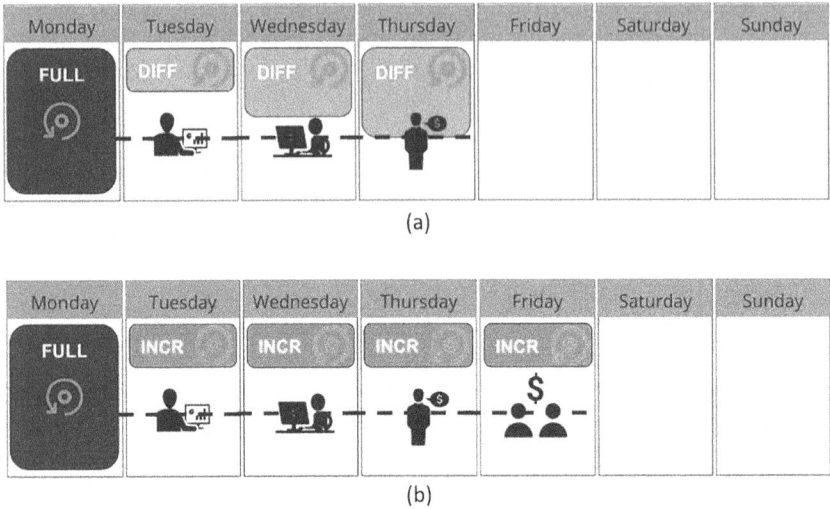

FIGURE 5.10 Backup strategies (a) show the differential backup and (b) show incremental backup.

Incremental—this backup agenda copies documents for every day because the ultimate backup became made into its very own archive; if the overall backup is taken each Sunday, Monday's incremental will most effectively include adjustments made on Monday, Tuesday's incremental will most effectively include adjustments made on Tuesday, and so forth. This approach has the first-rate overall performance and smaller back-up windows; however, to be able to repair from backup, every next backup needs to be restored in order. Figure 5.10 shows the difference between incremental and differential backup strategies.

5.6.6 CYBER SECURITY AWARENESS TRAINING

A security awareness training program is a formal learning process and an employee's basic understanding of computer security training and awareness. Security awareness and training programs train employees.

- How do you identify security threats?
- How do you reconcile trust and verification?
- How are your company's security policies being followed?
- How are safety procedures followed?

The importance of properly reporting growing information security concerns lies in awareness programs that enable employees to quickly understand security threats or vulnerabilities and report them to the security team. Therefore, it is important to have a culture of cybersecurity awareness and enforcement so that work-related computers and mobile devices are protected as a first line of defense. By educating and educating employees and stakeholders, workplaces understand what they need to do to reduce cyber risk and prevent cybercriminals from infiltrating corporate information assets. Digital attacks are constantly on the rise, so you need clarity on what your cyber defense strategy should be. Thanks to regular information security training, good security practices combined with basic network security best practices will give you a better understanding of IT governance issues. Simply because employee behavior leads to cybersecurity incidents. As they say, it's humans who make mistakes, and when we make mistakes, miss best practices, or simply fall prey to scams, it's right for humans. When security fails, what matters is an issue that is addressed in security awareness training. We often talk about people being the weakest link in the chain. But they are our strongest if we know what role we should play in fighting cyberattacks. We can be a line of defense for a strong cybersecurity incident response plan that can keep companies vigilant against cyber threats thanks to well-informed employees and digital networks.

Various research reports indicate that lack of cyber protection and social engineering implementation, insider threats, and advanced and persistent threats are the leading causes of cyber security threats. Minor differences between the trained and untrained workforce: if your employees are unaware of this, it could be the following:

Malware victims: "Inattentive or unsuspecting" employees can fall victim to phishing emails, and internal employees can become part of malware attacks. Email accounts for 94% of malware deliveries.

Be a victim of weak passwords: Weak or reused passwords can compromise the security of your data. The impact of weak passwords can be understood from the fact that one uses eight passwords for an average of 2.7 accounts for both personal and business use. Additionally, not secure passwords are another valid reason to compromise credentials for successful password spraying (a type of brute force attack).

Victim of social engineering: Social engineering scams are on the rise, and nearly 98% of his cyberattacks are socially engineered. Over 40% of IT professionals experienced social engineering tactics in 2019. Lack of cybersecurity compliance audits on cybersecurity procedures and risk management could be seen as a breach of the data security requirements proposed by the regulations: Payment Card Industry (PCI), Data Security Standards (DSS), Sarbanes Oxley (SOX), General Data Protection Regulation (GDPR), and CCPA, etc.

"Cyber protection way of life" amongst personnel desires to be developed. An organization attempt to include cybersecurity high-quality practices, for work-associated activities, to

- understand a Phishing electronic mail with malicious hyperlinks or malware;
- now no longer fall prey to faux login pages;
- have a stable password to log in to laptop systems;
- to have a grasp of encryption, etc.

5.6.7 Physical Security

At its core, physical security is about protecting facilities, people, and assets from real-world threats. This includes physical deterrence, intruder detection, and response to those threats. Although environmental events may be involved, the term is typically used to prevent people, whether outside parties or potential inside threats, from accessing areas or assets they should not be entering.

Physical attacks can break into secure data centers, enter restricted areas of buildings, or use devices that are inaccessible. Attackers can steal or damage critical IT resources such as servers and storage media, access critical end devices for business-critical applications, steal information over USB, load malware onto systems, and more.

There are likely to be tight controls at the outermost perimeter that should be able to keep outside threats out, while internal measures around access can reduce the likelihood of an inside attacker (or at least flag unusual behavior).

Mechanisms and methods of slowing or preventing threats are as follows:

Protective barriers—Mechanisms that provide physical barriers are often the ultimate security level. These include walls, fences, and reinforced windows.

Physical access control—A mechanism for defining secure areas by controlling traffic passing through specific entry points.

Mechanical—Turnstiles, gates, doors with locks (including electronic locks).

Human trap—A vestibule consisting of a small space between two sets of doors, such as a doorway.

Anti-theft devices—Combination safes, cable/locks for devices, and cable alarms may also have detection capabilities.

Closed circuit television (CCTV)—A control mechanism consisting of a video camera. However, it delays the monitoring of video data, which is typically only used to respond to incidents and gather evidence, so being visible to potential attackers can be considered a deterrent.

Some deterrents create psychological barriers to protecting assets by convincing attackers that an attack is unlikely to succeed.

Environmental considerations: In some cases, passive environmental design techniques are the best deterrent. For example, brighter areas with more open visibility are less likely to attract attackers. A common strategy is to place valuable objects where intruders must traverse large spaces to reach them, making them easier to detect.

5.7 CASE STUDY: APPLYING SECURITY CONTROLS IN A BANKING SYSTEM

A renowned bank in Saudi Arabia has recently hired an information security analyst to improve its security posture, find its vulnerabilities, and take actions to both detect and prevent their exploitation. The main responsibility of this position is to identify, detect, protect, and respond to the possible threats to the security of the organization's mission.

The bank is also looking to comply with regulatory requirements from the central bank that cover four core areas:

- Cybersecurity Leadership and Governance
- Cyber Security Risk Management and Compliance
- Cyber Security Operations and Technology
- Third-Party Cyber Security

The bank has 54 branches across different cities and more than 75 ATMs. It also provides online digital banking services. The core banking application executes in the data centers managed by the bank's IT department. However, the disaster recovery site is located in another city and is provided by a third-party service provider as a service.

The bank has also utilized the services of an outsourced SOC service provider. IBM Qradar is being used to forward all logs to SOC. Connection with SOC and all remote sites is through VPN.

The security analyst actions are explained in terms of core phases explained later.

5.7.1 GAP ASSESSMENT

One of the first steps the security analyst will take is to conduct a gap assessment to determine the current state of cyber security of the bank and compare it with the regulatory controls. [33, 34] The gap assessment will help identify the administrative controls in place addressing the requirements of the standard/framework such as policies, procedures, processes, guidelines, and governance. Along with other controls, the gap report will also contain a list of technical as well as physical controls that are already implemented in the bank.

To conduct a comprehensive gap assessment, the security analyst interviewed key stakeholders like the compliance department, IT department, and risk department. He also met business owners to understand core business processes. He also requested earlier audit reports and gap sheets. The gap sheet gave an idea about missing or inefficient controls. The gap assessment looks something like the next illustration.

The GAP dashboard is shown in Figure 5.11 and Figure 5.12 for the analysis of leadership and compliance in addition to risk assessment compliance.

Analysis of each control status and cyber security analysis are added. The dashboard ensures that management understands the current position and delivers support for remediation efforts. A detailed gap report with evidence forms the cornerstone of any effective cybersecurity program.

5.7.2 ROAD MAP DEVELOPMENT

Once gaps are identified, an analyst can plan for the mitigation of gaps. An effective road map requires input from control owners and business heads. Security analysts

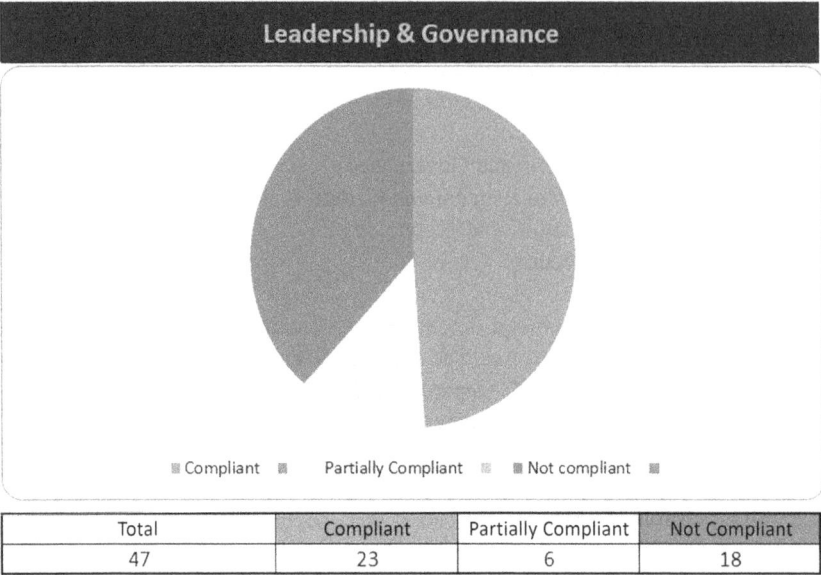

Total	Compliant	Partially Compliant	Not Compliant
47	23	6	18

FIGURE 5.11 The leadership and governance compliance in the bank. The analysis shows compliance of 50% in the leadership and governance.

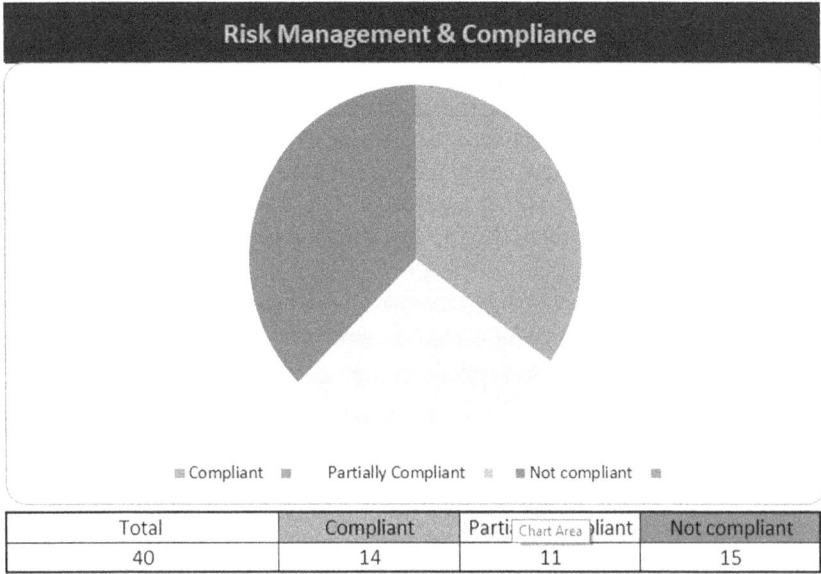

Total	Compliant	Partially Compliant	Not compliant
40	14	11	15

FIGURE 5.12 Risk management and compliance in the bank. The analysis shows a compliance of about 33% in risk management.

identified the level of risks against missing controls and threat scenarios to prioritize the implementation of the control. He defined three levels of risks, that is, high, medium, and low based on the impact on confidentiality, integrity, and availability. The road map got approval from senior management for implementation.

Security analysts took actions based on control requirements from each domain.

5.7.2.1 Cybersecurity Leadership and Governance

The analyst identified that the bank developed policies and procedures six months ago and got them approved by a cyber security steering committee. The committee is primarily responsible for governing cyber security in the bank. The committee however does not meet every quarter, and hence critical decisions are delayed. The analyst also identified that projects are initiated without discussing with the cyber security department, resulting in risks to infrastructure. The analyst added a project initiation process on the ITSM tool to ensure the cyber security department is onboarded at the start of a project.

The analyst also noted that although it's documented, controls are not reviewed at periodic intervals. This results in discrepancies in control implementation. To ensure effective significant controls, and review activity, he developed a checklist and questions to be asked of control owners. For technical devices like firewalls and IPS/IDS, he took guidance from manufacturer baseline guides. Table 5.1 shows the table used in leadership and governance compliance measurements.

TABLE 5.1

Part of the Analysis Table Used for Measuring the Complaint of Leadership and Governance in the Bank

Control code	Control name	Compliance status
3.1.1–1	A cyber security committee should be established and mandated by the board.	Compliant
3.1.1–2	The cyber security committee should be headed by an independent senior manager from a control function.	Compliant
3.1.1–3	The following positions should be represented in the cyber security committee: a. senior managers from all relevant departments (e.g., COO, CIO, compliance officer, heads of relevant business departments). b. Chief information security officer (CISO). c. Internal audit may attend as an "observer."	Compliant
3.1.1–4	A cyber security committee charter should be developed, approved, and reflect: a. committee objectives. b. roles and responsibilities. c. the minimum number of meeting participants. d. meeting frequency (minimum every quarter).	Compliant

(Continued)

TABLE 5.1 (*Continued*)
Part of the Analysis Table Used for Measuring the Complaint of Leadership and Governance in the Bank

Control code	Control name	Compliance status
3.1.1–5	A cyber security function should be established.	Compliant
3.1.1–6	The cyber security function should be independent of the information technology function. To avoid any conflict of interest, the cyber security function and information technology function should have separate reporting lines, budgets, and staff evaluations. The board of the member organization should allocate a sufficient budget to execute the required cybersecurity activities.	Compliant

5.7.2.2   Cyber Security Risk Management and Compliance

The bank conducts risk management against a cyber security risk management framework. The risk register was complete; however no follow-up on treatment was found. The analyst identified that the risk register is kept isolated in Cyber Security Department without discussing it with the risk owners.

This resulted in open risks with ineffective risk treatment and monitoring. The analyst decides to contact risk owners to ensure they implemented controls to mitigate risks. For example, a risk was identified that due to lack of network segregation, DDoS attacks can propagate fast. The analyst asked the network admins to segregate the network based on the criticality of risks and asked for evidence of completion.

5.7.2.3   Cyber Security Operations and Technology

The core areas that were reviewed in this section are as follows:

SERVERS: The servers used by a bank are mostly MS Servers or Ubuntu. The core system uses Red Hat. The servers are placed in isolated VLANs which are segregated by the CISCO firewall. All the servers have updated McAfee antivirus and are integrated through PAM (Privileged Access Management) solution. Table 5.2 shows the table used in operations and technology compliance measurements.

The security analyst identified that servers operating systems are not updated due to a lack of patch management process. The security analyst recommended ManageEngine patch management solution to identify new updates and apply them. The analyst also found the vulnerability scans of servers before moving to the production environment do not exist. This results in vulnerable systems placed in the production environment. To ensure the security of the production environment, the security analyst added the role of the cybersecurity department in change approval. Whenever any change in the production environment is required, cyber security approval is mandatory.

APPLICATIONS: The bank does not develop any in-house applications. All applications are procured by a competitive bidding process. The applications which

TABLE 5.2

Part of the Analysis Table Used for Measuring the Complaint of Operations and Technology in the Bank

Control code	Control name	Compliance status
3.3.4–1	The cyber security architecture should be defined, approved, and implemented.	Compliant
3.3.4–2	Compliance with the cyber security architecture should be monitored.	Compliant
3.3.4–3.a	The cyber security architecture should include: a. A strategic outline of cyber security capabilities and controls based on the business requirements;	Partially Compliant
3.3.4–3.b	b. approval of the defined cyber security architecture;	Compliant
3.3.4–3.c	c. the requirement of having qualified cyber security architects;	Not compliant
3.3.4–3.d	d. design principles for developing cyber security controls and applying cyber security requirements (i.e., the security-by-design principle);	Not compliant
3.3.4–3.e	e. periodic review of the cyber security architecture.	Compliant
3.3.5–1	The identity and access management policy, including the responsibilities and accountabilities, should be defined, approved, and implemented.	Partially Compliant
3.3.5–2	Compliance with the identity and access policy should be monitored.	Compliant

are external facing have strong security controls. All the data in transit is protected through HTTPS connections. The central core database is in Oracle and is encrypted using TDE (transparent data encryption). Vulnerability scans are conducted every quarter for all applications.

The analyst however reported that penetration testing of applications is not conducted. This resulted in serious issues which are not identified by vulnerability scans. Weaknesses such as malicious file uploads, weak passwords, and SQL injection were identified once penetration testing was conducted.

NETWORK: The bank has a complex network with multiple VLANS. A boundary firewall is from CISCO while the firewalls for each department are from Juniper. During the review of firewalls, it was identified that firewall rulesets are not reviewed causing performance issues. The analyst develops a yearly schedule for the review of firewalls.

5.7.2.4 Third-Party Cyber Security

The bank avails services from multiple third-party partners who provide technical support for the bank infrastructure. The bank reviews the cyber security posture of vendors based on filled questionnaires. The analyst identified that vendors are not ranked into categories based on the criticality of the services they provide. The analyst divided the vendors into three categories and required varying levels of control for them. This resulted in significant risk reduction from threats imported from vendor systems.

One of the core issues that analysts identified in the whole bank was the lack of cyber security monitoring. This was resulting in deficient controls.

5.7.3 Comparison of Different Gap Assessment Approaches

The gap assessment described in this paper is a compliance-based gap assessment. It focuses on identifying the differences between the organization's current security controls and the requirements of a specific standard or framework, such as policies, procedures, processes, guidelines, and governance. It also includes evaluating technical and physical controls that are already implemented in the organization. To conduct the assessment, the security analyst interviews key stakeholders, meets with business owners, and reviews earlier audit reports and gap sheets to identify missing or inefficient controls. This approach is commonly used to ensure that an organization's security controls meet the requirements of relevant regulations and standards.

There are several different approaches that can be used to conduct a gap assessment, including:

- Compliance-based: This approach focuses on determining whether the organization's security controls meet the requirements of relevant regulations and standards. [33]
- Risk-based: This approach focuses on identifying the organization's most significant security risks and determining whether the existing controls are sufficient to mitigate those risks. [34]
- Control-based: This approach focuses on evaluating the effectiveness of the organization's existing security controls and making recommendations for improvements. [35]
- Threat-based: This approach focuses on identifying the threats that the organization is most likely to face and determining whether the existing controls are sufficient to protect against those threats. [36]

Each approach has its own set of advantages and disadvantages, and the choice of which approach to use will depend on the specific needs of the organization.

A general explanation of the strengths and weaknesses of each approach, which may help you determine which approach is best suited for your specific use case.

Compliance-based: Strength: This approach can provide a clear and objective way to determine whether an organization is meeting the requirements of relevant regulations and standards.

Weakness: This approach may not address all potential security risks and may not identify all areas where security controls are lacking.

Risk-based: Strength: This approach can help organizations identify and prioritize their most significant security risks.

Weakness: This approach may not provide a comprehensive assessment of all security controls and may not identify all areas where security controls are lacking.

Control-based: Strength: This approach can provide a comprehensive assessment of an organization's existing security controls and identify areas for improvement.

Weakness: This approach may not take into account all potential security risks and may not identify all areas where security controls are lacking.

Threat-based: Strength: This approach can help organizations identify the threats they are most likely to face and determine whether their existing controls are sufficient to protect against those threats.

Weakness: This approach may not provide a comprehensive assessment of all security controls and may not identify all areas where security controls are lacking.

It is important to note that for best results, it is often recommended to use a combination of these approaches.

5.8 CONCLUSION

The chapter discussed the need for cybersecurity controls in place to prevent unauthorized access and undesired activities within an organizational network. Various vulnerabilities and types of attacks were highlighted to emphasize the importance of having security controls in place. It was concluded that proper security controls and employee security awareness training can help prevent these vulnerabilities and make enterprise networks more robust and secure against evolving cybersecurity incidents. A risk assessment was conducted to identify risks and apply controls to reduce them. The three main types of security controls discussed were administrative, technological, and physical controls. Administrative controls represent the organization's rules, processes, and procedures that guide the activities of workers, stakeholders, and senior management. Technical controls include firewalls, intrusion prevention systems, cryptographic controls, VPNs, and DMZs. Physical controls include CCTV, anti-theft systems, and physical entrance restrictions. It was emphasized that to address cyber resilience, cybersecurity should be integrated across the organizational life cycle to safeguard the company and adapt to new threats.

REFERENCES

1. Le, D. N., Kumar, R., Mishra, B. K., Chatterjee, J. M., & Khari, M. (Eds.). (2019). *Cyber Security in Parallel and Distributed Computing: Concepts, Techniques, Applications and Case Studies*. John Wiley & Sons.
2. Nayyar, A., Rameshwar, R., & Solanki, A. R. U. N. (2020). Internet of Things (IoT) and the digital business environment: A standpoint inclusive cyberspace, cyber-crimes, and cybersecurity. In *The Evolution of Business in the Cyber Age* (pp. 111–152). Apple Academic Press.
3. Lloyd, G. (2020). The business benefits of cyber security for SMEs. *Computer Fraud & Security*, 2020(2), 14–17.
4. Viegas, V., & Kuyucu, O. (2022). IT security technical controls. In *IT Security Controls* (pp. 103–172). Apress.
5. Ham, J. V. D. (2021). Toward a better understanding of "Cybersecurity". *Digital Threats: Research and Practice*, 2(3), 1–3.
6. Ding, D., et al. (2018). A survey on security control and attack detection for industrial cyber-physical systems. *Neurocomputing*, 275, 1674–1683.

7. Ali, S., et al. (2018). *Cyber Security for Cyber-Physical Systems*. Springer.
8. Hosam, O., & BinYuan, F. (2022). A comprehensive analysis of trusted execution environments. 2022 8th International Conference on Information Technology Trends (ITT). IEEE.
9. Kohnke, A., Shoemaker, D., & Sigler, K. E. (2016). *The Complete Guide to Cybersecurity Risks and Controls*. CRC Press.
10. Hosam, O. (2022). An earthquake query system based on hidden Markov models. *International Journal of Embedded Systems*, 15(2), 149–157.
11. Hosam, O. (2022). Intelligent risk management using artificial intelligence. 2022 Advances in Science and Engineering Technology International Conferences (ASET), 2022, pp. 1–9. doi:10.1109/ASET53988.2022.9734861
12. Hu, J., Liang, W., Hosam, O., Hsieh, M. Y., & Su, X. (2021). 5GSS: A framework for 5G-secure-smart healthcare monitoring. *Connection Science*, 1–23.
13. Liu, B., Xiao, L., Long, J., Tang, M., & Hosam, O. (2020). Secure digital certificate-based data access control scheme in Blockchain. *IEEE Access*, 8, 91751–91760.
14. Antunes, M., Maximiliano, M., & Gomes, R. (2022). A customizable web platform to manage standards compliance of information security and cybersecurity auditing. *Procedia Computer Science*, 196, 36–43.
15. Lei, Q., Xiao, L., Hosam, O., & Luo, H. (2020). A novel watermarking algorithm based on characteristics model of local fragmentary images. *International Journal of Embedded Systems*, 12(1), 11–21.
16. Hany, M. F., Youssef, B. A., Darwish, S. M., & Hosam, O. (2019, October). Intelligent watermarking system based on soft computing. International Conference on Advanced Intelligent Systems and Informatics. Springer, Cham, pp. 24–34.
17. Huang, W., Li, R., Xu, J., Huang, Y., & Hosam, O. (2019). Intellectual property protection for FPGA designs using the public key cryptography. *Advances in Mechanical Engineering*, 11(4), 1687814019836838.
18. Hosam, Osama. (2018). Hiding Bitcoins in Steganographic fractals. 2018 IEEE International Symposium on Signal Processing and Information Technology (ISSPIT). IEEE.
19. Hosam, Osama. (2019). Toxic comments identification in Arabic social media. *International Journal of Computer Information Systems and Industrial Management Applications*, 11, 219–226.
20. Hosam, O., & Hammad Ahmad, M. (2019). Hybrid design for cloud data security using combination of AES, ECC and LSB steganography. *International Journal of Computational Science and Engineering (IJCSE)*, 19(2).
21. Zhu, W., Hosam, O., & Zheng, X. (2019). A secure hierarchical community detection algorithm. *International Journal of Computational Science and Engineering (IJCSE)*, 19(2).
22. Hosam, O., & Ben Halima, N. (2016). Adaptive block-based pixel value differencing steganography. *Security Communications Networks*. doi:10.1002/sec.1676
23. Hosam, O. (2015). Colored texture classification with support vector machine and wavelet multiresolution analysis. The IEEE Symposium on Signal Processing and Information Technology (ISSPIT'05) December 2015, Abu Dabhi, UAE.
24. Rahalkar, S. (2018). *Network Vulnerability Assessment: Identify Security Loopholes in Your Network's Infrastructure*. Packt Publishing.
25. Halima, N. B., Hosam, O. (2016). Bag of words based surveillance system using support vector machines. *International Journal of Security and Its Applications*, 10(4), 331–346. http://dx.doi.org/10.14257/ijsia.2016.10.4.30
26. Hosam, O., & Halima, N. B. (2015). A hybrid Roi-Embedding based watermarking technique using DWT and DCT transforms. *Journal of Theoretical and Applied Information Technology*, 81(3), November.
27. Schreider, T. (2019). *Building an Effective Cybersecurity Program*. Rothstein Publishing.

28. Husák, M., et al. (2018). Survey of attack projection, prediction, and forecasting in cyber security. *IEEE Communications Surveys & Tutorials*, 21(1), 640–660.
29. Halima, N. B., Hosam, O. (2015). Embedding image ROI watermark into median DCT coefficients. *IRECOS*, 10(6).
30. Liang, Wei, Zhang, Dafang, You, Zhiqiang, Li, Wenwei, & Hosam, Osama. (2013). A survey of techniques for VLSI IP protection. *Information Technology Journal*. doi:10.3923/itj.2013, published June 12, 2013.
31. Hosam, Osama. (2013). Side-informed image watermarking scheme based on dither modulation in the frequency domain. *The Open Signal Processing Journal*, 5, 1–6.
32. Hosam, Osama, Sheta, W. M., Youssef, Bayumy A. B., & Abdou, M. A. (2012). Public watermarking scheme for 3Ds laser scanned archeological models. The Seventeenth IEEE Symposium on Computers and Communications (ISCC'12). July 2012, Cappadocia, Turkey.
33. Zhou, J., Ma, J., & Jia, W. (2016). A gap analysis method for evaluating the security of cloud services. *Journal of Computer Science and Technology*, 31(4), 779–792.
34. Ramachandran, K., & Suresh, S. (2015). A gap analysis of ISO 27001: 2013 implementation in small and medium enterprises. *International Journal of Advanced Research in Computer Science and Software Engineering*, 5(6), 538–543.
35. Saponas, E. S., Caralli, R. A., & Goodall, J. R. (2013). A control-based approach to conducting a security gap analysis. *Journal of Computer Science and Technology*, 28(1), 1–11.
36. DiGiambattista, E., & Gagliardi, M. J. (2011). A threat-based approach to conducting a security gap analysis. *Journal of Computer Science and Technology*, 26(1), 1–11.

6 A Comparative Study of Structureless Data Aggregation Protocols in Wireless Sensor Networks (WSNs)

Faizal N Vasaya, J Sandeep, and Libin Thomas

6.1 INTRODUCTION

WSNs have become an integral part of daily lives because of their varied applications in fields ranging from health care to military surveillance (Ian and Mehmet 2010). WSN contain sensors that act as data generating and forwarding nodes. A sensor node consists of sensing equipment along with a transceiver to transfer the sensed data. These nodes are deployed based on the application requirements, especially when the environment is critical to access (Jennifer, Biswanath and Dipak 2008).

WSNs can be applied in various areas like military operations (Simon et al. 2004), (Yick, Mukherjee and Ghosal 2005), natural disasters or human-caused disaster, health-related emergencies (Hande and Cem 2010), monitoring nuclear plants, etc. These applications can be classified as monitoring applications like monitoring agriculture, industrial process and production, etc. The application can be about tracking, like tracking wildlife, machines, vehicles and gadgets (Kay, Win and Meng 2005). The sophisticated and efficient communication protocols have realized the real potential of sensor networks. Generally, WSN are implemented in large numbers where the nodes are installed at random positions and sometimes installed with a predefined position in an structure form. A WSN node is a battery-operated component with constrained energy, computation, and memory capacities. These constraints are the key issues in establishing an efficient WSN with longer life span.

Current energy-efficient routing protocols provide a quick fix for designing and developing WSNs in different environments (Gaddafi et al. 2015). Existing protocols enable nodes deployment in a structured pattern or in a self-organizing manner. These self-organizing protocols are also referred to as virtually structured protocols for sensor networks.

Sensor nodes are usually deployed in harsh conditions with a battery-powered source of power supply that is often inadequate for the various operations and data collections processes that a sensor node needs to take care of (Sasirekha and Swamynathan 2017). As sensors are constrained by limited power, energy utilization

DOI: 10.1201/9781003343332-6

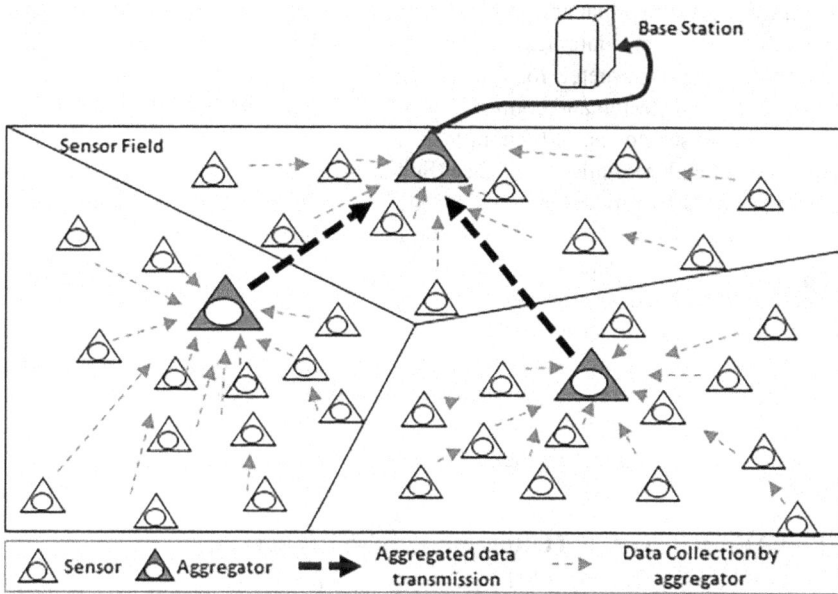

FIGURE 6.1   Data aggregation in wireless sensor networks.

is a critical concern in WSNs. Due to this reason, considerable significance has been given for research in the field of minimizing energy usage in WSNs. Efficient energy usage plays an important role in increasing network lifetime. The energy depletion of a particular node significantly affects the functioning of entire network. Hence energy should be economically used to increase network life span. Increase in network life span can be achieved through efficient and conscientious routing policies (Yongrui et al. 2015), (Ioannis, Wei and David 2007), smart sensor placement methodologies (Mustapha et al. 2014)(Carlos, Ivica and Luis 2009), and protocols that provide lofty data aggregation ratios (Feng and Jiangchuan 2011). Reduction in number of transmissions assists in increasing sensor lifetime and lowered bandwidth usage. This is possible by aggregating data from different sensor nodes and then forwarding it to sink. Figure 6.1 shows the network where aggregators collect the data from their nearest sensors. This collected data is then forwarded to base station for further processing. Data aggregation assists in collecting and fusing only useful information from the multiple sensors. These fused data is then efficiently routed to the base station. Increased aggregation ratios play an integral part in energy conservation.

Spatial convergence and the temporal convergence are the important criteria for improving data aggregation in the network. Spatial convergence focuses on establishing efficient routes from discrete sensor nodes to the sink whereas temporal convergence is responsible for efficient timing and control. There are many protocols developed and implemented for data aggregation in the network. These protocols can be divided into two categories as structured data aggregation protocols and structureless aggregation protocols. In Yadav and Yadav (2015), the authors have discussed about the design issues that are prevalent in wireless sensor networks based on structured and structureless data aggregation protocols. They identified that the field of energy conservation targeted for data collection and aggregation could be collaborated with data clustering and communication (Yadav and Yadav 2015). Energy consumption of sensor nodes, redundant transmission, node failures, etc., are some of the problems that largescale WSNs face (Subhashree and Tharini 2017). Energy consumption issue seems to be the one that requires the most amount of attention in the current situation as it is highly improbable for sensor nodes to have constant power supply when they are deployed at remote locations. In Dehkordi et al. (2020), the authors have reviewed various data aggregation techniques and protocols. Numerous data integration techniques were covered by the authors, and their focus was on modelling advanced designs for sensor networks using such updated protocols (Dehkordi et al. 2020). A review of hierarchical energy efficient routing protocols has been provided in Guleria and Verma (2019). The protocols have been categorized based on classical and swarm intelligence approaches.

6.2 VARIOUS STRUCTURED AND UNSTRUCTURED DATA AGGREGATION PROTOCOLS IN WSN

This section briefly discusses about some of the recent developments in structured and unstructured data aggregation protocols for WSN.

6.2.1 STRUCTURED DATA AGGREGATION

All the structures data aggregation algorithms are based on tree structure or based on clustering. In cluster-based data aggregation, the nodes of WSN are organized into different clusters based on performance and geographical features. Each cluster will have a head node which is responsible to aggregate data from the nodes in the same cluster. This aggregated data is then optimally forwarded to the base station. There are many data aggregation algorithms in this category that transfer the data through the optimal path to sink (Wendi, Anantha and Hari 2002), (Lingjun, Huazhong and Yun 2011) (Ossama and Sonia 2004). Compared to LEACH, DACP has an additional feature of data prediction which increases data aggregation ratios. HEED provides a solution to avoid two sensor nodes in the transmission range to become cluster head at the same time by considering its relative position (Ossama and Sonia 2004).

Tree-based approach works by constructing a minimum spanning tree structure that connects all the nodes and then transferring the aggregated data upstream until it reaches the sink. Mingxin Yang proposed a tree-based approach which reduces energy consumption by constructing minimum spanning tree and applying information entropy to find relationship among the sensed data (Mingxin 2015). Another tree-based data aggregation protocol was proposed (Fen et al. 2015) which deals with reducing the energy consumption of the network and increasing the lifetime of the path used for data aggregation. The approach makes use of tree structure and mixed integer linear programming for data aggregation. The protocol works in three modes: full, non-full, and hybrid partial aggregation modes (Fen et al. 2015).

6.2.2 STRUCTURELESS DATA AGGREGATION PROTOCOLS

A major drawback in structured data aggregation protocols is the cost involved in creation and maintenance of network structure. Structureless data aggregation is an approach that overcomes this overhead. Each node sends an anycast request To send (RTS) along with event data to figure out next hop to the base station. Next hops are selected based on the similarity of event data or its closeness to sink. The next hop sends back a Clear To Send (CTS) packet which begins the transmission.

Researchers have come up with many structureless data aggregation solutions. These solutions have its own pros and cons. The major motivation for this work was to analyze the design and working of different approaches and their impact on the performance of WSN. This review helps in identifying the research gaps to be filled up with systematic investigations and enhancements. The paper discusses some of the recent virtually structured data aggregation protocols with their advantages and overheads. Furthermore, this paper analyzes the protocols based on different performance measures.

6.2.2.1 Data Aware Forwarding and Random Scheduling

Data aggregation is an effective approach to save certain amount of resources in a wireless sensor network. This theory helps to reduce the number of transmissions which in turn saves energy. Another research conducted by Kai-Wei, Sha, and Prasun (2007) showed an efficient data aggregation protocol without explicit maintenance of the structure (Kai-Wei, Sha and Prasun 2007). The aggregation protocol developed in Kai-Wei, Sha, and Prasun (2007) achieved the advantages of early aggregation and fault tolerance. To achieve greater aggregation levels, it uses the concept of "gather before transmit" where a node closer to the sink is made to wait for a longer time before forwarding its data. This scheme covers both spatial and temporal data aggregation which led to higher aggregation ratios. The data gathering process begins when each node sending an anycast RTS to determine the next hop to sink. The node that receives this RTS serves as a candidate for the next hop. At this stage, it is assumed that all the nodes are aware of the geographic location of the one hop neighbors as well as the sink. To achieve higher aggregation ratios, the node which had the same data to forward or its closeness to the sink was given a higher priority to reply with clear-to-send (CTS). The randomized waiting (RW) scheme enables the protocol to reduce the number of transmissions. All the nodes that had a data to report had a random waiting time to start their transmission. Data aware anycast (DAA) follows media access control (MAC) layer anycasting, where a node with one hop distance from the sending node is selected. A major advantage of DAA+RW is that it does not require reconstruction of network structure, whereas in the case of structured approaches, a node failure may involve the overhead of network reconstruction.

6.2.2.2 Energy Aware Data Aggregation

In Nen-Chung et al. (2007), a grid-based routing protocol termed as energy aware data aggregation (EADA) is proposed. It proved that EADA algorithm efficiently reduces energy consumption as well as data collision in a sensor network. In the first phase, an M × N grid is formed whose cells are identified using XY coordinates. The next phase selects a gateway node. A gateway node is selected on the basis of the remaining residual energy. Whenever the gateway node detects an event, it informs to all the gateway nodes in the grid. Now, the mobile sink sends a request packet to the gateway node of the cell of the same grid where the sink is located. This cell's gateway node sends a query message to collect data from the interested gateway node where an event has occurred.

6.2.2.3 Real-Time Data Aggregation

The authors of Mohammad et al. (2011) proposed another structureless aggregation protocol called real-time data aggregation (RDAG). In this protocol, routing uses the real-time data and implements appropriate in predefined policies. This protocol makes use of geographical location of self and sink node. The protocol applies the concept of first in, first out (FIFO) queue to hold the packets at the aggregator nodes. The two major attributes used in the computation of routing are the packet's time-to-live (TTL) and the Euclidean distance from the sink. This protocol also measures estimated hop delay (EHD) for efficient data aggregation. It makes dynamic forwarding decision by sending packets only to those nodes which have similar kinds of packets waiting in the queue. A judicial waiting policy was proposed to achieve temporal convergence. This policy was applied to all the packets in the network by allocating each of them with a slack time in proportion to the destination hop count. Thus, their waiting time policy not only improved data aggregation but also helped during the periods of high contention without any synchronization within the nodes.

6.2.2.4 Data Aggregation Using Inverse Square and Survival Analysis

The authors of Vinayaga et al. (2011) proposed an efficient structureless approach for data aggregation based on inverse square law and survival analysis (Vinayaga et al. 2011). The approach was successful in reducing the number of transmissions along with an increase in packet delivery ratio. The authors defined a problem that DAA (Wendi, Anantha and Hari 2002) transmits CTS and RTS packet for every data transmission, which increases the workload. The protocol derives two step frameworks for data aggregation and forwarding. In the first step, it uses inverse-square law and survival analysis to detect their nearby neighbor nodes. Along with these two laws, it uses Kaplan Meier Estimator (KPE), which denotes the probability of a living node to survive. For delay calculation, it uses an enhanced random delay (ERD) where maximum delay value depends on the distance between a node and the sink. Further, it increases network efficiency by providing a fault tolerance mechanism (FTM), in which each node sends a beacon signal and starts the timer. If the node does not respond in the required time, it is considered to be failed. This failure message is also sent to other nearby nodes.

6.2.2.5 Energy-Efficient Data Aggregation Protocol
for Structureless Network

The authors of Chih-Min and Tzu-Ying (2014) came up with a structure-free and energy balanced (SFEB) protocol. The main aim was to reduce energy consumption by reducing the number of transmissions. It was assumed that SFEB works in a multi-hop network where each node is aware of the location as well as sink. This model works in two stages. In the first stage nodes are selected dynamically for the role of aggregators to collect data from the nearby nodes. Then the second stage helps in forwarding the collected data to the destination sink. To accomplish the task in phase one the entire network is divided into virtual parallelograms. After successful division, a primary aggregator (PA) is selected based on its nearness to the parallelogram corner. It also selects a secondary aggregator (SA), within the range of PA. In this protocol, the aggregators are selected in a hexagonal pattern, which reduces the number of aggregators further reducing the number of transmissions. The phase two deals with forwarding the collected data back to sink. This is attained by making the rear aggregators (RA) forward their data to the next nodes towards the sink.

6.2.2.6 Cycle-Based Data Aggregation Scheme

The authors of Yung-Kuei, Neng-Chung and Chih-Hung (2013) have proposed a grid-based algorithm for WSN. The model works on four stages. It constructs the grid structure for the network, identifying the cell heads for data collection, scheduling and cycle formation for data collection, and finally, transmission of collected data. To form the structure across the network, the entire sensor field is partitioned into $M * N$ cells, where each cell has an identity associated with it. Each cell in CBDAS consists of a head node (aggregator) and other nodes. In the next stage, cell heads are opted based on the energy. The third stage deals with construction of a cyclic chain to send data to its uplink as well as downlink node. This links up all the cell heads which result in a Hamiltonian cycle. In the data transmission stage, data is aggregated and transmitted to the sink node. The sink node selects a cycle head node from the set of head nodes based on the current residual energy. The sink node sends two packets to the cycle leader, which is recursively sent to the neighboring nodes. Finally, the head nodes transfer data to the cycle leader which forwards it to the sink node. The major disadvantage of this scheme is an overhead of maintaining a table for each sensor node to store its geographic location, cell ID, and current cell head. Extra packets are passed to form a cycle of head nodes which serve as the drawback to this scheme.

6.2.2.7 Wheel-Based Event Triggered Data Aggregation and Routing

Sutagundar and Manvi (2013) have come up with data aggregation and routing scheme for optimal routing in the network that uses wheel-based concept. This scheme works in seven phases. In the first phase, a dynamic wheel structure is constructed, and a path is discovered to the sink node. The second phase deals with the identification of secondary Aggregators (SA) along the spokes considering the factors like distance, energy, spoke angle, and signal strength. In the third phase, SA on the spoke aggregates the data from the neighboring nodes. The fourth phase deals with invoking event detection and classification process. In the fifth phase, boundary nodes aggregate data from the SAs along the spoke of the wheel. The sixth phase deals with maintaining the wheel structure. Finally, the aggregated data is forwarded to the sink along the selected path.

6.2.2.8 Virtual-Grid-Based Dynamic Routes Adjustment Schemes

Abdul et al. (2013) have designed a virtual-grid-based dynamic routes adjustment (VGDRA) scheme. The main aim was to minimize route reconstruction cost to optimally route data to the sink's new locations. Instead of using the grid construction approaches used in CBDAS (Yung-Kuei, Neng-Chung and Chih-Hung 2013) and dual-path-based data aggregation scheme (DPBDAS), this scheme uses heuristics used in LEACH (Wendi, Anantha and Hari 2002) and combines the features of other protocols (Manjeshwar and Agrawal 2001, 2000; Ghosh, Roy and Das 2009) to form a hybrid algorithm. In each cell, node closest to cells midpoint is selected as cell head. This approach of head selection consumes lesser energy as GPS locations are directly used for its selection. After the node head is selected, notifications are sent to neighboring nodes in the same cell as well as in the nearby nodes of the neighboring cell. This enables the protocol to identify cell adjacency, which forms a virtual backbone structure. Finally packets are forwarded to the sink through the optimal route. The major advantage of this scheme is that the routing information is stored only in the head node of each cell. Each time the base station changes its location, the routes to the sink are dynamically changed, which results in the optimal routes.

6.2.2.9 Dual-Path-Based Data Aggregation Scheme

Neng-Chung et al. (2014) proposed a multipath data aggregation scheme for grid-based network. This work is an enhancement to data aggregation scheme proposed by the same authors (Yung-Kuei, Neng-Chung and Chih-Hung 2013). The only difference between both the schemes is that the later selects two cell heads within a grid (cell head A and cell head B) and then connects all cell heads A and cell heads B resulting in a path for aggregated data transfer to the sink, all other procedure being same as the earlier version of the protocol.

6.2.2.10 Fermat-Point-Based Data Aggregating Routing Protocol

Kaushik, Pradip, and Sarmistha (2015) came up with an energy-efficient data aggregating protocol that uses Fermat point concept. The main aim was to come up with a distance vector protocol for data aggregation WSN. Advantage of this protocol is that it does not require any virtual structure like VGDRA (Abdul et al. 2013) and DPB- DAS (Neng-Chung et al. 2014). In this protocol, after the nodes have been deployed, a Fermat point is selected by forming a logical triangle/polygon from the source node to multiple sinks. This protocol used Minima algorithm to find a Fermat point between source and sinks (Ghosh, Roy and Das 2009). Now, the transmission begins by first transferring data from source to Fermat node (FN) and then from FN to sink. In this approach, the distance between the node and source impacts on the lifetime of network.

6.2.2.11 Directional-Segregation-Based Data Dissemination Protocol

Shubhra, Suraj, and Neeraj (2015) developed a vertical and horizontal segregation based data dissemination protocol. The major aim was to reduce the amount of energy consumed in informing the nodes about current location of the base station. This protocol consists of five phases starting from backbone creation, finding immediate hop nodes, identify the path based on tree structure, sharing base station location, and data distribution. In the backbone creation stage, the square sensor field will be divided into four equal parts. Again each of these four areas is divided into two equal parts by 45-degree angle from the center. All the nodes falling inside the horizontal and vertical strips are known as spine nodes. In the neighbor discovery stage, all the regular nodes will send packets to spine nodes for sharing locations. The next stage deals with the decision of nodes region discovery, which helps to identify the belongingness of regular nodes to a part of horizontal and vertical strips. Then the sink sends its location to the nearest node. This information is then sent to all the spine nodes that form the backbone. The regular node then communicates with nearest spine node to get the sink location information and then forward the data in a multi-hop manner.

6.2.2.12   Cluster-Chain Mobile Agent Routing

In Sasirekha and Swamynathan (2017), a routing protocol that can improve the existing network topologies has been proposed. This approach uses LEACH algorithm for building clusters of sensor nodes and identifying the cluster head for communication by header voting. The data collection with a power-efficient approach is used to reduce the high energy overhead. This is achieved by implementing long chain topologies with chain headers while avoiding periodic header voting (Sasirekha and Swamynathan 2017). Thus, the best of both have been utilized to implement CCMAR. However, the implementation faced security and fault tolerance related issues.

6.2.2.13   Energy-Efficient Data Aggregator Using Multiple Sink Node

The authors from Manishankar, Ranjitha, and Manoj (2017) prioritize the process of data collection in an energy-efficient manner and propose a method for the same. First it creates an efficient cluster within the network, and then data aggregation is done from the sensor node to the data node. LEACH has been utilized to create clusters of nodes within the network and also tries to improve the efficiency of data collection and forwarding process (Manishankar, Ranjitha and Manoj 2017).

6.3   ANALYSIS ON DESIGN FEATURES OF DIFFERENT DATA AGGREGATION PROTOCOLS

This section discusses on the structure of existing protocols. Design of protocol and additional features included were compared to relate with real-time requirements. Table 6.1 reviews the existing protocol structure, which also includes the type of packets being used. Table 6.2 presents the processes involved in the protocols with the overheads, and it also reflects the advantages.

6.4   FACTORS AFFECTING THE PROTOCOL'S PERFORMANCE

The efficiency of a data aggregation protocol depends on many factors. Some of the performance-based aspects are structure maintenance cost, mobility, routing table maintenance and consideration of sleep mode. Nodes in a wireless sensor network are randomly deployed in the sensor field. Each node needs to be aware of its geographical location. The location details are used for distance calculations while forming a virtual communication structure. A mobile node causes frequent update of location tables to attain maximum route efficiency. Node movement thus consumes energy to

TABLE 6.1

Review of Different Data Aggregation Protocols for Their Design

S.No	Existing methods	Structure	Processing phases involved	Type of packets
1	Data-Aware Anycast and Randomized waiting (DAA+RW).	Anycast based	Sending an anycast request to determine next hop to sink. Selecting a node for the next hop from the candidates who have received RTS.	RTS (Request to send), Class A (Same AID closer to sink), Class B (Same AID farther away from sink), Class C (Different AID closer to sink), CTS (Clear to send)
2	Energy-Aware Data Aggregation (EADA)	Virtual Grid	Randomized waiting grid construction. Gateway node selection. Event registration. Data collection	Even registration packet. Query message packet. Tree construct packets
3	Real-time Data Aggregation (RDAG)	No virtual shape formation	Real-time data aware routing. Judiciously waiting	Holding packets using FIFO queue. Time to Live (TTL). Estimated One hop Delay (EHD).
4	Data aggregation using Inverse Square and Survival analysis (DAISSA)	No virtual shape formation	Applying Inverse Square to identify distance. Survival Analysis for Event Density function. Kaplan Meier estimator to calculate the lifetime probability.	Beacon signal
5	Structure Free and Energy Balanced (SFEB) Data Aggregation	Virtual Parallelogram and sensor placement in hexagonal pattern.	Dynamically choosing the aggregators. Collecting and Forwarding it to sink node.	PA request (Primary Aggregator), SA request (Secondary Aggregator), RTS, CTS
6	Cycle Based Data Aggregation Scheme (CBDAS)	Virtual Grid	Grid Construction. Cell Head selection. Cycle Formation. Data transmission	Head electing. Head confirming. Path forming. Path reply. Token to Cycle Leader

7	Wheel based Event Triggered data aggregation and routing (WETdar)	Virtual Wheel Structure	Wheel Structure Construction. Identification of Secondary Aggregators (SA). Data aggregation by SA's. Event Detection and classification. Data aggregation from boundary nodes. Maintaining Wheel structure. Data dissemination to sink	Event triggering packet. Spoke generation packet
8	Virtual Grid Based Dynamic Routes Adjustment Scheme (VGDRA)	Virtual Grid	Grid construction. Cell head selection. Identification of optimal route to sink. Data dissemination	Sink location update packet. Route update packet
9	Dual-Path-Based Aggregation Scheme (DPBAS)	Virtual Grid	Grid Construction. Two cell head selection. Two cycle formation. Data transmission	Head electing, Head confirming, Path A forming, Path A reply, Path B forming, Path B reply. Token to Cycle Leader of A group. Token to Cycle Leader of B group
10	Fermat Point Based Energy-Efficient Data Aggregation routing Protocol (KPS).	No Structure	Node Deployment. Fermat point selection. Data transfer from source to FN. Aggregated data forwarding to sinks from FN.	Neighbor becon control packet
11	Vertical and Horizontal Segregation Based Data Dissemination Protocol (VHS-DDP)	Square division into Octants.	Backbone creation. Neighbor discovery. Tree construction. Sink location information. Data dissemination	NBR MSG, NBR REPLY, T MSG, HELLO MSG, HELLO REPLY, S MSG, SLQ, SLR

TABLE 6.2
Comparison of Various Protocols

Protocols	Aggregator node selection	Overheads	Assumptions	Type	Advantages
DAA+RW	Aggregation ID generated for each packet. Any node near to sink can serve as an aggregator	Increase in load because of CTS and RTS transmission which can reduce battery performance. Tree construction overhead	Node is aware about the geographic location of its one hop neighbours. Interference range is twice as transmission range. Nodes are time-synchronized. Static sensor nodes. Multi-hop communication. Aware of its geographic location. Location table of all the gateway nodes at the sink. Bidirectional Channels	Data aware anycasting at MAC layer	No overhead of constructing the structure. Reduced number of transmissions.
EADA	Cell head selection based on nodes energy	Event registration overhead. Increased rerouting overhead	Devices are aware of their geographic location. Limited energy of sensors. Multi-hop communication of nodes.	Grid-based	Reduced overhead of discovering new paths to sink. Increased network lifetime
RDAG	Dynamic routing of packets. Any node with similar packets can serve as aggregator.	Artificial Waiting policy. Estimation of end to end delay. Calculating the required velocity based on packet's TTL.	Failed node is a data forwarding node.	Structure less Dynamic forwarding.	No explicit maintenance of structure. Implementing artificial delays resulting in increase in temporal convergence
DAISSA	Dynamic forwarding of packets to the nodes with similar packets. Any node with greater survival probability can serve as aggregator.	Probability Value Calculation for each node	Failed node is a sensor node as well as aggregator node.	Structure less Dynamic forwarding.	Node failures are considered. High packet delivery ratio because of node failure avoidance mechanisms

SFEB	Primary Aggregator (PA) is selected based in the nearness of the parallelogram corners.	High Computation overhead. Longer Waiting in phase one may be in vain if aggregation levels are low	Each node is aware about its own as well as sinks geographic location. SFEB operates in a multi-hop network	Geometric shape based	Increase in event size has a little effect on aggregation ratio
CBDAS	Secondary Aggregator is selected within the range of PA Cell head based on remaining residual energy. Cycle leader selection based on distance from Base station.	Passing of extra packets to form a Hamiltonian cycle. Maintaining the cycle.	Static sensor nodes. Base station located far away. Location aware. Nodes are aware about Residual energy. Bidirectional channels.	Grid-based	Easy cyclic chain formation. Extended node lifetime
WETdar	Spoke Aggregators (SA) are along the spokes of the wheel. Aggregator selection is based on nodes energy and mobility.	Dynamic wheel structure maintenance. Calculation of Euclidean distance and spoke angle	Multi-ho communication with sink on event detection. Static nodes. Each sensor node has security features to perform inter agent communication.	Wheel Based	Two level aggregation increases aggregation ratio
VGDRA	Cell head selection based on nearness to the midpoint of the cell. Re-election considers midpoint closeness as well as residual energy.	Route updation at cell heads because of sink movements. Optimal route calculation based on sink location.	Random node deployment. Static sensor nodes and moving sink. Unlimited energy available for the mobile sink.	Grid-based	Optimal routing path is determined. Reduced reconstruction cost.
DPBAS	Cell head A and cell head B selection based on residual energy. Cycle leader A and Cycle leader B selection based in distance from base station.	Passage of extra packets to form cycles of head A and head B Token passing from base station for collection of aggregated data	Static sensor nodes. Base station located far away. Location aware. Nodes are aware about residual energy. Bidirectional channels.	Grid-based	Evenly distributed energy depletion. Dual path transmission

(Continued)

TABLE 6.2 (Continued)
Comparison of Various Protocols

Protocols	Aggregator node selection	Overheads	Assumptions	Type	Advantages
KPS	It's a Fermat Node located on the fermat point calculated by Minima algorithm.	Fermat point calculation. Multi-hop transfer to far away Fermat nodes. Confirming the residual energy above threshold of a Fermat node after each transmission.	Fermat point are near to the source node. Multiple sink are located at the edges of the network. Source node are aware about the position of Fermat node.	Random Distribution.	No structure maintenance overhead. Reduced disadvantage of greedy packet forwarding algorithm.
VHS-DDP	Spine nodes in the stripes forms the backbone.	Octants creation by dividing square into 8 parts. Large number of communication packets transferred between spine and regular nodes.	Location aware sensor nodes. Static nodes. Square-shaped sensor field.	Geometric Shape based.	Reduced number of transmissions to send sinks location.

TABLE 6.3

Comparison of Existing Data Aggregation Protocols

Protocols	Node movement	Location aware	Base station	Node failure reflection	Routing table	Sleeping nodes	Interference concern
DAA+RW	Dynamic	True	Static	Yes	At each node	Not implemented	Handled
EADA	Static	True	Mobile	No	Cell heads	Implemented	Not Handled
RDAG	Dynamic	True	Mobile	Yes	At each node	Not Implemented	Handled
DAISSA	Dynamic	False	Mobile	Yes	At each node	Not Implemented	Handled
SFEB	Dynamic	True	Static	Yes	PA and SA	Not Implemented	Handled
CBDAS	Static	True	Static	No	Cell heads	Implemented	Not Handled
WETdar	Static/ Dynamic	True	Static	Yes	Spoke Aggregator	Implemented	Handled
VGDRA	Static	True	Mobile	Yes	Cell heads	Not Implemented	Not Handled
DPBAS	Static	True	Static	No	Cell heads	Implemented	Not Handled
KPS	Static	True	Static	No	Fermat Point	Not Implemented	Not Handled
VHS-DDP	Static	True	Mobile	No	Spine nodes	Not Implemented	Not Handled

inform neighboring nodes about their current location by passing numerous control packets. Movement of base station causes a change in the network structure. This incurs a considerable usage of node energy as it transfers control packets to retrieve sink's current location. Factors such as energy depletion, link breakage, physical damage, and malicious attacks can cause a node to fail. A node failure might result in the breakdown of a part or an entire network. Many dynamic approaches have been proposed in the recent years to provide an alternative route in such situations. Studies have also proved that the dynamic structured approach has higher maintenance cost compared to that of static approaches. The amount of routing information stored at each node should be limited because of the memory restrictions in sensor nodes.

There are networks which generate data for transmission based on the occurrence of an event, whereas some networks generate data at frequent time interval. A pre-defined downtime can be set for the nodes which are ideal for times without any data to transmit in the network. This approach aids in increasing energy conservation as sleeping nodes consume very little battery power. A sensor network consists of nodes with short transmission ranges and limited bandwidth. Random node deployment increases the chances of intersection between transmission ranges of different nodes. This interference may cause packet collision that can result in invalid event data or complete loss of packets. Based on the previously discussed performance aspects, a study was conducted to analyze data aggregation protocols which are briefed in Table 6.3.

6.5   CONCLUSION

Optimizing energy consumption of the network in its data collection and forwarding process is one of the crucial aspects in the development of data aggregation protocols for WSN. This paper constructs a survey on the structureless data aggregation protocols which works with a virtual structure formation. Wireless sensor network has difference in requirement for its wide range of applications. Movement of nodes or a node failure can cause major changes in the network structure. This causes the reconstruction of network structure which consumes a considerable amount of battery power. A data aggregation protocol must consider these factors to attain appreciable aggregation ratios. Study shows wide diversity in the design of protocols that suits different requirements. Design and features of the existing approaches have been abstracted to visualize the research directions. Protocols were compared from their design prospectus that helps in understanding the paradigm for future requirements.

REFERENCES

Abdul W., Abdul H., Mohammad A., et al.: 'VGDRA: A virtual grid based dynamic routes adjustment scheme for mobile sink based wireless sensor networks', *Sensors Journal, IEEE*, Vol. 15, 2013, pp. 526–534.

B. Vinayaga S., Rajesh G., Khaja M., et al.: 'An efficient approach for data aggregation routing using survival analysis in wireless sensor networks', *International Journal of Wireless and Mobile Networks (IJWMN)*, Vol. 3, No. 2, April 2011, pp. 243–251.

Carlos E., Ivica K., Luis D.: 'A decision-making methodology for stochastic deployment of wireless sensor networks', UKSim 2009: 11th International Conference on Computer Modelling and Simulation, IEEE, 2009.

Chih-Min C., Tzu-Ying H.: 'Design of structure-free and energy-balanced data aggregation in wireless sensor networks', *Journal of Network and Computer Applications*, Vol. 37, 2014, Elsevier, pp. 229–239.

Dehkordi S.A., Farajzadeh K., Rezazadeh J., Farahbakhsh R., Sandrasegaran K., Dehkordi M.A., "A survey on data aggregation techniques in IoT sensor networks", *Wireless Networks*, Vol. 26, 2020, pp. 1243–1263.

Fen Z., Zhenzhong C., Song G., et al.: 'Maximizing lifetime of data-gathering trees with different aggregation modes in WSN', 2015 IEEE Global Communications Conference (GLOBECOM), IEEE, 2015, pp. 1–6.

Feng W., Jiangchuan L.: 'Networked wireless sensor data collection: Issues, challenges, and approaches', *IEEE Communications Surveys and Tutorials*, Vol. 13, No. 4, Fourth Quarter 2011.

Gaddafi A., Abdul H., Mohammad H., et al.: 'A comparative analysis of energy conservation approaches in hybrid wireless sensor networks data collection protocols', *Telecommunication Systems*, Vol. 61, Springer, 2015, pp. 159–179.

Ghosh K., Roy S., Das P.K.: 'An alternative approach to find the fermat point of a polygonal geographic region for energy efficient geocast routing protocols: Global minima scheme', NETCOM'09, First International Conference on Networks and Communications, IEEE, 2009, pp. 332–337.

Guleria K., Verma A.K., "Comprehensive review for energy efficient hierarchical routing protocols on wireless sensor networks", *Wireless Networks*, Vol. 25, 2019, pp. 1159–1183.

Hande A., Cem E.: 'Wireless sensor networks for healthcare: A survey', *Computer Networks*, Vol. 54, Elsevier B.V., 2010.

Ian F., Mehmet C.: *Wireless Sensor Networks*, Wiley, 2010.

Ioannis C., Wei L., David S.: 'Asymptotically optimal transmission policies for large-scale low-power wireless sensor networks', *IEEE/ACM Transactions on Networking*, Vol. 15, No. 1, February 2007.

Jennifer Y., Biswanath M., Dipak G.: 'Wireless sensor network survey', *Computer Networks*, Vol. 52, Elsevier B.V, 2008.

Kai-Wei F., Sha L., Prasun S.: 'Structure-free data aggregation in sensor networks', *IEEE Transactions on Mobile Computing*, Vol. 6, No. 8, August 2007, pp. 929–942.

Kaushik G., Pradip K., Sarmistha N., 'KPS: A Fermat point based energy efficient data Aggregating routing protocol for multi-sink wireless sensor networks', in *Advanced Computing and Systems for Security*, Springer 2015.

Kay S., Win N., Meng J.: 'Wireless sensor networks for industrial environments', Proceedings of the 2005 International Conference on Computational Intelligence for Modelling, Control and Automation, and International Conference on Intelligent Agents, Web Technologies and Internet Commerce (CIMCA-IAWTIC'05).

Lingjun M., Huazhong Z., Yun Z.: 'A data aggregation transfer protocol based on clustering and data prediction in wireless sensor networks', Wireless Communications, Networking and Mobile Computing (WiCOM), 2011 7th International Conference IEEE, 2011, pp. 1–5.

Manishankar S., Ranjitha P. R., Manoj K. T., "Energy efficient data aggregation in sensor network using multiple sink data node ", International Conference on Communication and Signal Processing, 2017, pp. 448–452.

Manjeshwar A., Agrawal D.P.: 'APTEEN: A hybrid protocol for efficient routing and comprehensive information retrieval in wireless sensor networks', Proceedings 16th International Parallel and Distributed Processing Symposium, IEEE, April 2001.

Manjeshwar A., Agrawal D.P., 'TEEN: A routing protocol for enhanced efficiency in wireless sensor networks', Proceedings 15th International Parallel and Distributed Processing Symposium, IEEE, April 2000, pp. 2009–2015.

Mingxin Y.: 'Constructing energy efficient data aggregation tree based on information entropy in wireless sensor networks', IEEE Advanced Information Technology, Electronic and Automation Control Conference (IAEAC), 2015, pp. 527–531.

Mohammad H., Hamed Y., Naser A., et al.: 'RDAG: A structure-free real-time data Aggregation protocol for wireless sensor networks', IEEE 17th International Conference on Embedded and Real-Time Computing Systems and Applications, IEEE, 2011, Vol. 1, pp. 51–60.

Mustapha R., Mohamed E., Mohamed A., et al.: 'A smart methodology for deterministic deployment of wireless sensor networks', Smart Communications in Network Technologies (SaCoNeT), 2014 International Conference, IEEE, 2014, pp. 1–6.

Nen-Chung W., Yung-Fa H., Jong-Shin C., Po-Chi Y.: 'Energy-aware data aggregation for grid-based wireless sensor networks with a mobile sink', *Wireless Personal Communications*, Vol. 43, Springer, 2007, pp. 1539–1551.

Neng-Chung W., Yung-Kuei C., Young-Long C., et al.: 'A dual-path-based data aggregation scheme for grid based wireless sensor networks', 2014 Eighth International Conference on Innovative Mobile and Internet Services in Ubiquitous Computing, IEEE, 2014.

Ossama Y., Sonia F.: 'HEED: A hybrid, energy-efficient, distributed clustering approach for ad hoc sensor networks', *IEEE Transactions On Mobile Computing*, Vol. 3, No. 4, October–December 2004, pp. 366–379.

Sasirekha S., Swamynathan S., "Cluster-chain mobile agent routing algorithm for effificient data aggregation in wireless sensor network", *Journal of Communications and Networks*, Vol. 19, No. 4, 2017.

Shubhra J., Suraj S., Neeraj B., 'A vertical and horizontal segregation based data dissemination protocol', in *Emerging Research in Computing, Information Communication and Applications*, Springer India 2015.

Simon G., Maroti M., Ledeczi A., et al.: 'Sensor network-based counter sniper system', Proceedings of the Second International Conference on Embedded Networked Sensor Systems (Sensys), Baltimore, MD, 2004.

Subhashree V.K, Tharini C., "An energy efficient routing and fault tolerant data aggregation algorithm for wireless sensor networks", *Journal of High Speed Networks*, Vol. 23, 2017, pp. 15–32.

Sutagundar A.V., Manvi S.S.: 'Wheel based event triggered data aggregation and routing in wireless sensor networks: Agent based approach', *Wireless Personal Communications*, Vol. 71, Springer, 2013, pp. 491–517.

Wendi B., Anantha P., Hari B.: 'An application-specific protocol architecture for wireless microsensor networks', *IEEE Transactions on Wireless Communications*, Vol. 1, No. 4, October 2002, pp. 1536–1276.

Yadav S., Yadav R.S, 'A review on energy efficient protocols in wireless sensor networks', *Wireless Networks*, Vol. 22, 2015, pp. 335–350. https://doi.org/10.1007/s11276-015-1025-x

Yick J., Mukherjee B., Ghosal D.: 'Analysis of a prediction-based mobility adaptive tracking algorithm', Proceedings of the IEEE Second International Conference on Broadband Networks (BROADNETS), Boston, 2005.

Yongrui C., Li G., Yulong X., et al.: 'Cross-layer design for energy-efficient reliable routing in wireless sensor networks', 2015 11th International Conference on Mobile Ad-hoc and Sensor Networks (MSN), IEEE, 2015.

Yung-Kuei C., Neng-Chung W., Chih-Hung H.: 'A cycle based data aggregation scheme for grid based wireless sensor networks', Innovative Mobile and Internet Services in Ubiquitous Computing (IMIS), Seventh International Conference, Sensors, 2013.

7 HR Analytics
A Model of Modern Human Resource Practices

Amit Kumar Srivastava, Shailja Dixit,
and Akansha Abhi Srivastava

Objective of Chapter

- This chapter with the help of content analysis and hermeneutics; a qualitative analysis approach critically evaluates the concept of big data in terms of human resource management which has its existence as HR analytics.
- Secondly it will explore HR analytics as a novel model of manpower practices by identifying its different dimensions at strategic and operational level.
- At the end, this chapter will provide concluding remarks to the practicing and budding managers about the role of human resource analytics to provide the competitive advantage to the companies by improving their business performance.

7.1 INTRODUCTION

If we will go into the past and analyze the history of management, then we can find that management models of the day were a hodgepodge that first came in existence and then disappear. Occasionally it has found to appear again. This was the time when human resource was just a domain where we consider the people who could not create any destruction/wealth in the organization. Many scholars have found two reasons behind the perception. Initially it was seen that many times HR people agreed that they were the expenses in their companies and nothing more, and in contrast to this some were opposing. Then it was observed that due to lack of a specific language, HR practitioners are fully unaware with the fact that what value they were generating for the organization. This was the main cause that they never talked about the value they were adding at their workplace, which provides a number of benefits to the company and its different stakeholders. All their terms they used in the human resource management were qualitative, subjective, and ambiguous.

They were only having simple stories in their hand in terms of response when they were supposed to provide evidence in terms of adding the value with their human resource services to the management. In case of their reply to the management on some questions like, "How is the morale of employee?" they would

DOI: 10.1201/9781003343332-7

generally reply, "It's good!"In case of a new question, "How good?" they frequently said, "Very good."

Then in case someone asked them, could you combine some other operations to increase the utility of performance indicators? It was enough to make them hopeless. With this hopelessness HR people got the solution that they should need to learn to express their ideas in quantitative terms and to make their responses more objective by using numbers to show their human resource activity in context of value addition for the whole organization. They realized that everywhere managers from other domains of management are using numbers to explain their business outcomes, and this was the main reason marketing, sales, finance in terms of operating expenses and profit, production in terms of time cycles, and volumes were the key indices of business of era. In their further research HR team get informed that productivity and quality was the key issue in a decade from 1970 to 1980. It was remarkably interesting for them to see the implementation of process quality as a solution to handle the aforementioned two issues successfully by the production department which was also relied on numbers to express degrees of change.

Then, in response to this analysis, HR experts have started to think about reengineering to quantify their process and results so that that will also be able show that human resource services were also involved in managing expense and generating the value by increasing the rate of capitalization of the company. Now seed of success was sown by the HR experts, which was ripened by the end of 1990. They have represented that HR function was undeniably adding measurable value throughout the organization. It was the great time when Saratoga Institute published the first benchmarks at the national level by applying a marketing model on HR function, which led to publication of human value management (HVM). By 2000, HR people were also fully equipped with advanced methodology to the question where they were talking about return on investment (ROI). Now it was the time when the whole field of human resource was shifted to the novel and more advanced statistical part of business management with the hope to touch the new heights of success in the process of value addition in the firm, and this has opened up the path of manpower analytics as a very new domain on the field of business analytics to complete the business process on new paradigm of global management practices in terms of quantification of HR data and information as well.

This drastic change was not a very easy task for HR team to adopt and that moment of time, but they were forced to adopt for making their survival in the long term and for providing efficiency to the management from their services. Without putting any confusion in their mind, the whole HR team has taken the decision to go in the depth of data science where they found big data analytics as the soul of this new science and started to learn the ways to implement it in the area of human resource management.

What analytics is in reality?

This question can be answered by combining the reality of arts and science (Figure 7.1). The arts provide an insight of how to look at the world, and the knowledge of sciences provides us an idea that how to do something. Whenever we will come in contact of the word analytics, our mind instantaneously think of statistics. In reality this approach towards analytics is not very much correct because critical

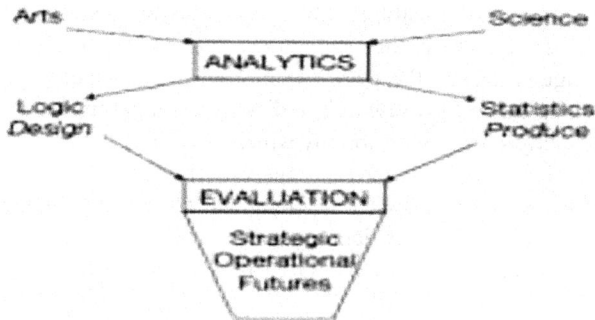

FIGURE 7.1 Reality of analytics.

(Source: Author)

evaluation of this term tells us that very first analytics is a mental framework with a logical progression and then a set of statistical framework which uses statistical tools and techniques.

In case of giving a solution to an organizational problem, it becomes important for managers to find out a logical structure to analyze the many variables which affects the human performance. Now, after identification of those variables, steps to apply statistical tools and techniques will come in the phase.

7.1.1 BIG DATA ANALYTICS

Today is the era of big data, and the traditional way of analyzing the data is not supporting the managers to manage a large quantity of data, which creates many problems among the managers in each domain to understand the complexity of data science. So it becomes important for them to learn the skills of this field and explore this new science upto the level of excellence so that they will be able to analyze the big data efficiently. Although the field of computer science and Internet technology has accepted Moore's law from many years, and now the problems of handling large data prevailing in the present scenario have been realized. Decision-makers must essentially gain significant insights from dynamism in different datasets in terms of frequent variations of data due to daily transactions with customer interactions and social networks. Such value can be obtained by using advanced analytics techniques.

It has been seen in many cases that today data is of many varieties and made up of a range of data types (Russom, 2011). These vast ranges of data have been realized with the unique features of colossal, multidimensional, assorted, multifaceted, amorphous, unfinished, and piercing. These features are the main reasons the result of statistical data analysis is varying (Jiang, Wang, & Zhao, 2014). Although it seems that big data provides an idea that more data is helpful in finding more useful information, but it is not a fact in every situation.

Hence, big data has variety and velocity, due to which traditional tools and techniques are not applicable on handling and conveying the correct value of a large quantity of data (Elgendy & Elragal, 2014). So how can we understand values of data? Few studies have found reality, soundness, worth, unpredictability, source, language, and ambiguity, and all these are grounds on which we can explain the values of dataset in the big data. Many research studies focus on the application of advanced technologies to analyze the data smoothly which also value the dataset.

Statistical methods are used for data analysis from decades, which help us to understand the value of data in different situations. The field of big data analytics is the best way of handling complex data, to analyze it more efficiently and accelerate the process of computation and decline the memory cost in analysis.

7.1.2 HUMAN RESOURCE MANAGEMENT AND ANALYTICS

As discussed previously, numbers are becoming a language of trade and commerce in every domain of management. In the present scenario, the field of analytics very thoroughly uses the descriptive, predictive, and prescriptive analyses derived from these numbers and useful for decision-makers to take decisions in companies. So companies are very much conscious at every level about exactness of decisions, which can be obtained only by their efficiency in using data analytics. In this way HR experts in the domain of human resource management must evaluate their human-related data very carefully so that it can be properly analyzed with human resource analytics to make effective decisions (Lochab, Kumar, & Tomar, 2018).

Hence, HR analytics seem to be a significant solution which can add value throughout the functioning personnel department and will cause improvement in the efficacy of other related functions by using logical and numerical explanations. This is the reason this new field of managing human resource data with analytics is becoming famous with different names like HR analytics or workforce analytics. By using data analysis in human resource management, HR professionals can move with the decisions which should be suitable for both employees and organization (Reena, Ansari, & Jayakrishnan, 2019). Here we can understand suitability in terms of how they can attract, preserve, and enhance the performance of the employees and company. It has been found in studies manpower analytics is an evidence-based approach, which helps in making rational decisions which will increase the intensity value added by HR practices on performance of the company (Singh, Upadhyay, & Srivastava, 2017). So workforce analytics has shifted the role of human resource management to a strategic partner, which was considered earlier as an operational partner (Malla, 2018). Now it becomes important for professionals that they should update themselves with the most recent advancements in the emerging field of manpower analytics. In this way this new discipline of analytics will be able to enhance the capability of leaders and managers to conduct their newly defined strategic goals through higher degree of workforce management. It is important for the experts in the domain of human resource management that they must be involved in multifaceted brainstorm on a regular basis to develop a comprehensive understanding of workforce contributions to strategic success of firms and ensure all aspects of their understanding should reflect in the workforce metrics while using analytics (Huselid, 2018).

By defining HR analytics in her research, Bassi in 2011 said that it can be defined in terms of systematic reporting on an array of HR metrics, and we can draw more simple solutions by using the predictive modeling and what-if analysis of scenarios.

Descriptive HR Analytics	Predictive HR Analytics	Prescriptive HR Analytics

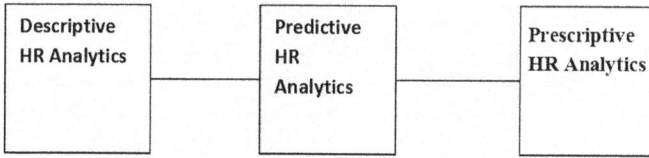

FIGURE 7.2 Components of HR analytics.

(Source: Author)

Her explanation includes the evidence-based approach in making human-related decisions, and she argued it is an evidence-based model which will help HR people in making better decisions in business. It is consisting of an array of tools and technologies, ranging from simple reporting of HR metrics to predictive modeling (Bassi, 2011). Finally, by critically analyzing Bassi's efforts, Mondare and his team in 2011 have rearticulated the term manpower analytics, that it's a tool to demonstrate the unswerving impact of people on business.

Generally, practitioners of data science have been observed that application of analytics has made a paradigm shift in business. Their expert view tells those analytics is the engine of business, and its intelligence is important for sustainable performance of the business (Fred & Kinange, 2015).

Based on in-depth analysis of Bassi's work, many scholars have emphasized manpower analytics requires different components of analytical course of actions, which includes descriptive, predictive, and prescriptive analytics (Jabir, Falih, & Rahmani, 2019).

7.1.2.1 Components of HR Analytics

Descriptive, predictive, and prescriptive human resource analytics as shown in Figure 7.2 are the three basic components of HR analytics.

Descriptive HR analytics—uses "mean median, variance, standard deviation," etc. It also describes the variables present in the dataset and answers questions like "what happened?" and "what is happening?"

Predictive HR analytics—uses "regressions, correlation and independent sample t-test etc. for finding out the predictive variables and making the predictive models to identify future trends, relationships, impacts, differences etc." (Jabir, Falih, & Rahmani, 2019). With a thorough analysis Jabir and his colleagues came on conclusion in 2019 that its major outcome is to answer the questions like what will happen? And why will it happen?

Prescriptive HR analytics—applies decision science, to make best use of scarce resources (Fred & Kinange, 2015), and it provides solution to the questions like what should be done? And why should it be done? (Jabir, Falih, & Rahmani, 2019).

7.1.3 ROLE OF MODERATING FACTORS IN HR ANALYTICS

It has also been found in the earlier studies it is important for the HR professionals to know the importance of different moderating factors. They play an important role in HR analytics for ensuring the smooth application of analytics in their organizations

(Kremer, 2018). Important moderating factors which are discussed in the study are problem identification, which means HR people should always be ready to identify organizational problems, know the data infrastructure. It means they must understand the reality that HR analytics requires the data that are accessible, accurate, and consistent across functions. Use of information technology means their analysis must be appropriate to advanced analysis and should focus on exploration, analysis, and modelling of data to implement the process of HR analytics more effectively. Analytically sound means HR professionals are supposed to be very critical in analytical skill to prepare the data, perform statistical process, and correspond the results in a meaningful manner and should have business focus, which means they must be very comprehensive throughout the process of implementing human resource analytics (Kremer, 2018).

7.1.4 HR ANALYTICS AND FUTURE DECISIONS

For promoting the HR analytics and its implementation, it has been observed that six points are required by human resource professionals to understand, which are formulation of an analytics policy regarding current trends and future requirements, detection of major outlay decisions on analytics software and personnel, recognize the related future issues, the way of utilizing the data, ensuring the cleansing of data, and minimizing the challenges related to validity of data (Levenson & Fink, 2017).

The newly developed area of HR analytics commits many benefits for companies, human resource professionals, researchers, and other stakeholders as well. This HR domain would be a source of improving the decisions in companies for its prosperous future. It would also assist the companies to solve the challenges by improving their workforce and services (Qadadeh & Mysore, 2022).

Decision-makers in the companies should use HR data analytics for any future changes. Different HR datasets can be utilized in conducting other important data analytics studies. All the output of the HR analytics process are also used in integrating the different datasets like emails, relationship, social media, etc. for adding the different dimensions to analysis (Qadadeh & Mysore, 2022).

7.2 LITERATURE REVIEW

7.2.1 HUMAN RESOURCE ANALYTICS: A TOOL OF UNLOCKING INSIGHTS FROM HR DATA

In a study it was found that the concept of big data is relevant in defining people analytics. It has been realized by the people from human resource management that use of analytics in their specific field make them able to provide the most accurate analytically driven and evidence-based decisions (Chattopadhyay, Biswas, & Mukherjee, 2017). Another study defines this new field of analytics as a modern tool of analyzing the existing situation and predicting the future results (Reddy & Lakshmikeerthi, 2017).

Hence, HR analysis is an exclusive blend of disengaging the insights from data, which will help in unveiling the findings to solve the human-related problems in the companies.

7.2.2 Human Resource Analytics and Integration of Data

A study on HR analytics involves integrating different internal functional data of human resource and external data of firm for using information technology to collect, manipulate, and report data (Sousa, 2018).

It has been observed that by using descriptive, visual, and statistical analyses of human-related data, people analytics provide information about human capital, organizational performance, which will be helpful for the organizations to find out the external economic benchmarks which establishes the business impact and enable them to make data-driven decisions (Marler & Boudreau, 2017).

7.2.3 Human Resource Analytics as a Process

It is realized by the HR professionals that people analytics is the progression of altering and managing HR-related data for the purpose of applying the more advanced analytical tools to solve the human-related problems in the companies (Kapoor & Sherif, 2012). Here we can say the whole procedure of people analytics is transforming the human resource management towards a new role of strategic business planning (Naula, 2015). Correspondingly, this process involves utilizing the HR data and linking it up with HR practices and policies for better understanding of employees' perception in terms of their involvement, contentment, efficiency, and recital (Reddy & Lakshmikeerthi, 2017).

7.2.4 Development of Human Resource Analytics

It has been found in a study that the journey of HR analytics has been started from fundamental aspect of mechanization to the era of digitalization followed by big data and then artificial intelligence (Heuvel & Bondarouk, 2017). This journey of human resource management was started in the 1980s with the computerization of few functions and managerial tasks. This has encouraged the corporate sectors to adopt the human resource information system (HRIS) at their places.

Analytics have poured an immense pleasure to enthusiasts of academics and research since the 1990s. People from different backgrounds have taken advantage of analytics. It has become one of the best ways to utilize their analytical skills in solving the human-related issues in industries. Now by understanding the amalgamation of HR and analytics, the business world has stepped onto a different dimension.

With the availability of high speed Internet in the year of 2000, many other corporate sectors have started the utilization of many new technologies, and it has given way to electronic HRM with the application of different analytical software on various dimensions of people management, talent acquisition, development, and retention (Dahlbom et al., 2020). In the present scenario, technological advancement in people management is placing HRM in the more strategic role of business organization (Meena & Parimalarani, 2019).

7.2.5 Adoption of Human Resource Analytics

Technology, organization, and people are the three main factors which have been observed to affect the adoption of HR analytics. A critical analysis of these factors tells that adoption of human resource analytics is still at a nascent stage resulting in more challenges than benefits (Mohammed & Quddus, 2019). Our analysis has found some common nuances in the adoption process of HRA: lack of analytical skills, availability of improper infrastructure, lack of understanding of its benefits and rationalized thinking to rely on traditional practices of HRM. With all these challenges, the main goal of analytics is to incarcerate, accumulate the data, and create reports (Marler & Boudreau, 2017). The digital era has allowed HR professionals to access large samples of data and analyze them in the aim to solve complex HR related challenges for better decision-making (Mohammed & Quddus, 2019). Also, implementing analytics require a strong visionary leader, with the right access to resources (Dahlbom et al., 2020).

7.2.6 Successful Implementation of Human Resource Analytics

Analytics has allowed HR professionals to access large samples of data and then to solve the complex human-related issues for better decision-making (Dahlbom et al., 2020). Our analysis shows that despite the high interest in data analytics and its benefits, many challenges have come at the time of using HR data for analytics. HR has just begun its journey from being descriptive to being prescriptive. Many organizations have claimed to adopt HR analytics as their HRM technique.

The idea of using data and big data to answer workforce-related questions improve HR functions and support strategic decision-making, which is still perceived as futuristic. The adoption of HRA in any organization is slow as the big data is itself lagging in many terms.

This is because using big data requires a shift in the processes used earlier, along with the shift in culture. This appears to be difficult for the organizations that are not ready to come out of their long established norms of conducting the HR management. Furthermore, to induce the HRA in any organization requires advancement in skills of the users or the employees of that organization.

So on the basis of this analysis, we can say that successful implementation of people analytics require a rigorous understanding of information technology to automate the data collection process, and most importantly it demands availability of analytically sound personnel in the human resource department (Marler & Boudreau, 2017). Hence, with this we come to a conclusion that HR people have to develop both business and analytical skills for the successful implementation of people analytics, which will help them to translate findings into solutions for improved business recital (Naula, 2015).

7.2.7 Results of Implementation of Human Resource Analytics

Now it is clear that proper implementation of analytics in human resource management help organizations to reach their strategic goals and desired outcomes (Ingham, 2011). It has been observed in a study that HR analytics help HR to play a more strategic role and have a more significant impact on the business (Lakshmi & Pratap, 2016). This implementation will improve various most wanted organizational outcomes like performance, level of people involvement, and their satisfaction.

A study states that HR professionals must focus on analytics by identifying the right measures and presenting them in a cohesive way to drive strategic business decisions (Ingham, 2011; Ben-Gal, 2019).

Hence, we can say that successful implantation of HR analytics create and maintain a balanced blend of different relevant capabilities in organizations' human capital, which completely changes the present way of perceiving and doing business. With analytically sound people, organizations will be aware that data and analytics provide them a way to achieve their business goals and objective more effectively. Thus, this implementation will help organizations to understand the fact incorporation of HR analysis is very significant in more informed and evidence-based decision-making to ensure organizational effectiveness.

7.2.8 NEW SCENARIOS OF HUMAN RESOURCE ANALYTICS

Now with the previous discussion we identified that HR analytics is an evidence-based approach to improve the process of decision-making w.r.t. human-related issues and functions in companies. It is for sure that this data-driven analytical dimension of human resource management adds value not only to the HR department but to the whole organization.

A study of qualitative interviews with nine leading Finnish companies revealed that top management did not have much faith in the future of HR analytics and had many doubts concerning the benefits it holds to the employees and the organization (Dahlbom et al., 2020).

So it is very important for HR professionals to better understand the impact and benefits of HR analytics to make clear among top management, in terms of cost-benefit analysis, before adopting analytics in the domain of people management (Ben-Gal, 2019). Consequently, their effort will determine the new scenario of HR analytics in the present arena of business. How the application, value, structure, and systems support of human resource analytics in 2025 might look like has been discussed in a study which argues that the main focus will be around fostering evidence-based decision-making and developing an analytical mindset specifically around all HR functions (Heuvel & Bondarouk, 2017).

7.3   PRESENT ERA OF ANALYTICS

It has been found in the research that an HR analytics system accumulates human related data from human resources information system (HRIS). A record of business and human performance data applies big data. Statistical analysis techniques provides measurement to these data for making the understanding of unexplained data patterns, relations, probabilities, and forecasting (Mohammed & Quddus, 2019).

Every business requires measurement in different terms, and it has been observed that things in business are generally measured in five ways. They are measured in cost, time, quantity, quality, and human inputs. It is important for managers and top level to understand the level of significance of these, that is, which is most important in the prevailing scenario. Understanding this depends on their approach of dealing with the organization.

If their approach accepts that all kinds of measurement involves an expense, then never will any of their actions create wastage of time and resources on evaluation of less important and insignificant value. As we all know, any kind of value we will get is part of knowledge of things, and it matters a lot in any setup. In reality "what matters in a business?" is not an HR question, while "what matters here is . . . that at what points in time and for what organizational purpose or value is significant?"

If they will try to move upto an optimum level of understanding, then they must go with five steps of analytics:

Step 1: Recording the operational domain of HR

It means critically recording the activities like recruitment, compensation, development, maintenance, and retainment. This measurement will create learning among HR professionals in a company regarding level of efficiency in their processes and provide an insight to them on how they can improve their human resource process to a higher level of excellence. This process improvement may ensure their success in creating the value by saving money and time and increasing the ratio of output to input. All these will also be reflected in satisfaction of employee and customer as well, and they all will be happier due to a less invasive process.

Step 2: Connecting to their organization's objectives

As we all know today's organizations are more concerned towards the objectives in terms of their quality, innovation, productivity, and services. All these criteria encompass the basic agenda of any organization. Higher management regularly involved in setting the targets across these process outcomes and reviewed it on a regular basis.

So all HR professionals are supposed to link the output of their workforce on these criteria, and they have to evaluate the result and check the alignment of results with quality, innovation, productivity, and services. HR professionals are also responsible to encourage the people for making the right perception regarding the value in their work.

Step 3: Benchmarking

HR professionals are required to prepare them and their team, which requires a lot of knowledge about organizations for the purpose of effective comparison in many domains that how one is affecting the other and the intensity of their impacts. Broadness of comparative data among the group of industry leads to marginal value because of the higher level of variance within that population. So more details will provide them significant values in all dimensions to achieve effective benchmarks to evaluate their performance in multiple ways.

Step 4: Using descriptive analytics to understand past behavior and outcomes

This is the initial step in the process of factual analysis. Descriptive analytics provides relationships among data without giving meaning to their patterns. Researchers believe that this process is exploratory, not predictive, and HR professionals are supposed to see the trends from the past, yet it is risky to extrapolate from the past into the future, considering the volatility and rapidly changing current markets.

Transactional Monitoring of People and Activity Reports	Performance Monitoring of Human Resources	Tying HR Metrics to Business with Business Analytics	Foretelling Effects with HR Metrics Analytics

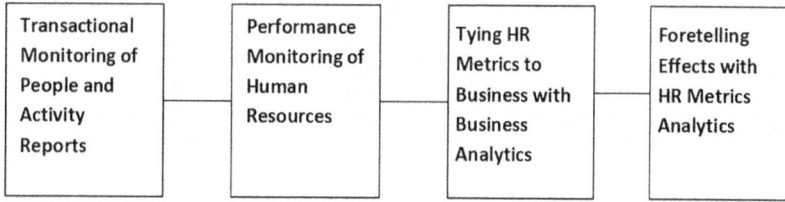

FIGURE 7.3 Process of HR analytics.

(Source: Author)

Step 5: Using prescriptive analytics to predict future likelihoods

This is a type of analysis which works on the reality of our current understanding by focusing on what we do not know, always comparing what happened earlier to what will happen later.

Predictive analytics describes the significance of the patterns observed in descriptive analysis. If we will try to take an example, then there is an observation that the banking sector uses this analysis in predicting the creditworthiness of borrowers, and the insurance sector is applying this analysis to predict the patterns of illness and mortality. It provides an idea to HR professionals of how they can utilize this predictive analysis in their process, and they can do it while making decisions about the expected return on their investments in hiring, training, and planning human assets on the HR matrix.

Metrics are the technical way of organizational management, and if professionals are able to speak with the same language of technicality, it will increase the impact of their efforts. Critical analysis of workforce intelligence reports of 2007 and 2008 tells a rationale for a management model that ensures the establishment of better communication between line managers and human resources professionals.

Hence, human resources analytics is an interpersonal tool to bring data from focus group study, survey, and in-depth analysis of different units at different levels (As shown in Figure 7.3), to paint an interconnected and actionable picture of current conditions and most likely futures. Human resource mapping has been observed in many cases for simply recording the inputs and output of the workforce. Accountants monitor income and expenses to explain the result of past decisions and investments to the management, and the new dimension of HR analytics prepares HR professionals to manage the human capital for tomorrow and gain sustainability in organizations.

Analysis of model

A critical review of this model (As shown in Figure 7.4) creates an explanation and understanding of human resource analytics in six different steps.

HR professionals have to come to a decision on human performance, training, recruitment, selection, turnover, absenteeism, employee health and safety, and other HR-related practices that need to be answered through HR analytics.

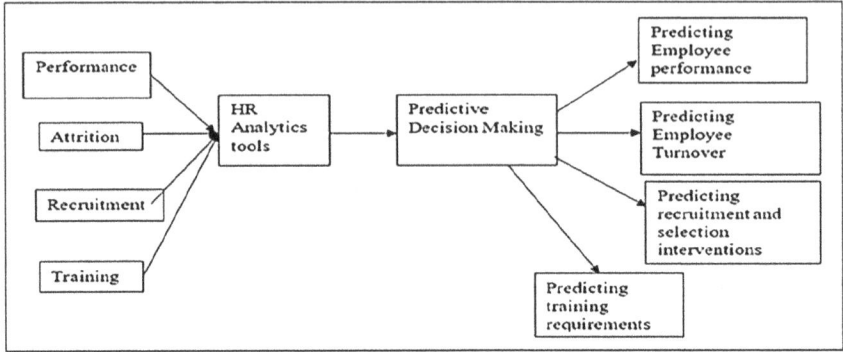

FIGURE 7.4   HR analytics model of predictive decision-making.

Source: (Mohammed & Quddus, 2019)

They have to make a decision on metrics for the variables which are supposed to be measured to derive the final conclusion w.r.t. identified objectives.

They are required to collect, organize, and store the data for identified metrics. It's advisable to collect data with the help of questionnaires, interviews, observations, etc.

After data collection they have to proceed to the process of analysis by utilizing the statistical tools like SPSS, Minitab, SAS, R, and Excel. During this step they also have to decide which technique they will use for data processing like descriptive statistics analysis or factor analysis or correlation analysis or regression analysis.

Then they have to derive the conclusions, and information needs to be presented using HR dashboards to the decision-makers.

At the end, the last step will provide them appropriate decisions relating to the identified HR objectives in a predictive manner.

Now the question arises, what fundamental the whole predictive model of human analytics works on which prepares HR professionals to maintain the sustainability of the organization and humans as well. What factors of the human resource process supports HR analytics process for predictive decision-making? With this intention an exhaustive review has been done which has provided the solution that the tool of analytics under HR domain comes to the predictive decisions by including some major operations of the human resource process like performance, attrition, recruitment, training, etc. (Mohammed & Quddus, 2019).

Mohammed postulated that all the data related to human resource factors in the organization, like human performance, their rate of attrition, recruitment, training, etc., are very much suitable for the purpose of analysis on HR analytics tools. In his research he has observed, data based on all these factors must support human resource analytics in predictive decisions by providing output which will predict employee performance, turnover, recruitment and selection interventions, and predicting training requirements from time to time. A critical review and in-depth analysis of this model by many scholars show that it is very significant in making effective and long-term decisions in the organization, which helps managers to predict behavioral patterns of people in the organizations in many terms like their rates of attrition, cost of development, and contribution of employee in every success and growth of the company.

Our study to find out whether HR analytics can be used as a modern tool of human resource practices in the contemporary era of business has explored a new dimension. This was proven from the study of Boudreau and Ramstad of 2004, which has given and is characterized by the LAMP frame work, that is, logic, analytics, measures, and process.

LAMP framework is considered the multistage processing model which appears to operationalize on HR scorecard. In many studies it has been found that researchers have a firm belief that every component of LAMP framework contributes to organizational effectiveness and efficiency (Bhattacharyya, 2017). HR scorecard is meant for linking people, strategy, and performance, which works as more frequent analytics element listed as newly drawn strategic tool among HR leaders (Becker, Huselid, & Ulrich, 2001).

So on the basis of people, strategy, and performance, the LAMP framework suggests that social influence and cognition would appear to be great potential to help, guide, and explain the cause-and-effect relationships between outcomes and moderators of human resource analytics (Cascio & Boudreau, 2010). It has also been suggested by many scholars that the leaders who are not from human resource management may make decisions with dominance of mental models and by retooling HR analysis on some other management disciplines such as operations, finance, and marketing (Boudreau, 2010; Rousseau & Boudreau, 2011).

7.4   BENEFITS OF HUMAN RESOURCE ANALYTICS

While it has been proven HR analytics is a powerful tool and model of modern human resource practices, it is important to know the reason why. Meaning, in what ways manpower analytics help in increasing the business impact:

1. Companies are regularly investing in human resources, so manpower analytics helps in redirecting all these investments of companies to more employee-centric initiatives. It covers initiatives which impact more critically on business metrics and outcomes instead of non measurable HR trends that promise to make employees happier, more engaged, and satisfied.
2. Companies are making more investments in expectation of tangible outcomes that benefit all internal and external stakeholders, so manpower analytics help companies to ensure the tangible outcomes with their investment.
3. As earlier in this chapter we have found that number is the language of analytics, then manpower analytics help companies to quantify the returns on the investments in all aspects of managing people via their impact on the top and bottom lines of organizations.

4. Manpower analytics help HR departments to become more accountable in their practices so that they can add more value in every part of bottom lines very precisely and in a quantifiable manner, which will be reflected in the performance of the company.
5. Manpower analytics bring a paradigm shift in the role of HR practitioners as it helps to involve them in any discussion of the company with full facts, figures, and logics by quantifying the impacts of their operational and strategic practices in outcomes of business.

While the professional discipline of manpower analytics is relatively new, by making more efforts on behavioral modeling, predictive modeling, impact analysis, cost-benefit analysis, and return on investment, the increasing challenges to HRM can be minimized.

HR people are supposed to focus their attention to build the capability internally to move beyond some common part of analysis like scorecards and dashboards to come on the advanced version of analytics, which help them to improve the organizational effectiveness by providing solid foundation of analytic and its competency among HR leadership team, generalists, and functional experts.

The conceptual framework of HR analytics consist of application of statistical science, research design, identifying meaningful questions, using appropriate data to solve the queries, implementing the scientific benchmarks to examine the outcome, and converting the outcomes into meaningful form for the use of business.

Manpower analytics is instrumental in process analytics, which is responsible for conducting cause and effect analysis on individual HR processes. It is also known as integrated analytics, which pulls multiple HR processes together to optimize different strategic issues such as career and succession planning.

Contemporary era of business is witnessing the use of diverse types of analytics to discover the drivers of tangible business outcomes, which can be observed very frequently in different settings in the form of predictive models. It is used by banking sectors to estimate the risk of customers and commercial credit. In the same way marketing demographics are utilized in predicting buying behaviors of the customer. These approaches make a well-informed, predictive assessment based on available data. The financial department in any business also uses the same pattern at the time of dealing with financial predictions and cost-benefit analysis. The major aim of doing this is to evaluate the past, understand the present, and to predict the future by analyzing the present facts and data.

In the same way practitioners in the field of human resource management also realized that HR analytics demonstrate the direct impact of human resource data on important business outcomes so it is the real truth that advanced data science is highly prioritized in the domain of modern HRM, and it makes human resource management a strategic business partner which helps in executing the value in every sphere of a company.

7.4.1   HRM: A STRATEGIC BUSINESS PARTNER

Manpower analytics is an upcoming field which can help human resource professionals to become a strategic partner in a business. In recent years, this field has gotten many conceptual achievements like HR metrics and scorecards, but the fast-changing pace of information technology has rolled up the needs of human resource professional towards the power of analytics to analyze the different HR measures and their significance.

This is the reason professionals should keep their focus on application of HR analytics in their operations. Successful implementation of analytics in workforce management enhances the ability of leaders and managers to achieve their operational and strategic objectives effectively. It tells every organization has to develop a comprehensive understanding of the way human resource management contributes in strategic success of a company. It is important for professionals to make an understanding with the concept and their insights must be reflected in the workforce metrics and analytics performed by HR professionals there (Huselid, 2018). This is the reason HR analytics is becoming a novel model of human resource practices.

David Ulrich, the father of modern human resource management, in an interview has given an idea on three major HR deliverables: administrative efficiency, functional excellence, and strategic alignment. He discussed all three deliverables can be instrumental for a firm to be successful if centrally focused by human resource professionals in their practices.

Administrative efficiency indicates HR should ensure administrative tasks like payroll, office space, role clarity, etc. are done efficiently, often through technology solutions.

Functional excellence tells HR should provide innovation and integration of their functional areas like hiring, training, motivating, standards, accountability, rewards, and communication.

Strategic alignment provides an idea that every organization has to have a strategy on where to play and how to win. HR professionals enable these strategies through talent, leadership, and organization.

Manpower or HR analytics is the combination of quantitative and qualitative data to get an insight on the strategic management of people in organizations. Previous literature shows that Dr. Jac Fitz-enz in 1978 has given the concept of metrics to measure the impact of human resource functions strategically on an organization's bottom lines.

To be a consistent strategic partner, manpower analysis can provide a key direction to actions, and hence, all HR functions are recognized as a strategic partner of every large and small company, which also accelerates the financial performance of a company and pushing it to new heights of success in terms of human satisfaction and financial prosperity as well.

7.4.2 HR ANALYTICS AND FINANCIAL PERFORMANCE OF COMPANIES

An empirical study has observed the chemistry of manpower analytics and financial performance of companies (Aral, Brynjolfsson, & Wu, 2012). Harris in 2011 has explained the cost savings on the processes of human resources are unlikely to result in business impact because administrative costs typically represent 3% of a company's selling, general, and administrative expenses. And no amount of savings wrung from reducing HR administrative expenses is likely to have any impact on business performance. Aral and his research team in 2012 have given some facts explaining the relationship between business performance and human resource analytics in their study. In exhaustive literature review we have observed two examples for the purpose of creating the overall understanding among the future researchers that how the domain of HR analytics is a new model for human resource practices in the contemporary era of business which tighten the knot of financial performance and human management. First example discusses the case of Lowes, a retail chain, and how the use of manpower analytics paved the way for Lowes towards employee engagement, performance of store, and manpower processes (Coco, 2011). In this case the application of manpower analytics provided the conclusion to Lowes that a highly engaged employee is responsible for 4% higher average sales per store. Six analytical tools of manpower analytics have been discussed in the case study, which explores how the use of analytics in manpower management increases the impact of business and using manpower analytics helped them in predicting employee performance by using their applicant database. And the second example discussed the case of Sysco and how it has established the significant links between work climate surveys, employee satisfaction, customer loyalty, and high revenue by applying HR analytics.

7.5 CONCLUSION

There are many other studies which provide a strong chemistry between manpower analytics and outcomes of business. Falletta, in his study of 2014, found that manpower analytics played a significant role in determining and implementing the human resource strategy. Lawler and his team in 2004 found in their survey over 100 Fortune 500 companies have suggested the role manpower analytics plays in measuring the relationship between human resource processes and business impact.

On the basis of all the discussions, at the end of this chapter we are moving to the conclusion that HR analytics is an outgrowth of and marriage between human resource metrics and general business analytics. And this outgrowth brings a lot of logics in the domain of manpower management. The only limitation of this marriage is that the business analytics demands a number of specific data from human resource management.

So human resource professionals are required to be very much equipped in data as a critical component of business intelligence for the purpose of applying descriptive, predictive, and perspective techniques of human resource analytics.

The outgrowth of business analytics and human resource management in the form of HR analytics has drastically changed the traditional human resources metrics which were confined almost to employee's issues to a much broader and more useful view of the metrics. It can draw on any or all data related to business intelligence for both support in the delivery of human resources services and influence the behavior at all levels of employees.

At the end, this chapter also concludes that human resource analytics turn HR metrics toward the future strategic and operational decisions regarding retention, readiness, leadership, and engagement, which speaks to what is likely to come tomorrow. Indeed, the area of HR analytics is the novel model of human resource practices in the contemporary era of business.

REFERENCES

Aral, S., Brynjolfsson, E., & Wu, L. (2012). Three-way complementarities: Performance pay, human resource analytics, and information technology. *Management Science*, 58(5), 913–931.

Bassi, L. (2011). Raging debates in HR analytics. *People and Strategy*, 34(2), 14.

Becker, B. E., Huselid, M. A., & Ulrich, D. (2001). *The HR Scorecard: Linking People, Strategy, and Performance*. Cambridge: Harvard Business Press.

Ben-Gal, H. C. (2019). An ROI-based review of HR analytics: Practical implementation tools. *Personnel Review*, 48(6), 1429–1448.

Bhattacharyya, D. K. (2017). *HR Analytics: Understanding Theories and Applications*. New Delhi: SAGE Publications.

Boudreau, J. W. (2010). *Retooling HR*. Boston, MA: Harvard Business Publishing.

Cascio, W., & Boudreau, J. (2010). *Investing in People: Financial Impact of Human Resource Initiatives*. London: Ft Press.

Chattopadhyay, D., Biswas, B. D., & Mukherjee, S. (2017). A new look at HR analytics. *GMJ*, 11(1), 41–51.

Coco, C. T. (2011). Connecting people investments and business outcomes at Lowe's: Using value linkage analytics to link employee engagement to business performance. *People and Strategy*, 34(2), 28.

Dahlbom, P., Siikanen, N., Sajasalo, P., & Jarvenpää, M. (2020). Big data and HR analytics in the digital era. *Baltic Journal of Management*, 15(1), 120–138.

Elgendy, N., & Elragal, A. (2014, July). Big data analytics: A literature review paper. In *Industrial Conference on Data Mining* (pp. 214–227). Cham: Springer.

Fred, M. O., & Kinange, U. M. (2015). Overview of HR analytics to maximize human capital investment. *International Journal of Advance Research and Innovative Ideas in Education*, 1(4), 118–122.

Heuvel, S., & Bondarouk, T. (2017). The rise (and fall?) of HR analytics: A study into the future application, value, structure, and system support. *Journal of Organizational Effectiveness: People and Performance*, 4(2), 127–148.

Huselid, M. A. (2018). The science and practice of workforce analytics: Introduction to the HRM special issue. *Human Resource Management*, 57(3), 679–684.

Ingham, J. (2011). Using a human capital scorecard as a framework for analytical discovery. *Strategic HR Review*, 10(2), 24–29.

Jabir, B., Falih, N., & Rahmani, K. (2019). HR analytics a roadmap for decision making: Case study. *Indonesian Journal of Electrical Engineering and Computer Science*, 15(2), 979–990.

Jiang, J., Wang, S., & Zhao, S. (2014). Does HRM facilitate employee creativity and organizational innovation? A study of Chinese firms. *The International Journal of Human Resource Management*, 23(19), 4025–404 Publisher: RoutledgeInforma Ltd Registered in England and Wales Registered Number: 1072954 Registeredoffice: Mortimer House, 37–41 Mortimer Street, London W1T 3JH, UK (5) (PDF) Does HRM Facilitate Employee Creativity and Organizational Innovation? A Study of Chinese Firms. Available from: https://www.researchgate.net/publication/241725532_Does_HRM_Facilitate_Employee_Creativity_and_Organizational_Innovation_A_Study_of_Chinese_Firms [accessed May 01 2024].

Kapoor, B., & Sherif, J. (2012). Human resources in an enriched environment of business intelligence. *Kybernetes*, 41(10), 1625–1637.

Kremer, K. (2018). HR analytics and its moderating factors. *Vezetéstudomány-Budapest Management Review*, 49(11), 62–68.

Lakshmi, M., & Pratap, S. (2016). HR analytics—a strategic approach to HR effectiveness. *International Journal of Human Resource Management and Research*, 6(3), 21–28.

Lawler III, E. E., Levenson, A., & Boudreau, J. W. (2004). HR metrics and analytics–uses and impacts. *Human Resource Planning Journal*, 27(4), 27–35.

Levenson, A., & Fink, A., (2017). Human capital analytics: Too much data and analysis, not enough models and business insights. *Journal of Organizational Effectiveness: People and Performance*, 4, 145–156.

Lochab, A., Kumar, S., & Tomar, H. (2018). Impact of human resource analytics on organizational performance: A review of literature using R-software. *International Journal of Management, Technology and Engineering*, 8, 1252–1261.

Malla, J. (2018). HR analytics center of excellence. *International Journal of Business, Management and Allied Sciences*, 5, 282–284.

Marler, J. H., & Boudreau, J. W. (2017). An evidence-based review of HR analytics. *International Journal of Human Resource Management*, 28(1), 3–26.

Meena, R., & Parimalarani, G. (2019). Human capital analytics: A game changer for HR professionals. *International Journal of Recent Technology and Engineering*, 8(2S11), 3963–3965.

Mohammed, D., & Quddus, A. (2019). HR analytics: A modern tool in HR for predictive decision making. *Journal of Management*, 6(3).

Naula, S. (2015). HR analytics: Its use, techniques and impact. *International Journal of Research in Commerce & Management*, 8, 47–52.

Qadadeh, W., & Mysore, V. (2022). HR analytics to understand employees' behavior against HR policies changes in the UAE. *London Journal of Research in Computer Science and Technology*, 22(3), 1–17.

Reddy, P. R., & Lakshmikeerthi, P. (2017). HR analytics'—an effective evidence based HRM tool. *International Journal of Business and Management Invention*, 6(7), 23–34.

Reena, R., Ansari, M. M., & Jayakrishnan, S. S. (2019). Emerging trends in human resource analytics in upcoming decade. *International Journal of Engineering Applied Sciences and Technology*, 4(8), 260–264.

Rousseau, D. M., & Boudreau, J. W. (2011). Sticky findings: Research evidence practitioners find useful. In *Useful Research: Advancing Theory and Practice* (pp. 269–287). San Francisco: Barrett-Koehler.

Russom, P. (2011). Big data analytics. *TDWI Best Practices Report, Fourth Quarter*, 19(4), 1–34.

Singh, P., Upadhyay, R. K., & Srivastava, M. (2017). The role of HR analytics in higher education institution. *International Journal of Engineering Sciences & Research Technology*, 6(7), 92–100.

Sousa, M. J. (2018). HR analytics models for effective decision making. 14th European Conference on Management, Leadership and Governance, ECMLG, pp. 256–263.

8 Technology and Design Advancements

Implementation of Cloud Computing Using a Service-Oriented Architecture (SOA) Model

Shaikha Alqassemi and Zakea Ilagure

8.1 INTRODUCTION

8.1.1 BACKGROUND

Cloud computing describes the delivery of computing and related communications services based on distantly located network-based resources without requiring a user to possess those resources. Usually, but not always, the network is on the Internet. The services the supplied resources cover include software, information, security, storage, and so on. For example, a person utilizes cloud-based resources when using mail services like watching a video on YouTube, storing files using Dropbox, shopping at Alibaba or Amazon, and so on. The roots of cloud computing can be traced back to the 1960s when John McCarthy proposed the idea of a computer network that could connect across the planet and offer computation as a public service. These ideas laid the groundwork for the development of cloud computing (Arutyunov, 2012).

8.1.2 HISTORY OF CLOUD COMPUTING

To understand what cloud computing is, it is necessary to first understand that this term (cloud computation) has two words, where "cloud" means the network utilized for service provisioning. According to Bairagi and Bang (2015), "the cloud" was first used in the telecom industry in 1966 during the invention of virtual private networks (VPN) services by Douglas Parkhill. Providers and clients could offer the same amount of bandwidth using VPN services with the help of a cable to transfer data (the hard wire data circuit). A reduced cost was incurred by rerouting the network traffic to account for fluctuating network use. More so, Ma (2012) argues that before the advent of VPNs, they only provided specialized point-to-point data connections that were inefficient in terms of bandwidth usage. With VPN services, however, it became

DOI: 10.1201/9781003343332-8

possible to efficiently balance network traffic. This concept has since been expanded by cloud computing to encompass servers and network infrastructure. Most organizations have embraced the use of cloud computing. For example, Amazon has implemented the Amazon Web Service, along with IBM and Google, which are currently running cloud computing projects.

Badger et al. (2012) argue that the term "cloud" is the Internet and related infrastructure. In a broad sense, cloud computing refers to on-demand utility computing for individuals with cloud access. According to Synergy Research Group (2019), cloud computing increases flexibility and automation process to the extent that customers are not afraid of software up-gradation, increase accessibility across the globe, and reduce hardware and maintenance cost.

Diagrams illustrating cloud-based services frequently show the cloud as nothing more than a hand-drawn outline on paper. For several decades, visualizations of networks like the Internet have frequently used forms that resemble clouds. These diagrams are a mainstay of textbooks and popular articles on data transformation networks (Sahu & Pateriya, 2013). The phrase "cloud computing" is a very recent invention and better understood by referring back to computing history and analyzing prior models of supplying services across communication networks, which is the forerunner of modern cloud computing (Taylor & Metzler, 2009).

Working "on the cloud" in the context of trading refers to files typically written and stored on a cloud server rather than the computer on which the application (app) was first installed. If the program is kept on distant servers, machines linked to the cloud can access it (Katzan, 2010). Since the user data and software are stored on remote servers, the program can operate concurrently on numerous linked machines. The program is accessible using a browser or app from any networked device (such as a tablet).

The Internet has prospered, and IT has grown in popularity during the past ten years. People use the Internet to find the necessary knowledge and information for business and social activities. The emphasis on cloud computing results from this reliance on the Internet. However, many industry analysts and businesses have voiced concerns about cloud computing, which is anticipated to mature in the following four to six years (Shiau et al., 2012). By 2020, most Internet users will "live largely in the cloud," according to a CNN survey of almost 900 Internet, technological, and social professionals (Shiau et al., 2012, p. 44).

From a strategic perspective, cloud computing presents unique difficulties. The biggest issue is that it does not function effectively with slow connections like dial-up services because it demands a consistent Internet connection and high speed. Even worse, if there is no Internet connection, nothing can be accessed, not even previously prepared documents. No work due to a dead Internet connection. Additionally, there is no assurance that the data will never be lost because once it is on the cloud, any hacker or virus with sufficient sophistication can access it.

Cloud computing maturity in every industry to comprehend their current situation, the capabilities they still need to build, and how long it will take to migrate to the cloud.

Maturity models have been developed to evaluate the level of cloud computing maturity in various industries, allowing them to understand their current status, identify any remaining capabilities they need to develop, and estimate the time required for migration to the cloud.

8.1.3 CHAPTER FOCUS AREAS

While not all businesses are fully aware of the benefits of cloud computing and may not be using it to its fullest potential, some companies have fully embraced it. Examples of these companies include Amazon and eBay Noon. Nearly 200 IT decision-makers and CIOs who are not yet using the cloud were polled. The results showed that 61% said they are delaying using cloud computing services until they are reasonably certain that their networks pose no substantial security threats (Micro, 2009). Another 27% stated that security concerns were a significant factor but did not cause them to postpone or abandon their plans to use new cloud computing solutions (Micro, 2009). Therefore, a decision to use cloud computing can be made, delayed, or limited based on security concerns or a lack of trust. These findings suggest that one possible reason for not applying cloud computing in some businesses in the trading sector is the lack of trust in cloud service providers to offer the same level of security or service that companies can provide themselves in business locations.

8.1.4 CHAPTER DELIVERABLES

The main aim of this study is to assess the implementation of cloud computing in business using the service-oriented architecture (SOA) model. The research-specific objectives guided the study to achieve the primary goal, which stated as follows:

1. To evaluate how cloud computing can be implemented to its fullest level at business
2. To discuss the pros and cons of implementing cloud computing in business
3. To discuss how the service-oriented architecture (SOA) model can evaluate the maturity level of cloud computing by assessing how this model is used to implement, measure, and enhance cloud computing in the business sector.

This chapter comprehensively examines cloud computing, including its history, capabilities, and the pros and cons of using it in a business setting. The history of cloud computing includes its evolution and the key players and innovations that have shaped the field. The power of cloud computing includes the ability to provide businesses with access to unlimited computing services and resources and the potential benefits and cost savings that result from embracing and implementing cloud computing.

This chapter does not cover the technical features of cloud computing, such as the underlying architecture, components, and technologies involved. Instead, it will concentrate on the more intensive effects and factors of using cloud computing in a professional setting.

The application of cloud computing in fields other than business, such as government, health care, and education, will also not be covered in this research. It will pay particular attention to the advantages, difficulties, and factors involved in employing cloud computing in a commercial setting.

8.1.5 Chapter Rationale and Value

The chapter assesses the maturity of cloud computing in the business sector. The development of a suitable model for business cloud computing implementation is an important step in creating customized models that consider the local context, thus enabling greater success. These processes can apply in other contexts. More broadly, this research will highlight the benefits of cloud computing for trading goals in general.

The chapter findings impact the readers with knowledge that can help them improve business operations. Understanding the implementation of cloud computing with the SOA model will help the readers to understand how businesses can achieve better scalability, ensure better integration with other systems, and improve resource utilization. The reader, through this study, can better understand technology. Investigating the implementation of cloud computing in business using SA models helps to understand the advantages and disadvantages of this technology and its application in real-world scenarios.

It further contributes to the existing body of knowledge on the ways in which cloud computing implementation can help businesses and organizations improve resource utilization, reduce IT infrastructure costs, and achieve cost savings. More so, the study provides knowledge on how businesspeople can improve the competitiveness of their businesses by implementing cloud computing. The study reviews how SOA can help businesses stay ahead of their competitors by providing more efficient and better customer service.

By understanding the challenges and benefits of implementing cloud computing in the singing SOA model, organizations can make informed decisions about their future technology investment and ensure they are ready to adapt to new technologies.

8.2 EVALUATING HOW CLOUD COMPUTING CAN BE IMPLEMENTED TO ITS FULLEST LEVEL FOR BUSINESS

8.2.1 General Overview of Cloud Computing

In recent years, interest in cloud computing has increased dramatically. This interest has mostly been sparked by the cloud's promise to lower prices and delivery times and develop new services and business models that were previously more difficult to implement (California Department of Technology, 2014). Different scholars have given different understandings of cloud computing. According to Slahor (2011), cloud computing involves storing and accessing data and applications on the Internet instead of a local computer's hard drive. It is a system of interconnected data that works together to minimize duplication. IT services and resources can be accessed through the Internet using web-based tools and programs. The data and software packages are stored in servers where the users can access, edit, and save changes (Buckman & Gold, 2012). Cloud computing, however, is dependent on Internet access for its functionality. The term "cloud computing" is derived from the metaphor of storing information in the "clouds" and being able to access it from any location, without being confined to a specific device. Armbrust et al. (2010) describe it as a computing approach that delivers computing services and resources through the Internet, allowing users to store data and run programs on remote servers, instead of on their personal device, for processing and access. It is believed that through cloud computing, users can access data and applications anytime and anywhere as long as they are Internet-connected. The providers and the customers have access to scalable and flexible infrastructure that allows businesses to scale their resources down and up as required.

According to Badger et al. (2011), scaling resources enables organizations to focus on their business projects and outsource IT and information tasks to cloud service providers. Over time, third-party companies have been offering cloud computing services to businesses, with various service models such as Software as a Service (SaaS), Platform as a Service (PaaS), and Infrastructure as a Service (IaaS) available for selection. These service models give businesses the flexibility to choose the level of support they require, with SaaS providing ready-to-use software applications, PaaS offering a platform to develop and run applications, and IaaS supplying basic IT infrastructure.

Cloud computing has significantly altered the IT sector and changed how businesses function. It has numerous advantages, including lower IT expenses, greater effectiveness and productivity, improved security, and improved cooperation and flexibility (Badger et al., 2011). Despite these advantages, Badger et al. (2012) provided that there are still worries about data privacy and security, so it's critical for businesses to pick trustworthy cloud service providers and have security measures in place to safeguard their data.

8.2.2 Cloud Computing in Early Stages

In the past, companies used many machines to provide the necessary power and access to multiple assets simultaneously. A primitive version of cloud computing as we know it today was employed in the past when humans used rows and rows of magnetic tape machines. Because they were the only locations with enough space in the 1950s, the massive machines were deployed in schools, government agencies, and large enterprises (The Are of service, 2009). Over time, as technology improved, the status of the cloud changed too. The rise of cloud computing occurred in the 1960s when people were using mainframes. It started with "Utility computing" by scientist John McCarthy, which became a main business need for companies such as IBM **(The Are of service, 2009)**. The main goal of utility computing was to break down computing power into daily business needs, such as how electricity and telephone companies operate services for customers. In the late 1990s, companies started promoting the concept that "the network is the computer." Combined with the introduction of the Internet in the mid-90s, the computer became a prime method for sharing information. Amazon started to invest in servers to sell its products and became the first company built on the technical innovation that began after the dot com bubble. Then, the idea of indexing the Internet arose, and Yahoo and Google pushed networking to what we now call "cloud computing." So search engines were the first version of cloud computing, organizing huge amounts of information over servers (Sclater, 2010). The history of cloud computing has shown the parallel stages it went through to reach what we have today. This improvement shows a real need for this technology, and people are getting used to it. Throughout the research, the researcher has investigated the possibility of raising the maturity level of the cloud and how they can enjoy the full benefits.

8.2.3 THE NEED FOR CLOUD COMPUTING

Modern company operations depend on cloud computing, and its significance is increasing. The Synergy Research Group recently conducted a study that projects that the worldwide cloud computing market will reach $500 billion by 2023. (Synergy Research Group, 2019). This expansion underlines the growing demand for cloud computing among companies in various industries. Organizations need specialized staff to install, configure, test, run, secure, and update IT programs and hardware. It is difficult for all business types to get the needed programs, requiring significant effort and time. With cloud computing, these issues are eliminated, as they no longer manage hardware and software, with the responsibility now on the service provider. Traditional business applications have been complex and costly for the past 30 years. However, cloud computing has undergone significant changes that have made it more accessible and cost-effective, and it has evolved into a vital component of contemporary IT solutions. The various stages of cloud computing's development over the years demonstrate its importance in modern IT solutions.

Lin et al. (2009) assert that the need for cost-effectiveness in the organization's operation is one of the key drivers of cloud computing adaptation. By outsourcing information and technology functions to a cloud service provider, businesses can save on software, hardware, and staffing costs. Organizations believe that cloud computing provides scalability, which allows them to easily scale down and up their IT resources as required. This allows them to increase their operation's efficiency and reduce costs. The organization's desire to access specialist IT resources is a major driver of the need for cloud computing. Organizations may access the resources they require without the fear of their management and software thanks the cloud computing, which eliminates the need to hire and maintain expensive and time-consuming specialists' IT workforces (Lin et al., 2009). This allows businesses to focus on their core strengths and stay competitive in the fast-paced business environment.

8.3   THE ADVANTAGES AND DISADVANTAGES OF IMPLEMENTING CLOUD COMPUTING

8.3.1 ADVANTAGES OF CLOUD COMPUTING

According to Sclater (2010), using cloud computing services allows businesses to save money on software, maintenance costs, and hardware. Instead of investing in expensive IT infrastructures, businesses can hire resources on an as-needed basis and pay for the required infrastructure. More so, email servers such as Microsoft Azure, Amazon Web Services (AWS), and Google Cloud Platform (GCD) are some of the popular cloud computing service providers that offer a cost-effective solution for businesses of all sizes. In addition, with cloud computing, organizations can pay per use rather than paying for an annual package.

Moreover, Sclater (2010) posits that there is no need for complicated setup or configuration because cloud computing services are often delivered over the Internet. This may lead to a quicker adoption procedure and enable businesses to quickly profit from cloud computing.

According to Sclater (2010), cloud computing enables users to operate from any location and on any device because it is web-based and accessible from anywhere with an Internet connection. It also offers 24/7 access to vital apps and data. Collaboration and productivity may both benefit from increased availability and accessibility.

Companies such as Airbnb, Uber, Amazon, and Netflix use cloud computing services to handle high traffic levels during peak seasons and support their rapidly growing businesses. Therefore, cloud computing can allow organization to easily scale their computing resources up and down as required. Scalability allows the company to respond to changing business environments and shrink or grow its infrastructure (Mell & Grance, 2011). Ambrust et al. (2010) assert that the organization may implement cloud computing to acquire multiple copies of data stored in different locations from the cloud service providers. This minimizes the risk of losing data, increases reliability, and ensures less downtime as they are connected to a team dedicated to monitoring and maintaining their systems. Online services such as Microsoft Office 365 and Salesforce, which are hosted in the cloud, provide reliable access to critical data and applications for business (Badger et al., 2011).

Badger et al. (2012) posit that cloud computing service providers have invested heavily in data security and employed a qualified team of experts to maintain and monitor their systems and provide a higher level of security than any other organization could achieve. Therefore, the organization would need to implement cloud storage solutions, such as Google Drive and Dropbox, that uses advanced security measures to ensure the privacy and confidentiality of sensitive information and protect users' data.

8.3.2 DISADVANTAGES/CHALLENGES OF CLOUD COMPUTING

Cloud computing is a flexible platform that allows users to carry out various activities such as accessing online storage, using business applications, developing custom software, and creating virtual network environments. Over time, the usage of cloud services has risen dramatically, with huge amounts of data stored in the cloud. Nevertheless, with this rise in usage comes a corresponding rise in the threat of data breaches as hackers try to take advantage of security weaknesses in the cloud's design Kazim and Zhu (2015). Kuyoro et al. (2011) posit that despite the advantages of cloud computing, many businesses are reluctant to use it because of worries about security and other associated issues. The growth of cloud computing is significantly hampered by security because it can be unsettling to entrust important data to another organization. Customers should be aware of the potential dangers associated with data breaches in the cloud environment.

Given the data is not stored on a single physical entity, and the enterprise does not technically own the data, this can raise security concerns, as data can be lost or attacked by criminal bodies (Sclater, 2010). Storing sensitive data in the cloud can result in privacy and security issues, as this information may be vulnerable to data breaches, unauthorized access, and cyber-attacks. In recent years, a data breach has impacted over 100 million customers, a good example of a cloud computing-related security issue (Barona & Anita, 2017).

Data ownership is a complicated matter in the realm of cloud computing. The question of who has the right to access, control, and manage data stored in the cloud is a complex one. According to Rittinghouse and Ransome (2017), cloud services providers control the data as they oversee and maintain the systems and IT infrastructures used to hold data. As a result of giving the vendor these tasks, companies may not have direct ownership over their data files and the contents involved. Organizations may have a legal and regulatory obligation that compels them to control their data or privacy concerns regarding the storage and processing of their data by a third party.

Companies such as Google and Microsoft allow institutions to offer their cloud services to their customers, which raises the risk an institution can be "locked-in" to the services of a single provider, thereby limiting service offerings. Moreover, providers are now charging fees when they transfer their services to another provider (Sclater, 2010).

Lock-in, a situation where an organization is tied to a specific cloud service provider and cannot switch to a different provider without facing technical difficulties or significant costs, is a significant challenge in the cloud computing context. This challenge mostly occurs when an organization is heavily invested and customizes its systems and applications to work with a particular cloud platform. Cloud service providers like Microsoft and Google are vulnerable to the risk of lock-in as they offer cloud services by institutions that may only offer services from a single provider. This limits the institution's options for service offerings, reducing flexibility, increasing costs, and limiting institutional innovativeness (Opara-Martins et al., 2016).

Data recovery issues are a rampant challenge in the cloud computing context. Data recovery can be described as returning data, applications, and systems to their default condition in the event of a disaster, such as a natural disaster, cyberattack, or hardware failure. Although data storage on the cloud might be advantageous for disaster recovery since it can be spread over several servers and places, enhancing accessibility and availability, establishing the precise location of backups in the cloud can be more difficult than with conventional hosting, which can make disaster recovery more difficult (Gorelik, 2013). While this increases access, if data is lost, it can be difficult to define the exact location of backups, unlike with traditional hosting.

8.4 IMPLEMENTING THE SERVICE-ORIENTED ARCHITECTURE (SOA) MODEL

8.4.1 THREE TYPES OF CLOUD COMPUTING

There are three major types of cloud computing services. First, Infrastructure as a Service (IaaS) provides virtualized computing services such as storage, serveries, and servers over the Internet. In this case, customers can rent and pay for what they use. Second, Platform as a Service (PaaS) includes services and tools for application development, testing, and deployment and provide a platform for developing, running, and managing application without the fear of infrastructure management. The last type of cloud computing is a Software as a Service (SaaS), which provides access to a software application on a subscription basis over the Internet. Customers need not worry about installation, upgrading, or maintenance. For example, office productivity applications, customer relationship management (CRM), and email.

Various servers may host different resources at their data centers using one of the three cloud computing types. There are three major types of cloud computing services, which are important to know. For further understanding of the three types, refer to Skiba (2011).

8.4.2 Deployment Methods

Cloud computing can be implemented in four ways, based on the desired level of ownership, scale, and access. These are as follows:

8.4.2.1 Public Clouds

A public cloud is one that is run and administered at data centers owned by the service provider and utilized by various clients (Goyal, 2014). As a result, it is neither intended for exclusive usage nor resides in a company's internal firewall. Gorelik (2013) posits that the public cloud is the most popular deployment method as it is publicly accessible to anyone who wants to sign up and use it (Gorelik, 2013). This allows applications from different organizations to be mixed together on one cloud's server, storage system, and network. Gorelik (2013) conducted a Trend Micro survey involving 1,200 businesses with at least 500 employees across six countries. The results indicated that 93% of the respondents indicated that they at least use one type of public cloud service provider, with 38% believing that the providers were not meeting their business and IT needs. While 55% indicated that they were worried about unencrypted share storage, 43% indicated that security issues are the main challenge, and 85% indicated that using their own data centers would be much better. The use of the public cloud subjects the organization to reduced control, transparency, and visibility of infrastructure issues and the need to comply with various regulatory and compliance requirements. Examples of public cloud vendors include the Azure Service Platform from Microsoft, the Blue Cloud of IBM, the Elastic Compute Cloud (EC2) from Amazon, and App-Engine from Google.

8.4.2.2 Private Clouds

A private cloud is an exclusive cloud infrastructure created for the use of a single organization and can be both more expensive and require more investment compared to a public cloud. Some businesses opt to create their own private clouds to have complete control, which requires a significant investment in purchasing, building, and managing the infrastructure. An outsourced private cloud, on the other hand, is still more expensive than a public cloud but requires less investment in creating and managing the cloud. However, outsourced private cloud is more expensive than public clouds, and therefore, it is said to negate the economic benefits that come with outsourcing IT infrastructure, which is a significant motivation for many firms to shift to a public cloud. However, some businesses have attempted to create private clouds in contrast to public cloud services because they offer more control, privacy, customizable flexibility, stability, and security. Examples of private cloud toolkits include Amazon's Virtual Private Cloud, HPs Cloud-Start, and IBM's Web Sphere Cloud-Burst Appliance.

8.4.2.3   Hybrid Clouds

A hybrid cloud is a combination of both public and private cloud models (Gorelik, 2013). It allows an organization to modify access, privacy, and other settings for different data components, depending on need, cost, logistics, etc. Hybrid clouds help the business to enjoy benefits for both public and private clouds as outlined previously, but also incur related drawbacks. Every business is trying to combine the best of both worlds through hybrid clouds. Mission-critical or internally focused apps may be installed on the private cloud in some businesses, whereas other applications may have hosted on public clouds (Goyal, 2014). One reason for the adoption of hybrid cloud is the practice of "cloud bursting," where a private cloud application is temporarily deployed on a public cloud to handle an increase in demand. Goyal (2014) explains that the hybrid cloud is seen as a transitional step for companies moving towards greater usage of public clouds, while still retaining some critical functions on private clouds.

8.4.2.4   Community Clouds

A community cloud is similar to a hybrid cloud, but with a distinct difference. It is the least commonly used option; it is a cloud shared by several organizations (Gorelik, 2013) so that different users can access different elements of datasets from different organizations. Community cloud is a type of cloud computing that addresses the security, regulatory compliance, and service level concerns of a specific group of users. These users share a multi-tenant infrastructure, similar to that of a public cloud, but the concerns of the group are addressed by the community cloud for the benefit of the group. Goyal (2014) highlights that a community cloud is a combination of the advantages of both public and private clouds, and examples of community cloud solutions include CFN services' Global Financial Services Cloud, Adobe's Creative Cloud, United Health Group's Optum Health Cloud, and International Game Technology's IGT Cloud. Understanding the cloud deployment models can tell how cloud computing can help an organization to identify which model suits its needs based on its size, provided services, and ownership to enhance the way of learning, especially for the business sector.

8.4.3   Benefits to the Field of Business

Cloud computing has brought many benefits to the field of business, such as the following:

8.4.3.1   Lower Costs

Cloud users are not required to have high-performance or high-memory computers to run applications as the data is not stored on physical hardware (Suicimezov, 2015). Applications can be executed from various devices like personal computers, mobile phones, and tablets, as long as they have an Internet connection, with little setup required.

8.4.3.2   Improved Performance

As cloud computing solutions have most processes and applications in the cloud (Suicimezov, 2015), updates to the cloud will cascade down to the applications, minimizing service disruptions.

8.4.3.3   Greater Compatibility across Users

Given files and documents viewed online, devices can be of almost any brand or operating system and still be compatible with the cloud software (Suicimezov, 2015).

8.4.3.4   Greater Collaboration

As employees and customers can view pictures and videos online, they can interact with each other freely, and edit and share documents, without needing to be in the same physical space (Suicimezov, 2015).

8.4.4   CHALLENGES TO THE FIELD OF BUSINESS

The challenges are listed here:

1. Standardizing the format of the content is crucial in ensuring compatibility between software and hardware. Compatibility with hardware refers to the ability to connect different devices without the need for specialized hardware or software. On the other hand, software compatibility refers to the capability to use an application on various computers without altering its format. The various formats available have different characteristics, such as PDF, ePub, gpj, and HTML, and different nations may employ different forms. As a result, customers who make online purchases may not have access to content from other merchants or nations.
2. Reliability is another major challenge faced by cloud services. Service reliability is the likelihood that a cloud service will be completed within a specified amount of time. The lack of dependability can result in service failures and inconvenience. For example, in the case of an e-Textbook service outage, students and teachers may not be able to access the learning materials or resources they have saved, which could result in device failure, service failure, or power outage. This can lead to students being unable to function if there is no alternative solution to address the service outage. The service also requires the uploading and downloading of large files, which can result in high bandwidth consumption, leading to bandwidth usage and service quality issues, especially when dealing with rich multimedia films and images that are very appealing.
3. Space is a key factor in cloud computing (Banciu & Cirnu, 2014). Since cloud computing is used by businesses and clients to store, save, and transfer their files, it requires a significant amount of device storage. However,

cloud computing does not use traditional storage components such as motherboards, hard drives, or flash memory.

4. The limited life of ownership is another important issue in cloud computing. The life of ownership refers to the time a person retains ownership of a product after purchasing or renting it. If the life of ownership is less than the required duration for the business life of the organization and customers, various problems may arise (Lee et al., 2013). For instance, if the ownership life is not long enough, the data entered into the website or application by the users may be lost.

8.4.5 GOOGLE

One of the earliest reported incidents of cloud hacking occurred in March 2009, when Google disclosed a breach in connection with its documents and spreadsheet products (Bender, 2012). The company sent a letter to some users informing them that some of their documents, as identified in the letter, may have been hacked into without their knowledge. This breach was unique compared to most previous publicized breaches as it occurred remotely on a Google server rather than on the user's device or service provider (Bender, 2012). This incident serves as a reminder of the lower level of data integrity that might be associated with cloud computing, which should be considered when opting for cloud storage. It is important to note that while US law imposes specific privacy restrictions on some industries, such as healthcare or financial services, most industries are subject to generic privacy and security laws (Bender, 2012). Implementing stronger rules, regulations, obligations, and protections can help enhance the security of cloud storage.

8.4.6 UP IN THE AIR OVER CLOUD COMPUTING

Many organizations' CIOs believe that businesses of all sizes will eventually move to cloud computing, but more cautious experts say that cloud computing is just a technology trend and will never replace on-premise computing (Otey, 2009). In December 2008, a publication in SQL magazine featured an article discussing decision-makers' opinions regarding cloud technology. The article noted that despite the cost-saving benefits of cloud computing, there was still skepticism among these individuals due to concerns over availability, reliability, and security (Otey, 2009). The article highlights the mixed feelings held by decision-makers towards the cloud and how issues with availability, reliability, and security can offset the cost-saving benefits of this technology. In addition, respondents to a reader survey noted that they are generally skeptical about the cloud and would either avoid or limit its use.

8.4.7 EVALUATION OF OPPORTUNITIES AND CHALLENGES

While the challenges and risks associated with cloud computing cannot be denied, overall, it appears that its benefits outweigh these risks when it comes to the business sector. There seems to be a lack of case studies in the business sector that suggest a significant failure of cloud computing. Moreover, protections can be put in place to minimize these risks. For example, the University of Hong Kong announced,

"For the protection of data privacy, sensitive data usually has to be encrypted before outsourcing, which makes effective database utilization a very challenging task" (NewsRx, 2015).

One more supported example shows that cloud computing not only benefits the business sector but also solves problems in the education sector. A study at the British University of Egypt (BUE) discovered the numerous benefits of web-based solutions in the education sector. The research found that these solutions addressed several issues, such as overcrowded classrooms, the high cost of educational materials, transportation difficulties, the need for further education and specialized training, interaction with the global education community, and the advancement of national education (El-Seoud et al., 2013). The project's findings at BUE demonstrate the potential for web-based solutions to positively impact education in various ways, from reducing overcrowding and costs to facilitating interaction and growth within the field. So as some institutions are using the cloud to increase the benefits, other institutions adopted the cloud only to solve problems or minimize the risk, such as (BUE). BUE has found that money should be invested to make staff more aware of this technology and capable of using it to maximal advantage.

The previous section has discussed the opportunities and challenges with examples of where the cloud has been implemented, and it can be seen that there are more opportunities than challenges. Hence, it approves that cloud computing can be beneficial in the field of business. In addition, there are no clear examples found on the challenges of cloud computing in business, which tells that the cloud has added value to the field of business.

8.4.8 Cloud Computing Maturity Model

8.4.8.1 Service-Oriented Architecture (SOA) Maturity Model

This model was tested on cloud computing implementation at HCT, so in this research, the same model has been explained clearly to be tested in the business sector in the future.

The SOA Maturity Model is a tool that helps organizations adopt cloud technology by determining critical capabilities necessary for successful adoption. This model outlines four key concepts: capabilities, domains, maturity, and adoption (Bob Hensle, 2013). According to some researchers, the model consists of five maturity levels, although there is disagreement among researchers, and some suggest that there are six levels. These levels are further defined and explained. The higher the level of maturity is, the greater the overall success of the cloud computing strategy.

8.4.8.2 Initial Services

This is the first level of maturity and focuses on establishing new functionality; it can be seen as the initial learning phase. Simply, it creates connections between different applications/services using a service interface (Meier, 2006).

8.4.8.3 Architected Services

This level aims to reusability of services and to define standards for the enterprise SOA. It is only achieved when new functions are implemented in multiple projects to serve different services (Meier, 2006).

8.4.8.4   Business Service and Collaborative Services

This level has two strategies. Strategy A aims to optimize internal business processes and creates a link between business and technology, allowing users to access applications online. On the other hand, Strategy B focuses on enhancing processes with external partners and aims to improve collaboration with these entities. These two strategies work together to facilitate better overall business performance and improvement (Meier, 2006) and the establishment of collaborative services to connect internal and external services.

8.4.8.5   Measured Business Service

This level is achieved by successfully implementing both strategies A and B. At this stage, the performance of services can be monitored and evaluated in real time, allowing for a transition from reactive to proactive business implementation (Meier, 2006). This level also requires that all cloud activities are recorded and logged to accurately measure and assess the functioning of cloud services. By measuring and tracking cloud service performance in real time, organizations can make informed decisions and optimize their cloud operations to drive business success.

8.4.9   Optimized Business Services

The highest maturity level of the service oriented architecture (SOA) maturity model, as described by Meier (2006), is focused on the automation of business processes. This level endeavors to optimize business operations by enabling event-driven automation, allowing real-time responses to changing conditions, and facilitating a shift from a reactive to a proactive approach. The SOA maturity model, as explained by Bob Hensle (2013), helps to guide organizations through the cloud adoption process by identifying crucial capabilities and defining the key concepts of maturity, adoption, domains, and capabilities.

The model is organized into eight domains, encapsulating more than 80 capabilities that help organizations assess and manage their cloud computing practices. The domains are used to classify and categorize the related capabilities and provide a road map for improvement in areas deemed important. The assessment of maturity and adoption of appropriate governance serves as key indicator of the overall success of a cloud computing strategy. The SOA Maturity Model provides a clear path for organizations to follow in order to effectively adopt and utilize cloud computing.

8.4.10   Domains in Cloud Computing

By implementing the following domains in a cloud computing service, businesses can address all areas of the cloud and get the full benefits.

8.4.10.1   Business Strategy

This category of capabilities is crucial in providing the necessary infrastructure to support the cloud initiative. It includes business incentives, expected outcomes,

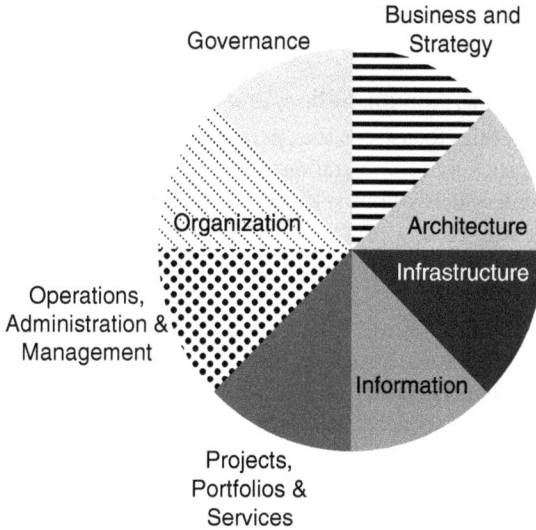

FIGURE 8.1 SOA maturity model domains.

guiding principles, estimated costs, and funding. These elements work together to drive the success of the cloud initiative. As shown in Figure 8.1, the pie chart illustrates the various domains of the Service-Oriented Architecture (SOA) maturity model, categorized into eight key areas. These domains represent critical aspects for assessing and enhancing SOA maturity within an organization.

8.4.10.2 Architecture

This category of capabilities focuses on the overall architecture of the cloud initiative and provides guidelines for different stakeholders to ensure compliance. Key elements of cloud architecture include resource pooling, interoperability, and self-service, which are essential for a well-functioning cloud system (Bob Hensle, 2013).

8.4.10.3 Infrastructure

This category of abilities focuses on the technical foundation of the cloud initiative, including the service infrastructure and tools. Key areas of emphasis include shared services, the process of providing resources, and the packaging of models (Bob Hensle, 2013).

8.4.10.4 Information

This category of capabilities is focused on the information elements of cloud computing, including metadata management, customer entitlements, and data durability. These elements ensure that the cloud initiative can effectively manage and protect its data (Bob Hensle, 2013).

8.4.10.5 Projects, Portfolios, and Services

This category of capabilities focuses on planning and developing cloud services and managing the portfolio of required services. It is an important aspect of the cloud initiative, as it helps to ensure that the right services are delivered to meet customer needs (Bob Hensle, 2013).

8.4.10.6 Operations, Administration, and Management

This category of abilities focuses on the operational aspects of the cloud environment after deployment, such as administration and management. It plays a critical role in ensuring the seamless delivery of self-service functions and efficient change management (Bob Hensle, 2013).

8.4.10.7 Organization

This category of abilities aims to enhance organizational efficiency by establishing an appropriate organizational structure, building skills, and securing executive support and authority. This is crucial for ensuring the necessary support and resources for the cloud initiative to succeed (Bob Hensle, 2013).

8.4.10.8 Governance

This category of capabilities focuses on the management structures and procedures that support and manage the cloud initiative. It includes elements such as policy management, risk management, and auditing capabilities, which help to ensure that the cloud environment is well-governed and secure (Bob Hensle, 2013).

8.4.10.8.1 SAO Maturity Model

Despite the benefits that this model could bring to an organization, there are still some weaknesses that have been captured in the implementation of such a model. For example, the level of maturity in each domain and the capabilities within them can vary greatly from one organization to another. Some organizations may be more advanced in certain domains, while others may have made more progress in certain capabilities within a domain. As a result, it is important to assess the level of maturity within each domain and across domains to identify any gaps or areas for improvement. This model, although not comprehensive, can be adapted and utilized within the specific context of an organization's setup. Measuring the maturity of each domain and capability helps organizations understand their strengths and weaknesses and plan accordingly for future improvements.

8.5 CONCLUSION

This paper has addressed the main elements that are required to answer the research questions that have been mentioned earlier. It gave a full description of what cloud computing means, its types and advantages, and disadvantages. In addition, a maturity model was explained in detail in order to measure and implement cloud computing in business to its extent. Cloud computing can save vast values of data and save budget and time. This can be achieved by following the right model with regard to all business aspects. More research needs to be done in this field, considering confidentiality and assurance of the data. No doubt that such research will bring unlimited benefits to entrepreneurs and web developers.

REFERENCES

Armbrust, M., Fox, A., Griffith, R., Joseph, A. D., Katz, R., Konwinski, A., . . . Zaharia, M. (2010). A view of cloud computing. *Communications of the ACM, 53*(4), 50–58.

Arutyunov, V. V. (2012). Cloud computing: Its history of development, modern state, and future considerations. *Scholarly Journals, 39*(3), 173–178.

Badger, M. L., Grance, T., Patt-Corner, R., & Voas, J. M. (2011). Cloud computing synopsis and recommendations (draft), nist special publication 800–146. *Recommendations of the National Institute of Standards and Technology, Technical Report, 12.*

Badger, M. L., Grance, T., Patt-Corner, R., & Voas, J. M. (2012). *Cloud computing synopsis and recommendations.* National Institute of Standards & Technology.

Bairagi, S. I., & Bang, A. O. (2015, March). Cloud computing: History, architecture, security issues. In *National conference convergence* (Vol. 2015, p. 28). IEEE Computer and Reliability Societies.

Banciu, D., & Cirnu, C. E. (2014). Cloud challenges for the e-learning process. *Romanian Cyber Security Journal, 2014,* 278–734.

Barona, R., & Anita, E. M. (2017, April). A survey on data breach challenges in cloud computing security: Issues and threats. In *2017 International conference on circuit, power and computing technologies (ICCPCT)* (pp. 1–8). IEEE.

Bender, D. (2012, October). *Privacy and security issues in cloud computing.* IEEE.

Bob Hensle, M. D. (2013, September). *SOA maturity model—guiding and accelerating SOA success.* Redwood Shores, CA 94065.

Buckman, J., & Gold, S. (2012). Privacy and data security under cloud computing. *Privacy and security issues in cloud computing, 88*(2), 10–22. (AACRAO).

California Department of Technology. (2014, January 2). *Cloud Computing (CC) Reference Architectur e (RA).* California Department of Technology.

El-Seoud, M. S., El-Sofany, H. F., Taj-Eddin, I. A., Nosseir, A., & El-Khouly, M. M. (2013). Implementation of web-based education in egypt through cloud computing technologies and its effect on higher education. *Higher Education Studies, 3*(3), 62–76.

Gorelik, E. (2013, January). *Cloud computing models.* 02142. Addison-Wesley Educational Publishers Inc.

Goyal, S. (2014). Public vs private vs hybrid vs community-cloud computing: A critical review. *International Journal of Computer Network and Information Security, 6*(3), 20–29.

Katzan, H. J. (2010). The education value of cloud computing. *ProQuest,* 37–42.

Kazim, M., & Zhu, S. Y. (2015). A survey on top security threats in cloud computing. *International Journal of Advanced Computer Science and Applications, 6*(3), 109–113.

Kuyoro, S. O., Ibikunle, F., & Awodele, O. (2011). Cloud computing security issues and challenges. *International Journal of Computer Networks (IJCN), 3*(5), 247–255.

Lee, H. J., Messom, C., & Kok-Lim, A. Y. (2013). Can an electronic textbooks be part of K-12 education? Challenges, technological solutions and open issues. *TOJET: The Turkish Online Journal of Educational Technology, 12*(1).

Lin, G., Fu, D., Zhu, J., & Dasmalchi, G. (2009). Cloud computing: IT as a service. *IT Professional Magazine, 11*(2), 10–13.

Ma, S. (2012). A review on cloud computing development. *Journal of Networks, 7*(2), 305.

Meier, F. (2006, August 30). Service oriented architecture maturity models: A guide to SOA adoption? *Högskolevägen, 541 28 Skövde.*

Mell, P., & Grance, T. (2011). *The NIST definition of cloud computing.* NIST.

NewsRx. (2015, April 30). *Data encryption; Research data from city university of Hong Kong update understanding of data encryption (L-EncDB: A lightweight framework for privacy-preserving data queries in cloud computing),* p. 242. MDPI.

Opara-Martins, J., Sahandi, R., & Tian, F. (2016). Critical analysis of vendor lock-in and its impact on cloud computing migration: A business perspective. *Journal of Cloud Computing*, *5*, 1–18.

Otey, M. (2009, March). Cloud computing: SQL mag readers weigh in. *SQL Server Magazine*, 7.

Rittinghouse, J. W., & Ransome, J. F. (2017). *Cloud computing: Implementation, management, and security*. CRC Press.

Sahu, Y., & Pateriya, R. K. (2013). Cloud computing overview with load balancing techniques. *International Journal of Computer Applications*, *65*(24), 40–44.

Sclater, N. (2010, September). *Cloud computing in education*. 8 Kedrova St, Bldg. 3.

Shiau, W.-L., Chao, H.-C., & Chou, C.-P. (2012). An innovation of an academic cloud computing service. *Scholarly Journals*, *5*, 938–943.

Skiba, D. J. (2011, August). Are you computing in the clouds? understanding cloud computing. *Nursing Education Perspectives*, 266–268.

Slahor, S. (2011). What is cloud computing? *Law & Order*, *59*(8), 10.

Suicimezov, N. (2015). *Cloud computing—an innovative solution for E-learning platforms*. Carol I National Defence University Publishing House.

Synergy Research Group. (2019). *Global cloud computing market to reach $500bn by 2023*. Retrieved from. https://srgresearch.com/press_release/global-cloud-computing-market-to-reach-500bn-by-2023/

Taylor, S., & Metzler, J. (2009, May 19). *What is cloud computing?: Possible solution in the cloud for fitting various vendor offerings together into a synergistic system*. Senior Technology E.

The Are of Service. (2009, May 12). *A complete guide to cloud computing*. The Are of Service.

Yuan-Shun Dai, B. Y. (2010, February 18). *Cloud service reliability: Modeling and analysis*. IEEE Press.

9 Redefining Traditional Education Using Augmented Reality and Virtual Reality

Kavitha Rajamohan, Manu K S,
Anju Kalluvelil Janardhanan,
and Sangeetha Rangasamy

9.1 INTRODUCTION

The education system has undergone a profound transformation during the last decade to meet the growing demands of the industry, society, and economy. Now as a part of COVID-19 recovery, the education system must move forward to reinvigorate. All levels of education will need to cultivate creativity, collaboration, and innovative capacities as there is no "quick fix" formula for education. The key to future education is taking what has been learned from the COVID-19 crisis and leveraging the lessons for systematic reform that predates it to address the challenges. Technological innovations can design creative ways to teach, learn, and assess students. Technology enhanced education (TEE) combines digital technologies (Law et al., 2016) and uses different strategies to facilitate teacher-student interaction in online education and other types of Internet-supported mediums of learning (Tsai, 2017; Yu, 2022). It challenged the traditional teaching and learning models to meet the expectation of 21st-century learners. A number of immersive technologies such as augmented reality (AR) and virtual reality (VR) are revolutionizing the educational system, changing the traditional methods of teaching and learning. The technological growth in both hardware and software has made AR and VR a game-changer in different fields, including education. The development of AR and VR in its way reach as an invaluable teaching tool for visualized learning where there is only contextual concept or abstract ideas of the real word problems. A fully computer-generated display or a modified interaction with real-world objects are the core concepts that define VR and AR. The concept of VR is based on creating a simulated and interactive environment through the use of computer-generated three-dimensional (3D) images of objects, spaces, and events. To take part in the immersive environment, the user needs a head-mounted display (HMD). In turn, a composite view of two existing parallel worlds (real and computer-generated) is provided by AR that enables users to display additional digital information and enhance their experience of reality (Pregowska et al., 2021). Images, videos, sounds, or text can all be used to convey this information. The study by Freina and Ott (2015) reported that subjects like computer science, medicine, paramedicine, engineering, construction, etc. in the United States and United Kingdom tertiary level frequently used AR and VR as they required spatial knowledge and safety. Chiang et al. (2014) and Wang et al. (2014) found that utilizing AR applications to promote inquiry-based learning could boost student motivation and engagement in the learning process. Especially in healthcare education, AR and VR can help students, patients, and healthcare supporters to understand the behavior of critical health issues such as stroke and an autism spectrum disorder. According to Market Research Future (2022), the compound annual growth rate (CAGR) of AR and VR in the education market is expected to grow at 18.2% from 2022 to 2027. Further, almost 30% of Americans now use AR on a monthly basis. The Metaverse market might be worth between $8 trillion and $13 trillion by 2030. (Lee et al., 2022). The University of Michigan provides funding and guidance for implementing AR/VR projects, introducing various courses and innovative AR/VR labs across disciplines as a part of the university's Center for Academic Innovation. (University of Michigan Emerging Technologies Group, 2023). All full-time Harvard students can use the AR/VR Studio at the Harvard Innovation Labs. The studio offers open sessions and guided seminars to allow students to explore the potential of this technology. It has a variety of AR/VR devices as well as content production tools and software (Dick, 2021). The Colorado State University has put up a new virtual reality (VR) lab. The facility's headsets run customized software that enables students to have simultaneous immersive learning experiences and view medical imaging in a shared virtual environment (Lang, 2019). In this context, it is essential to understand the applications and challenges of using AR and VR for current education scenarios. This chapter proposes to throw light on paradigm shift in TEE, the technological advances in education, and the application of AR and VR in education from the perspective of teaching and learning in the new normal. Further, it also discusses the trends, benefits, risks, challenges of AR and VR applications and the way forward.

9.2 MOVING AHEAD WITH EDUCATION 4.0

Throughout the last 50 years, technology has developed in a seamless process driven by human, social, and economic needs. Technology connected with the Internet is transforming more and more areas of human life. Besides web services 1.0, 2.0, and 3.0, the process of these changes involves education. There has been a shift in education's goal to accommodate students' needs. Miranda et al. (2021) identified education 1.0, 2.0, 3.0, and 4.0 as the related development stages.

9.2.1 EDUCATION 1.0: DOWNLOAD EDUCATION

As Western Universities and the Industrial Revolution developed in the late 18th century, classrooms that aimed to promote instructional approaches were born. Aside from books, teachers were "sage on the stage" with a primary source of knowledge. A teacher-centered model of education such as Gurukul in India is a classic example of what Education 1.0 looks like. It describes the learning environment of an authoritarian leader where students are passive recipients of concepts or ideas taught by the teacher within a physical classroom. There was unidirectional dissemination

TABLE 9.1
Transition from Education 1.0 to Education 4.0

	Education 1.0 (Download Education)	Education 2.0 (Open Access Education)	Education 3.0 (Knowledge Producing Education)	Education 4.0 (Innovation Producing Education)
Period	Late 18th century	Early 20th century	Late 20th century	Present
Teacher Role	Sage	Guide, information source	Orchestrator, curator, and collaborator	Mentor, coach, collaborator, reference
Student Role	Largely passive	Emerging active	Active, initial independence	Active, high independence
Approach	Teacher-centered	High teacher importance	Co-constructed	Mostly student centered
Learning Outcome	Grades, graduation degree	License to professional practicing	Prepared for practice and scenario analysis	Training of key competencies
Enablers	Mechanical printing, ballpoint pen, typewriter	First computer	Computer and widespread use of Internet	ICTs tools and platforms powered by IoT
Information Source	Standard texts	Adopted text, open-source materials	Text, case studies	Online sources
Facilities	Classroom	Blended lab, classroom	Blended and flexible physical shared spaces	Cyber and physical spaces both shared and individual
Technology	Mechanical system	Mass production	Internet access	Connectivity, digitalization, virtualization

Source: Germain, 2019; Miranda et al., 2021

of knowledge from the teacher to the student. Rather than seeing students as unique beings, they are receivers of knowledge. It is the teacher's role to determine which topics are most important for students to learn without considering the student's interests and needs. All students are seen as one, and the same level of education is provided to all. This one-size-fits-all instructivism approach focused on the 3 Rs—reading, responding, and regurgitating (Tan et al., 2018). Education 1.0 does not offer students much creative freedom as students are only obliged to follow instructions and assessments to ensure that students do not engage in an open-ended learning process. As shown in Table 9.1, the table outlines the evolution of educational paradigms from Education 1.0 to Education 4.0, highlighting key changes across several dimensions.

9.2.2 EDUCATION 2.0: OPEN ACCESS EDUCATION

In a world filled with ambiguity and uncertainty, instructivism did not work in the early 20th century. In search of ways to improve the learning process, educators realized students learn best when they gain real-life experience. So as a "guide on the side," teachers act as facilitators and encourage students to interact with their peers and create interesting learning experiences. Teacher-to-student, student-student, and student-to-content relationships were considered an essential part of this learning process. This constructivist approach focused on the 4 Cs—communicating, contributing, collaborating, and co-creating (Tan et al., 2018). With Web 2.0, a gradual blurring of lines occurred between the creator and the recipient because of the new use of resources. This technological breakthrough led to Education 2.0 where learning shifted with technology, and teachers became secondary sources of information because of the Internet and mobile. For instance, wikis, discussion forums, chats, instant messaging, personal websites, blogs, emails, and social media platforms facilitated collaboration, social learning and co-creation of knowledge among students. Students are active learners and develop their knowledge and skills through activities such as projects that allow them to explore the problems, communicate with one another, and find potential solutions.

9.2.3 EDUCATION 3.0: KNOWLEDGE PRODUCING EDUCATION

Semantic web and artificial intelligence (AI) are milestones that distinguish Web 2.0 and Web 3.0. Since the emergence of web services 3.0, the education system has been transformed as knowledge was freely and easily available. A new generation of education took the shape of 3.0 in the early 21st century, which was based on new technologies. Compared to the previous eras of education, Education 3.0 shows a radical shift to personalized learning that was self-directed, interest-based, problem-solving, innovative, and creative. Educators became learning designers and leaders of collaborative knowledge creation. Students became co-developers, co-researchers, authors, drivers, and assessors of learning experiences. Students could choose their learning objectives and seek the expertise of their teachers to decide what they want to learn. They also interacted with other experts within their learning communities to discuss the new ideas and concepts they are learning. Thus, they became active participants in the learning process, and student-centered education was the goal. This connectivist approach focused on the 3 Cs—connecting, collectives, and curating (Tan et al., 2018). As opposed to the previous eras of education, student initiative is a key feature of Education 3.0 because they seek knowledge out of their interest rather than to get a formal education. In the second half of the 20th century, interactive whiteboards replace chalkboards in the classroom. As the digital age progressed, learning management systems (LMS) such as Moodle, Canvas, Blackboard, etc. were developed to assist students in learning. Learning happened anywhere. Diversity in networks paved the way for the fading of the common classroom style. The flipped learning approach emphasizes greater student-teacher interaction by combining physical and online learning.

9.2.4 EDUCATION 4.0: INNOVATION PRODUCING EDUCATION

The growth of online learning has sped up over the past decade, and it has been incorporated into formal education systems. To lay the foundation for modern education, the education environment should accommodate the changes in the industry. Industry 4.0 resulted in the evolution of Education 4.0. Remote learning, personalized learning, bring your own device (BYOD) learning, project based learning (PBL), and practical learning are the key concepts of Education 4.0. Using digital media as a tool for the integration of cultural, educational, and social activities, it aims to ensure sustainable development of people. Today's educational institutions are at the forefront of meeting students' future needs with the help of advanced technologies such as artificial intelligence (AI), augmented reality (AR), virtual reality (VR), etc. in the classroom. Education 4.0 depends on the availability of the content. Educational materials have to be provided to students whenever and wherever they need them. With TEE, students learn at their own pace and develop their skills on their own terms. By integrating a heterogeneous group of students with a dynamic learning system that continuously interacts with them and adapts the content to their specific needs, the students can be active in their learning process. With AR students may experience a blend of virtual and real content in an immersive learning environment (Klopfer & Sheldon, 2010). Modern-day neurosurgical practice paradigms are poised to embrace VR and AR technologies (Godzik et al., 2021). Due to the COVID-19 pandemic, students and teachers experimented with technology more and more, accelerating the adoption of Education 4.0. Appropriate use of information and communication technologies (ICT) tools such as AR and VR will improve the quality of education (Raja & Lakshmi Priya, 2022).

9.3 TECHNOLOGICAL ADVANCES IN EDUCATION

In this 21st century, effective teaching is defined not only by a broad dimension of knowledge but also includes the capability of the teacher to grab students' attention and encourage their learning interests. Technological advancements such as artificial intelligence (AI), gamification, augmented reality (AR), virtual reality (VR), and mixed reality (MR) have been additively used in the education field to promote student learning and increase their motivation. There are different ways to design AR and VR technologies to better equip the teachers to facilitate teaching and learning experience better. This might vary from 3D imagery, 3D animations, 3D videos, 360-degree videos, 360-degree photosphere, and others. Perhaps, AR/VR technologies give users a real experience with virtual elements.

9.3.1 ARTIFICIAL INTELLIGENCE (AI)

AI is also called a technological game changer since it created a revolution in education in terms of innovation and knowledge gain. This game changer has the potential to address the challenges as well as enhance the teaching and learning process. AI is not used as a replacement for the teacher but assists teachers to understand students' individual needs according to their learning capabilities. AI also helps teachers in grading and provides immediate feedback. In the current scenario, AI is used for classroom management, lesson planning, assessment, scheduling class and exam, campus maintenance, safety, and security. In higher education AI is used for many purposes like lecture transcription preparation, chatbots-assisted enrollment, online discussion boards, academic research in a connected campus environment, student success metric analysis, and many more. In recent years, many AI-influenced technologies were introduced in education. ThinkerMath—math tutoring program for personalized learning, Jill Watson—AI-enabled virtual teaching assistant, Brainly—social media site for doubt clarification, Nuance—speech recognition transcribers to assist writing, Cognii—virtual assistant for higher education and corporate training are a few examples. Software-assisted AI learning was introduced by technological companies, namely, Palitt—which creates customized lecture series, and Cram101—which summarizes textbooks into a smart guide and generates practice questions (University of San Diego, 2022).

9.3.2 GAMIFICATION

In education, gamification is an approach that uses gaming elements in the learning environment to increase the level of classroom engagement for students' skill development. The main goal of gamified education is to engage students, provide optimized learning, support behavior challenges, and socialize. The authors (Smiderle et al., 2020) explored the features of gamification on students' learning, performance, and commitment based on their personality behavior in a web-based learning environment. The study result shows that the outcome of gamification is based on the precise characteristics of the students. A systematic review (Dichev & Dicheva, 2017) is structured based on mechanisms, subjects, the learning activities, and the study goals, which are purely based on gamification. This review also discusses the involvement of gamification in education by categorizing it as (1) learner-centric: based on learners and their perception, (2) platform centric: based on gaming elements and gamified platforms. Even the development of gamification in education is still moving forward, soon gamified learning would become a recognized instructional approach.

9.3.3 AUGMENTED REALITY (AR)

AR is created in a real-world environment overlaid with virtual objects. In the past decade, AR has been one of the actively growing technologies, especially in education, which gained momentum worldwide. In this digital era, students of the new generation have more skills in technology, which made the educational institution and teachers adopt new technology for the teaching and learning process. According to the literature (Garzón, 2021), the University of North Carolina developed the first AR for teaching 3D anatomy in 1995. The application of AR in education is divided into three generations, namely, hardware-based AR (1995–2010), application-based AR (2010–2020), and web-based AR (2020 onwards). In the first generation, AR is characterized as expensive and complex head-mounted, heads-up, and handheld displays that are used to teach engineering, health science, and natural sciences. Due to this high cost and low usability, AR was unpopular in the education domain. The outbreak of the second generation of AR overcame the economic drawback by deploying AR applications on mobile and the development of game engines, SDKs (software development kit), and libraries related to AR that enables an easy development environment. This resulted in the five directions of AR application in education such as skills training, object modelling, discovery-based learning, AR books, and AR gaming. The third generation brings AR and AI together to provide intelligent solutions for everyday life problems (Garzón, 2021).

9.3.4 VIRTUAL REALITY (VR)

Virtual reality is a computer-simulated software, which is an immersion in 3D and allows interaction with virtual objects in the virtual environment. Many schools started adopting this VR because it allows students to get the experience of the destination from their classrooms. The education tool developed using VR provides interactive content made using images and videos that enable the student to explore 360 degrees of a scene. VR gives students a chance to learn through immersive experience by allowing them to flourish their imagination by sitting in their classroom. The most common way of using VR in school education is virtual field trips where all students can enjoy a trip to museums, monuments, or other countries with no excuses like expenses, disability, or travel issues. Despite many advantages, the usage of VR technology in education needs to be taken care seriously while developing since VR-world cannot fully reproduce the real world (Paszkiewicz et al., 2021).

9.3.5 MIXED REALITY (MR)

MR is created in a real-world environment with a virtual object where anyone can interact with the virtual object. MR is an upgraded version of AR and VR. It is developed based on a real-time environment that interactively synchronizes with virtual objects. In the past few years, MR has become one of the booming technologies in education since it transforms the way teachers teach and students learn. A Microsoft product, HoloLences-2, uses four different MR technologies such as HoloStudy, HoloTour, HoloHuman, and Lifeliqe (Woods, 2020). HoloStudy is an MR learning tool that simplifies the learning of complicated subjects in a revolutionary way. HoloTour takes the user virtually to different time periods and locations as field trips. HoloHuman provides a chance to get a deeper knowledge of human anatomy as holographic illustrations of the human body.

9.4 APPLICATIONS OF AUGMENTED REALITY IN EDUCATION

Education of the younger generation plays a vital role in the growth and future of any country. Generally in the traditional teaching method, which is also called teacher-centric learning, the teacher will teach and students will follow the lecture where there is no opportunity for creativity and thinking. But nowadays most educational institutions follow student-centric learning where the teachers are the facilitators for the students to understand the concept and help them to think beyond based on their learning capacity. AR is a computer technology with great potential and pedagogy that provides innovative methods for the teaching and learning of education. This technology allows the students to see the real world environment and with virtual objects which enhances the learning methods from traditional to technological. This technology contributes a lot to different levels of education such as primary education, higher education, engineering, medical science, architect, and special education, etc.

9.4.1 AR FOR PRIMARY EDUCATION

In recent days, experimental learning in education is a new approach, especially in primary education. Generally, primary education begins at age 6 and lasts for a minimum of four years when the students are getting prepared for the next level of education. At the same time, the brain development of this age group is super fast, which is made to introduce the elements of sense methods like touch, hearing, and sight. AR allows the students to see the real-world environment with augmented sensory input like video and graphics with a special sound. A system of AR for teaching (SMART) (Bistaman et al., 2018) is designed to teach types of animals and modes of transportation to second-grade students. This system comprises a webcam, computer, projector, and AR marker. This SMART system created a positive impact on students in terms of collaboration and motivation among students during teaching and learning. A computer simulation, Alien Contact! (Bistaman et al., 2018), was designed using an AR game to teach language arts, math, and scientific skills. To play this AR game, students need to form a team of four and take up different roles, such as FBI agent, cryptologist, hacker, and chemist. The team has to solve the puzzle of the AR game by collaboratively sharing ideas among them.

9.4.2 AR FOR HIGHER EDUCATION

The innovation and faster technological changes impact directly the education sector and amplified the assurance of a dynamic and competitive educational system. AR in education shows a different direction to the learning process by creating a fun experience for the young, energetic, restless group of higher education students through a visual simulation. AR has been used for different purposes in higher education. Many AR applications have been developed to enable learners to have a better understanding of scientific concepts such as electromagnetism, movement of Earth around the Sun, food digestion, etc. NetAR (Criollo-C et al., 2021) was developed for mobile devices to learn the Network model and static routing. This app was developed to learn the networking concept visually by avoiding AR devices. A game-based AR, ChronoOps (Godoy Jr., 2020) has been developed for language learners. This game creates a location-based (for different languages) match that has situational scenes with encouraging respondents to improve the language skills of the player/learner. Integrating AR/VR in higher education proved to have a positive effect on the learning process with increased student engagement, elevated interaction, and improved authenticity.

9.4.3 AR FOR STEM

A meta-discipline STEM (science-technology-engineering-mathematics) is an emerging area in education for social scientific development. The educators and the instructors continuously seek an innovative instructional approach to create STEM professionals. AR-technology-incorporated teaching shows the way to significant learning and in-depth understanding of STEM concepts. It also helps to get mandatory skills, such as problem-solving, teamwork, and communication. A systematic review (Mystakidis et al., 2021) of ten years of published work on AR contribution in STEM fields is analyzed and provides the seminal information that is listed here.

(1) AR received predominantly a greater appreciation in engineering-related subjects for better understanding.
(2) Desktop computers, mobile devices, AR wearable devices (AR glasses), etc. are used as development devices.
(3) Three types, namely, marker-based, location-based, and markerless augmentation are used for creating real-world environments.
(4) Three augmentation techniques, namely (a) augmentation of the specialized lab equipment, (b) physical objects augmentation, and (c) course handbook/instructional sheets using augmentation.
(5) Identified five instructional approach across the STEM spectrum, namely (a) experiential, (b) cooperative/collaborative, (c) presentation, (d) activity-based, and (e) discovery (scientific inquiry).
(6) Five different instructional techniques used in STEM studies are (a) instruction through simulation, (b) project, (c) observation, (d) problem-solving, and (e) question-answer.
(7) Taxonomy of instructional methods.

From the aforementioned points, one can conclude that usage of AR in STEM education has brought both teaching-learning; however, the learning curves and adapting to usability features are still challenging elements.

9.5 APPLICATIONS OF VIRTUAL REALITY IN EDUCATION

VR technology is revolutionizing the entire learning process by creating real learning environments, especially training in dangerous, costly, and complex environments. VR is not just about replacing traditional learning methods, it enhances the overall learning effectiveness by saving money and reducing training time and errors. The Global VR in the education market is expected to increase from $8.66 billion in 2022 to $32.94 billion in 2026 at a CAGR of 39.7%. The VR system in education mainly comprises hardware, software, and related solutions. The hardware components such as VR heads, head-mounted displays (HMDs), and projectors are used to create interactive experiences by projecting the VR content (Business Wire, 2022). Virtual reality (VR) is a multisensory immersive environment that comprises different sensors in the VR headset that captures users' movements and provides a real-life experience via HMDs. This gives a feeling of being physically present and creates virtual interaction to users (Deloitte, 2016). Figure 9.1 depicts when a trainer can decide on designing VR technology in the learning process based on the level of information required and complexity involved. VR is more suitable for complex environments, and the learner requires higher order thinking to understand practically with help of simulation models. If the learner requires simple information, then the traditional methods of classroom teaching using books and case study methods can be more

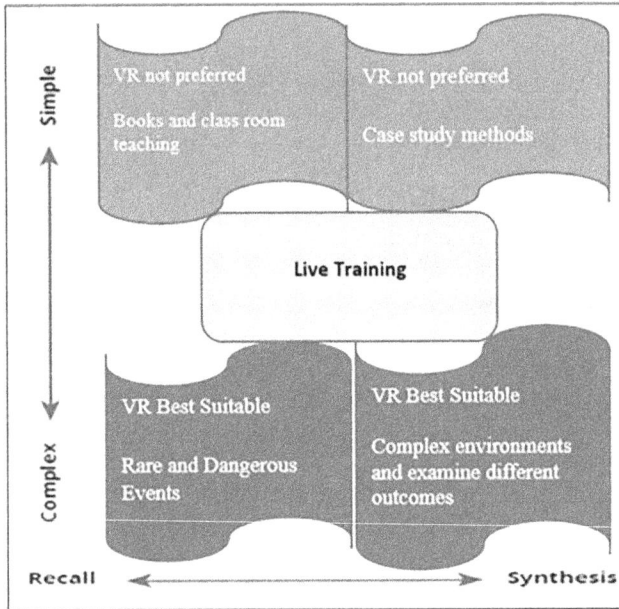

FIGURE 9.1   When and how to use VR—a decision framework.

Source: Deloitte Analysis

effective (Deloitte Insight Contributors, 2018). The application VR in education contributed to the development of the teaching and learning process. Figure 9.2 shows the application of VR in medical and aircraft training.

9.5.1 HIGHER EDUCATION: MEDICAL

VR technologies are extensively used in complex biological systems and biomedical applications such as molecular data visualization and medical and surgical training. Neuroscience-related research found that VR-based content delivery systems provide better learning experiences with emotional engagement than 2D and 360° learning platforms. Some of the VR applications used in biomedicine are as follows: (1) clinical assessments where the patients and doctors can enter a virtual environment for interactions and suggestions; (2) ConfocalVR application is used to visualize the complex 3D cell structure and molecular distributions; (3) neuroscientists use the TeraVR application to visualize complex neurons in brains (Venkatesan et al., 2021).

9.5.2 AEROSPACE ENGINEERING: HIGHER EDUCATION

Virtual avionics procedure trainer (VAPT) provides low-cost training to flight crews in avionics—an electronic system used in modern aircraft, resource management, and basic flight and cockpit operations. As shown in Figure 9.2, the VR-based avionics system simulators such as flight simulator and VR engine simulation create a

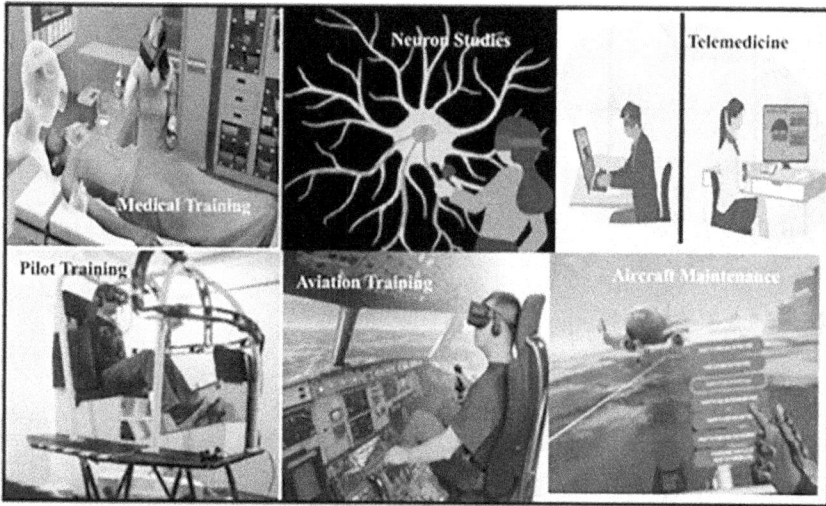

FIGURE 9.2 Applications of VR in medical and aircraft services.

Source: Medical Training (Andres, 2022), Neuron Studies, Telemedicine (Venkatesan et al., 2021), Pilot Training (Morozova, 2018), Aircraft Maintenance (Lozé, 2019)

realistic training experience for crew members (Choudhury, 2022). VR technologies can reduce maintenance training time by 75% in aviation mechanics (Zazulia, 2019).

9.5.3 SCIENCE STUDENTS

VR technology provides a real-world immersive experience that makes learning more realistic and interactive. The fully immersion form of VR means a realistic feeling of being present in the real world by the movement of one's body and being able to interact and control actions in a simulation. Further, they found that VR-based platforms enhance the interest and motivation of learning science concepts by developing scientific attitudes in science students (Sarpal & Nangia, 2022).

9.5.4 PRIMARY SCHOOLS

Some of the government schools in India are procuring VR headsets for educational purposes and implementing VR-based teaching and learning. Students experience the real world through VR headsets. For example, as shown in Figure 9.3, students can see, interact, and learn with a virtual simulation of any animal, planets, human anatomy structures, science concepts, and rockets (Choudhury, 2022).

9.5.5 VIRTUAL FIELD TRIP FOR EDUCATION: SCHOOLS AND HIGHER EDUCATION

Immersive VR platforms provide students to experience real-world locations and their surroundings by sitting inside the classroom with help of VR headsets. The various applications of VR in education are as follows:

FIGURE 9.3   VR-based learning in Indian schools.

Source: Choudhury (2022), InGage (2018), Marr (2021)

FIGURE 9.4   Virtual field trips.

Source: Connections, 2022; Peter and Philip, 2019

- Provides an opportunity for experiential learning to the students than classical methods of simply reading and writing.
- VR platforms completely engage the students with their full attention by providing memorable experiences.
- VR platforms inspire students to involve, learn, and understand the concepts within the classroom.
- The immersive experience of VR technology encourages students' out-of-the-box thinking and improves their creativity and imagination (VR for Education—the Future of Education, 2022).

Harvard University used virtual field trip to teach "The Pyramids of Giza: Archaeology, History" concept of Egyptology course to their graduate students. Figure 9.4 shows that the students were in visualization classrooms to have a virtual experience of the largest Egyptian pyramids, and they connected live with students of Zhejiang University in China (Mohammad, 2017). Further, the students can experience forests with animals (Pettai, 2022).

9.6   TRENDS, BENEFITS, RISKS, AND CHALLENGES OF IMPLEMENTING AR AND VR IN EDUCATION

Teachers could easily teach abstract concepts to students with the help of AR. The overall classroom experience can be enhanced by implementing the interaction and experimentation technologies of AR. Thus, teaching them new skills, inspiring their minds to explore and seek newer academic interests. AR provides an enhanced version of the surroundings by layering digital content on top of the graphic representation of the real world. A combination of three salient features, namely, real and virtual worlds, real-time interaction, and accurate 3D registration of virtual and real objects was proposed by Parveau and Adda (2018).

AR technologies could be used for enabling the following:

- learning content in 3D perspective;
- ubiquitous, collaborative, and situated learning;
- presence, immediacy, and immersion of learners senses;
- envisioning the invisible; and
- associating formal and informal learning.

A multidimensional space for learning is facilitated by AR and VR. Thus, enabling students to interact, visualize, and immerse themselves into the subject and explore it completely. The immersive feature of VR provides an opportunity to the learners to explore and learn at their own place and space. The implementation of this feature and these technologies has been done in majority of the disciplines and advance studies—maths, science, and history in school level and health care, engineering, and architecture.

The fundamental aspects of AR and VR are "immersion" (from reality to virtual), "ubiquity" (stationary to omnipresent), and "multiplicity" (single user to potentially everyone). The ease and widespread implementation of AR and VR technologies to desktop computers, handheld devices (mobile, tabs, and laptops), head-mounted displays, immersion technologies, and so on. These are supported by the novel assistive tools, the educational values of AR, which are not solely based on technology but also on the educational system requirements. Support to engage learners in authentic exploration is provided by virtual elements like texts, videos, pictures, and graphics. Sequestered subject data can be made available to the students with the help of AR technologies.

Although there are a lot of strengths and benefits in adopting a newer technology in the education system to provide better learning experiences, the risks and challenges involved in such technologies cannot be overemphasized. There are inherent risks in applying AR and VR in three-dimensional ways, namely, security, privacy, and safety. Most cases of data breaches result in integrity issues. The authenticity of the apps and sites used by the teachers may not be guaranteed. Allowing access to technology implementation in the devices might give access to confidential information stored in the devices too. The well-being of an individual is challenged resulting in cyber sickness due to the overuse of devices. Various stakeholders in the education system can enjoy the benefits of AR and VR if the following challenges are handled well. Teachers must be equipped with the skills required to handle AR and VR. Students must

Trends
1. Combination of real and virtual worlds
2. Real time interaction, and
3. Accurate 3D registration of virtual and real objects

Benefits
1. Content in 3D format
2. Ubiquitous, collaborative and multiplicity
3. Learners' senses of presence, immediacy, and immersion.
4. Visualizing the invisible, and
5. Bridging formal and informal learning

Risks
1. Security – Integrity Challenges
2. Privacy – access to confidential data
3. Safety – Wellbeing of the user impacted

Challenges
1. Teachers – AR training
2. Students – Cognitive overload
3. Teaching – Technical and scientific competency
4. Infrastructure – Incompatibility

FIGURE 9.5   Trends, benefits, risks, and challenges of implementing AR and VR in education.

Source: Authors

manage the cognitive overload to enjoy the enhanced learning environment. A better teaching environment can be provided if technical and scientific teaching tools are made available to overcome the incompatibility issues in the environment. Further, Fernandez (2017) highlighted the significant barriers of use of AR and VR technologies in education such as matching the training content with current curriculum, technological training to teachers, and how to make use of this AR/VR technology within the teaching system. Some more challenges like focusing on learning outcomes when using AR. AR classrooms should be a collaborative learning environment. AR must be understood, accessible, and time-constrained to be overcome ("Augmented Reality in Teaching: Key Challenges and How to Overcome them," 2023). Figure 9.5 depicts trends, benefits, risks, and challenges of implementing AR and VR in education.

9.7 CONCLUSION

Evolution in ICT transformed all the aspects of human life to make it comfortable, safe, and secure. IoT, AI, and ML have introduced a new prefix, "Smart," to most aspects, starting from smart phones, smart homes, smart education, smart watches, smart classes, smart boards, and so on. In the same way, the education sector is also continuously evolving, constantly instilling appropriate skills and knowledge to enrich and progress young minds to meet up with the technological changes brought into all other sectors. The intervention of ICT has transformed all the processes of the education sector. Perhaps teaching pedagogy continues to evolve to make more innovative and simple ways in which these lessons are presented. However, it is noteworthy that these tools are not meant to replace traditional education systems, curriculums, or workers, but rather as catalysts to enhance them. Integrating AR and VR technologies in teaching young learners provides an exciting opportunity for teachers in enriching the student experience. COVID-19 brought sudden changes to the education system. The transition from classroom teaching to remote or online teaching was possible due to the availability of existing tools and techniques like smart education. There are shreds of evidence to prove that the intervention of AR and VR entices students to remain interested, engaged, and have fun learning (Lee,

2012; Huang et al., 2019). Better class engagement in K-12 schools can be assured due to the real-time exposure to the content and context explained through AR and VR. To make use of the power of AR and VR, it is important to create an appropriate plan with the education system. Students, teachers, and policymakers should work together to practice adaptive leadership, responsive teaching, and generative assessment for a better future using AR and VR applications. By doing so, educational institutions will become stronger, smarter, and more resilient. Smart industries and smart education are the need of the hour. At the same time, the goal of the education system should be to make an equal opportunity of learning and as well as learning environment available, accessible, and affordable for everyone on the planet.

REFERENCES

"$32.9 Billion Worldwide Virtual Reality in Education Industry to 2031—Identify Growth Segments for Investment—Researchandmarkets.Com. 2022." *Business Wire*. www.businesswire.com/news/home/20220609005644/en/32.9-Billion-Worldwide-Virtual-Reality-in-Education-Industry-to-2031-Identify-Growth-Segments-for-Investment—ResearchAndMarkets.com.

43 Examples of Artificial Intelligence in Education. 2022. *University of San Diego*. https://onlinedegrees.sandiego.edu/artificial-intelligence-education/.

Andres, M. 2022, March 12. "How to Use Virtual Reality in Medical Training in 2022." *I. Program-Ace*. https://program-ace.com/blog/virtual-reality-in-medical-training/.

Augmented Reality in Teaching: Key Challenges and How to Overcome Them. 2023. "THE Campus Learn, Share, Connect." www.timeshighereducation.com/campus/augmented-reality-teaching-key-challenges-and-how-overcome-them.

Bistaman, Izwan Nurli, Syed Zulkarnain Idrus, and Salleh Abd Rashid. 2018. "The Use of Augmented Reality Technology for Primary School Education in Perlis, Malaysia." *Journal of Physics: Conference Series* 1019: 012064. doi:10.1088/1742-6596/1019/1/012064.

Chiang, Tosti H. C., Stephen J. H. Yang, & Gwo-Jen Hwang. 2014. "An Augmented Reality-Based Mobile Learning System to Improve Students' Learning Achievements and Motivations in Natural Science Inquiry Activities." *Journal of Educational Technology & Society* 17 (4): 352–365. www.jstor.org/stable/jeductechsoci.17.4.352.

Choudhury, Moumita Deb. 2022. "Virtual Reality Headsets Enter Computer Labs in Indian Schools, Colleges." *Mint*. www.livemint.com/technology/tech-news/virtual-reality-headsets-enter-computer-labs-in-indian-schools-colleges-11645766029704.html.

Criollo-C, Santiago, David Abad-Vásquez, Marjan Martic-Nieto, Fausto Andrés Velásquez-G, Jorge-Luis Pérez-Medina, and Sergio Luján-Mora. 2021. *Towards a New Learning Experience through a Mobile Application with Augmented Reality in Engineering Education*. MDPI. Multidisciplinary Digital Publishing Institute. www.mdpi.com/2076-3417/11/11/4921.

Deloitte. 2016, July 21. "The Very Real Growth of Virtual Reality." *ME PoV*. https://www2.deloitte.com/xe/en/pages/about-deloitte/articles/treading-water/the-very-real-growth-of-virtual-reality.html.

Deloitte Insight Contributors. 2018. "Real Learning Virtual World [White paper]." https://www2.deloitte.com/content/dam/insights/us/articles/4683_real-learning-virtual-world/4683_real-learning-in-a-virtual-world.pdf.

Dichev, Christo, and Darina Dicheva. 2017. *Gamifying Education: What Is Known, What Is Believed and What Remains Uncertain: A Critical Review—International Journal of Educational Technology in Higher Education*. SpringerOpen. Springer International Publishing. https://doi.org/10.1186/s41239-017-0042-5.

Dick, Ellysse. 2021. "The Promise of Immersive Learning: Augmented and Virtual Reality's Potential in Education." *ITIF*. https://itif.org/publications/2021/08/30/promise-immersive-learning-augmented-and-virtual-reality-potential/.

Fernandez, Manuel. 2017. "Augmented-Virtual Reality: How to Improve Education Systems." *ScholarWorks*. Accessed June 16. https://scholarworks.waldenu.edu/hlrc/vol7/iss1/3/.

Freina, L., and M. Ott. 2015, April. "A Literature Review on Immersive Virtual Reality in Education: State of the Art And Perspectives." *The International Scientific Conference Elearning and Software for Education* 1 (133): 10–1007.

Garzón, Juan. 2021. *An Overview of Twenty-Five Years of Augmented Reality in Education*. MDPI. Multidisciplinary Digital Publishing Institute. https://doi.org/10.3390/mti5070037.

Germain, Marie-Line. 2019. *Integrating Service-Learning and Consulting in Distance Education*. Emerald Publishing Limited.

Godoy Jr., C. H. 2020, June 18. "Augmented Reality for Education: A Review." *International Journal of Innovative Science and Research Technology* 5 (6): 39–45. https://doi.org/10.38124/ijisrt20jun256.

Godzik, J., S. H. Farber, T. Urakov, J. Steinberger, L. J. Knipscher, R. B. Ehredt, L. M. Tumialan, and J. S. Uribe. 2021. " 'Disruptive Technology' in Spine Surgery and Education: Virtual and Augmented Reality." *Operative Neurosurgery* 21 (1): S85–S93.

Huang, K. T., C. Ball, J. Francis, R. Ratan, J. Boumis, and J. Fordham. 2019. "Augmented versus Virtual Reality in Education: An Exploratory Study Examining Science Knowledge Retention When Using Augmented Reality/Virtual Reality Mobile Applications." *Cyberpsychology, Behavior and Social Networking* 22 (2): 105–110. Accessed June 16. https://pubmed.ncbi.nlm.nih.gov/30657334/.

InGage. 2018, October 16. "Govt Schools Adopt Tech in Teaching." www.myingage.com/blog/govt-schools-adopt-tech-in-teaching/.

Klopfer, J., and E. Sheldon. 2010. "Augmenting Your Own Reality: Student Authoring of Science-Based Augmented Reality Games." *New Directions for Youth Development* 2010 (128): 85–94. U.S. National Library of Medicine. Accessed June 16. https://pubmed.ncbi.nlm.nih.gov/21240956/.

Lang, B. 2019, October 15. "Colorado State University Has Deployed a 100 Headset VR Lab for Biomedical Education." *Road to VR*. https://www.roadtovr.com/colorado-state-university-immersive-reality-training-lab-vr/.

Law, N., D. S. Niederhauser, R. Christensen, and L. Shear. 2016. "A Multilevel System of Quality Technology-Enhanced Learning and Teaching Indicators." *Journal of Educational Technology & Society* 19 (3): 72–83.

Lee, Kangdon. 2012. "Augmented Reality in Education and Training—Techtrends." *SpringerLink. Springer US*. https://link.springer.com/article/10.1007/s11528-012-0559-3.

Lee, Nicol Turner, Rashawn Ray, Samantha Lai, and Brooke Tanner. 2022. "Ensuring Equitable Access to AR/VR in Higher Education." *Brookings. Brookings*. www.brookings.edu/blog/techtank/2022/09/06/ensuring-equitable-access-to-ar-vr-in-higher-education/.

A Literature Review on Immersive Virtual Reality in Education . . .—CNR. 2023. Accessed June 16. www.itd.cnr.it/download/eLSE%202015%20Freina%20Ott%20Paper.pdf.

Lozé, Sébastien. 2019. "Beyond the Manual: VR Training on Aircraft Maintenance." *Unreal Engine*. www.unrealengine.com/en-US/spotlights/beyond-the-manual-vr-training-on-aircraft-maintenance.

Market Research Future. 2022. "AR and VR in Education Market Anticipated to Grow at a CAGR of 18.2% during 2022 to 2027—Report by Market Research Future (MRFR)." *GlobeNewswire News Room. Market Research Future*. www.globenewswire.com/en/news-release/2022/06/07/2458167/0/en/AR-and-VR-in-Education-Market-Anticipated-to-Grow-at-a-CAGR-of-18–2-During-2022-to-2027-Report-by-Market-Research-Future-MRFR.html.

Marr, Bernard. 2021. "10 Best Examples of VR and AR in Education." *Forbes Magazine*. www. forbes.com/sites/bernardmarr/2021/07/23/10-best-examples-of-vr-and-ar-in-education/ ?sh=7eb3b6ee1f48.

Miranda, Jhonattan, Christelle Navarrete, Julieta Noguez, José-Martin Molina-Espinosa, María-Soledad Ramírez-Montoya, Sergio A. Navarro-Tuch, Martín-Rogelio Bustamante-Bello, José-Bernardo Rosas-Fernández, and Arturo Molina. 2021. "The Core Components of Education 4.0 in Higher Education: Three Case Studies in Engineering Education." *Computers & Electrical Engineering* 93: 107278. doi:10.1016/j. compeleceng.2021.107278.

Mohammad, A. K. 2017. Virtual Reality in Avionics Design [Shenyang Aerospace University]. https://www.researchgate.net/publication/354035366_Virtual_Reality_In_ Avionics_Design_VR_In_Avionics.

Morozova, Anastasia. 2018. "Using Virtual Reality to Prepare People with the Most Dangerous Jobs for High-Risk Situations." *Jasoren. Jasoren.* https://jasoren.com/using-virtual-reality-to-prepare-people-with-the-most-dangerous-jobs-for-high-risk-situations/.

Mystakidis, Stylianos, Athanasios Christopoulos, and Nikolaos Pellas. 2021. "A Systematic Mapping Review of Augmented Reality Applications to Support STEM Learning in Higher Education." *Education and Information Technologies* 27 (2): 1883–1927. doi:10.1007/s10639-021-10682-1.

Parveau, M. and M. Adda. 2018. 3iVClass: "A New Classification Method for Virtual, Augmented and Mixed Realities." *Procedia Computer Science* 141: 263–270.

Paszkiewicz, Andrzej, Mateusz Salach, Paweł Dymora, Marek Bolanowski, Grzegorz Budzik, and Przemysław Kubiak. 2021. "Methodology of Implementing Virtual Reality in Education for Industry 4.0." *Sustainability* 13 (9): 5049. doi:10.3390/su13095049.

Peter, D. M. and J. K. Philip. January 8, 2019. "How Harvard University Uses Rumii Virtual Reality Software to Explore The Giza Pyramids with Zhejiang University in China." *Doghead Simulations.* https://www.dogheadsimulations.com/doghead-blog/2019/1/7/ rumii-customer-spotlight-how-harvard-university-uses-virtual-reality-to-explore-the-pyramids.

Pettai, S. 2022, May 23. "3 Use Cases of Virtual Reality in Higher Education." *UX Connections*. https://www.uxconnections.com/3-use-cases-of-virtual-reality-in-higher-education/.

Pregowska, Agnieszka, Karol Masztalerz, Magdalena Garlińska, and Magdalena Osial. 2021. "A Worldwide Journey through Distance Education—from the Post Office to Virtual, Augmented and Mixed Realities, and Education during the COVID-19 Pandemic." *Education Sciences* 11 (3): 118. doi:10.3390/educsci11030118.

Raja, M., and G. G. Lakshmi Priya. 2022. "Using Virtual Reality and Augmented Reality with ICT Tools for Enhancing Quality in the Changing Academic Environment in COVID-19 Pandemic: An Empirical Study." *Technologies, Artificial Intelligence and the Future of Learning Post-COVID-19*: 467–482. doi:10.1007/978-3-030-93921-2_26.

Sarpal, S. and A. Nangia. 2022. "Trends in Use of Virtual Reality (VR) Technology in Science Education: A Systematic Review." *Indian Journal of Educational Technology* 4 (2): 225–242.

Smiderle, Rodrigo, Sandro José Rigo, Leonardo B. Marques, Jorge Arthur Peçanha de Miranda Coelho, and Patricia A. Jaques. 2020. "The Impact of Gamification on Students' Learning, Engagement and Behavior Based on Their Personality Traits." *Smart Learning Environments* 7 (1). doi:10.1186/s40561-019-0098-x.

Tan, Sin Ying, Dhiya Al-Jumeily, Jamila Mustafina, Abir Hussain, Anne Broderick, and Henry Forsyth. 2018. "Rethinking Our Education to Face the New Industry Era." *EDULEARN Proceedings*. doi:10.21125/edulearn.2018.1564.

Tsai, Chin-Chung. 2017. "Conceptions of Learning in Technology-Enhanced Learning Environments." *Asian Association of Open Universities Journal* 12 (2): 184–205. doi:10.1108/aaouj-12-2017-0038.

University of Michigan Emerging Technologies Group. (2023, March 22). "Stephanie O'Malley, Author at X-Reality." *X-Reality*. ttps://xr.engin.umich.edu/author/somalley/.

Venkatesan, M., H. Mohan, J. R. Ryan, C. M. Schürch, G. P. Nolan, D. H. Frakes, and A. F. Coskun. 2021. "Virtual and Augmented Reality for Biomedical Applications." *Cell Reports Medicine* 2 (100348): 1–13.

VR for Education—the Future of Education. 2022. "Immersion VR." https://immersionvr. co.uk/about-360vr/vr-for-education/.

Wang, Hung-Yuan, Henry Been-Lirn Duh, Nai Li, Tzung-Jin Lin, and Chin-Chung Tsai. 2014. "An Investigation of University Students' Collaborative Inquiry Learning Behaviors in an Augmented Reality Simulation and a Traditional Simulation." *Journal of Science Education and Technology*. Springer. 233 Spring Street, New York, NY 10013. Tel: 800–777–4643; Tel: 212–460–1500; Fax: 212–348–4505; e-mail: Service-ny@springer. com; Web site: www.springerlink.com. https://eric.ed.gov/?id=EJ1041092.

Woods, J. 2020, November 18. "Mixed Reality Classrooms: The New Era of Education." *XR Today*. https://www.xrtoday.com/mixed-reality/mixed-reality-classrooms-the-new-era-of-education/.

Yu, Mengting. 2022. "Technology-Enhanced Education." In *The Wiley Handbook of Sustainability in Higher Education Learning and Teaching*, 133–151. doi:10.1002/97811198 52858.ch7.

Zazulia, Nick. 2019. *Virtual Reality Could Cut Maintenance Training Time 75 Percent*. Avionics International. www.aviationtoday.com/2019/07/18/virtual-mro-training-could-be-a-massive-time-saver/.

10 AI-Based Automatic Detection of IP Network Performance in Telecommunication

Shamoona Imtiaz, Oliver Popov, and Jaume Rius i Riu

10.1 INTRODUCTION

Until the 21st century, engineers urged to persuade fast communication in network performance. Then quick file transfer was the most desirable network service (Olifer and Olifer 2005). Later, at the beginning of the 21st century, the inception of real-time communication endorsed the demand for low latency and jitter. However, with the convergence of computer and telecommunication networks, integrated services digital networks (ISDN) innovated telephony services into intelligent networks (Olifer and Olifer 2005). This way, intelligent networks could prioritize the network traffic such that sometimes FTP (file transfer protocol), traffic can get higher priority and sometimes VoIP (Voice Over IP) is of utmost importance. This has been achieved with many different approaches like software defined networking (Feamster, Rexford, and Zegura 2014), knowledge defined networking (Mestres et al. 2017), and knowledge plane of the Internet (Clark et al. 2003). The idea was to pull the intelligence out of network devices and keep it somewhere in the control plane. Such approaches allow shaping the network traffic based on dynamic requirements so that a network service does not need to starve due to its static budget allowance, yet other services are not utilizing their full budget. However, existing network solutions may not always be able to handle the dynamic network conditions due to exponential traffic growth while maintaining service level agreement (SLA).

As Morais and Pedro said, IoT and 5G will impact the volume and dynamicity of the network traffic (Morais and Pedro 2018), so as for the telecommunication industry. Therefore, in 2007, to address the ongoing challenges for network operators, Netrounds' founders introduced software as a service (SaaS) solution (software defined networking—SDN based). The solution offers active network analytics and a service assurance (Netrounds 2019b). As a default solution, the web graphical user interface (GUI) of Netrounds Control Center (NCC) runs on-demand tests configured on each Netrounds probe and generates aggregated results of network KPIs (such as threshold, delay, and packet loss) and SLA monitoring metrics. These probes continuously generate data at the finest granularity. Whereas programmable REST

DOI: 10.1201/9781003343332-10

API (Representational State Transfer Performance Metric Application Programming Interface) allows the orchestration of automated tests and monitors.

Besides all the efficient network monitoring services offered by Netrounds, it does not provide a relational picture of this sheer amount of data. Automation of these test results, prediction of IP network behavior, and timely anomaly detection can minimize the violation of SLA. Therefore, applying machine learning algorithms on Netrounds metrics is a decent approach to visualize the relational picture of network telemetry, which otherwise may not be visible to the human eye. Similar approaches have already been used for visualization, prediction, and anomaly detection of network performance (Skënduli, Biba, and Ceci 2018) (Sun et al. 2018) (Morais and Pedro 2018).

Furthermore, such knowledge enables telecom operators to maintain business continuity by forecasting any disruption or degradation in the network performance. For compliance and legalities, business continuity is the result of good management and governance of critical processes of an organization. It is an integral part of an organization's risk management process (ENISA 2023). Moreover, according to ISO 22301 "Business Continuity Management" is a resilient approach to the critical infrastructure of organizations (For Standardization 2019). An important aspect of business continuity is information technology, which comes under the discipline of IT Disaster Recovery (ITDR), such that

> In the event of an incident the plans and systems in place should ensure a resump tion of service within the agreed Service Level Agreements (SLAs) ensuring compliance and customer satisfaction as well as aiding in Business Continuity.
>
> (ENISA 2023)

Therefore, for companies, it is just not only the analytics they are interested in. Today, it is imperative to go after the reasons why things do not behave normally. Thus, the aim is to visualize, predict, and detect anomalies of network performance with respect to "Delay average (ms)" and "Lost packets." The scope of the problem is illustrated in Figure 10.1, which covers the points from "IPEdge" through "Distribution Routers" to the "Edge."

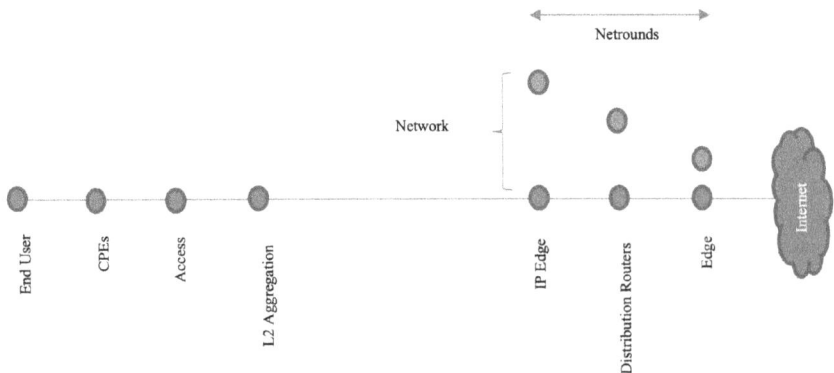

FIGURE 10.1   Telecom network infrastructure.

10.2 BACKGROUND

10.2.1 NETROUNDS

Netrounds is a programmable test and service assurance solution which offers traffic generating test agents. These software-based test agents can provide the measurements for Internet Performance, IPTV and OTT Video, Voice, Security, Network Performance, Wi-Fi, Mobile Radio, and Remote Packet Inspection (Netrounds 2019b).

The solution comes from and is based on three different venues such as "Datacenter and Cloud," "SDN-WAN and Dynamic VPN," and "IP Core and Mobile Backhaul Performance" for service activation testing, quality monitoring, and troubleshooting. For each solution, Netrounds probes monitor and analyze the protocols by deploying automated tests and monitors for target KPIs. Measurement can be of type test or monitor (Netrounds 2019d). Test measurement is a finite set of sequential steps over a specified period only, whereas monitor measurement executes indefinitely. These measurements can be retrieved using "Netrounds Control Center Web GUI," "Netrounds REST API," or "Netrounds NETCONF and YANG API." In this study, the focus is on the "Netrounds REST API" with the "read" function, as shown in Figure 10.2. With the inherent capability, REST API can be used in connection to any protocol, including HTTP/HTTPS.

10.2.2 MACHINE LEARNING OVERVIEW

Artificial intelligence and machine learning are the same is a misconception. They are similar but separate in the perimeter. Machine learning is learning through experiences, whereas artificial intelligence is a broader concept of machine learning which carries out human tasks by the acquisition and application of knowledge (GeeksforGeeks 2023). The process is used to learn, predict, decide, remember,

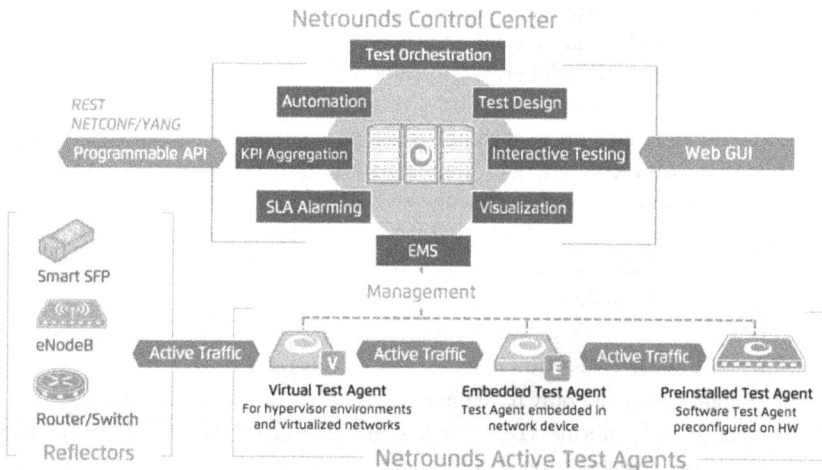

FIGURE 10.2 Netrounds Control Center (Netrounds 2019b).

analyze, and recognize the data. Whereas machine learning is limited to seeking the logical relationships between observed data when there is no known or clear mathematical relationship. Machine learning is categorized into four, that is, supervised machine learning, unsupervised machine learning, reinforcement learning, and federated learning.

10.2.2.1 Supervised Machine Learning Models

In supervised learning, features are labeled, and output can be predicted either as a class label (classification) or continuous value (regression) (Al-Rubaie and Chang 2019). The data selected for the training of the model is called features, and the predicted output is called a class label. This is the most common category in use. The two most popular models are regression and classification. For the predictions phase of this research, supervised learning was deemed appropriate as a modeling technique. The selected models are discussed here.

- Regression—Regression can be interpreted as a task of predicting the next value based on the previous values (Cios et al. 2007). It is a statistical analysis when the data values grow gradually or are continuous in nature. The equation of the regression line is described as follows:

$$y = m \times x + c$$

where y is a dependent variable, x is an independent variable, m is the slope of the line, and c is the coefficient of the line.

- Decision Tree—A decision tree is one of the most popular classification models which organizes its knowledge in the form of nodes. The top node is the root which is followed by the internal/decision nodes and reaches out to the leaf nodes (Skënduli, Biba, and Ceci 2018). Branching begins with Boolean conditions on each node, and each branch is a decision. Decision tree is easy to interpret and visualize but is prone to overfitting the noise. They show high variance even for a small variation.
- Random Forest—Random forest is a collection of decision trees where each decision tree votes and the majority of voting is regarded as the final decision tree to reduce the chance of overfitting. During the modeling process, it randomizes the sampling of training data as well as randomizes the subset of features to split the nodes (Koehrsen 2018b). While training, each sample learns from a random sample, which means some samples may be used more than once. The main aim is to lower the variance, but it comes at the cost of increased bias.

10.2.2.2 Anomaly Detection

Usually, anomalies do not only have their characteristics, but they are also defined in comparison to what is normal (Dunning and Friedman 2014). Therefore, an anomaly is something that deviates from normal and expected patterns.

- Types of Anomalies—In general, anomalies can be global or local. Global anomalies are easy to locate because they deviate from the instances in the denser area. To see the local anomalies, one needs to leave the global view and zoom in on specific dense clusters. These are mostly hard to detect and can only be detected when compared to the close neighbor (Goldstein and Uchida 2016). Therefore, anomalies can be classified into three types concerning data instances: point anomaly, collective anomaly, and contextual anomaly.

  Point anomaly refers to a data point that is too far from the rest of the data points (Samuelsson 2016). This research detects this kind of anomaly. Collective anomaly is where a set of related data instances helps to detect the anomaly. The dataset requires a relationship between the data items such as sequential data, spatial data, and graph data. Although each point itself may not be anomalous within the collection (Ahmed 2018). In contrast, a contextual anomaly is an abnormality that becomes an anomaly in a specific context only. This is also known as a conditional anomaly.

- Learning Methods of Anomaly Detection—There are three main approaches for anomaly detection, that is, unsupervised, supervised, and semi-supervised (Goldstein and Uchida 2016). In unsupervised anomaly detection, all the data is passed into the model regardless of separating the test and train data, shown in Figure 10.3(a). In this approach, decisions are made based on the intrinsic properties of the data. In supervised anomaly detection, data is divided into test and train labelled sets, and the training set does include anomalous data points during the training of the classifier, as shown in Figure 10.3(b). The trained classifier is later used for detection. Since anomalies are normally infrequent as compared to the normal data so the class distribution can be uneven. Inequality can be a problem for supervised anomaly detection algorithms because their performance reduces with the uneven class distribution. Lastly, semi-supervised is like a supervised approach where test and train data are used but without any anomalous data in the train set, as shown in Figure 10.3(c). In this work, unsupervised anomaly detection is performed.

- Methods of Anomaly Detection—There are five main methods of anomaly detection. These methods are distance-based, density-based,

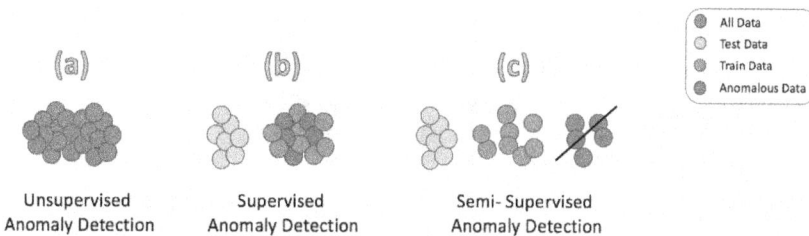

FIGURE 10.3 Data classification for anomaly detection approaches.

boundary-based, partition-based, and property-based. Since the focus of the research was point anomalies and contextual anomalies, partition-based and property-based were not included in the experiment.

In a distance-based method, distance is measured between the data instances and their neighbors. Among many different distance measures, Euclidean and Mahalanobis are the two most common ones [19]. The example algorithms are kth-nearest neighbor (KNN); k-means clustering and regression hyperplane distance. The kth-nearest neighbor is a famous algorithm for anomaly detection and was used for global anomaly detection.

In a density-based method, normal data instances make up a dense area, and anomalies fall sparsely in the area. Some popular examples are the local outlier factor (LOF), kernel density estimate (KDE), and DBSCAN (Huang 2018). In this research, the LOF was performed to detect global anomalies.

The boundary-based method defines a boundary around normal data instances, and anything outside that region is considered as an anomaly. The most common methods are k-centers, one-class support vector machine, and elliptic data description (Huang 2018). One of the methods used in this research was one class support vector machine.

10.2.2.3   Unsupervised Anomaly Detection Models

Unsupervised machine learning groups similar information together and uses the technique of clustering (Al-Rubaie and Chang 2019). Usually, it is used when there is no known output label but input labels are known. Therefore, the unsupervised method was used for anomaly detection.

- kth-Nearest Neighbor—This model is used to detect global anomalies. It classifies the data points based on their neighbors' placement. k is the number of nearest neighbors used for majority voting, and its value depends on the nature of the dataset. It is important to choose an optimal value of k; the recommendation is to keep it odd. If k is too small, then there is a chance of bias and noise, and if it is large, then it may take ages to classify because it is a lazy learner.
- One-Class Support Vector Machine—One-class support vector machine (SVM) trains mostly on normal data so is used when there is a lot of normal data (Azure 2019). It infers the properties of normal datasets, then anything different than normal is treated as anomalous.
- Local Outlier Factor—This is the density-based method which is rather famous for the detection of local anomalies but could also be used for global anomalies. If the local anomalies are not of interest, then this model can generate a lot of false alarms.

10.2.3   Related Work

Due to intensive data collection behavior, machine learning and artificial intelligence have received huge attraction in the digital world to perform business activities based on the inferred knowledge from the captured data. Industries like telecommunication, financial services, agriculture, stock market, health care, retail, automotive, government, transportation, oil and gas, and information and cyber security have shown extreme interest in machine learning applications ranging from financial solutions to network performance evaluation and cyber threat detection.

The concept of DevOps had been also welcomed in the network community, especially for SDN and APIs. APIs are universal plugs for communication regardless of the corresponding operating system. APIs can act as a filter for active testing in the multilayered architecture (Encyclopedia 2023). According to Kim, Kim, and Ko, "SDN decouples the control and network plane"; therefore network policies can be managed dynamically (Kim, Kim, and Ko 2014). Many researches have shown that SDN has enabled network stakeholders to monitor the network policies with reduced manual and physical effort (Zhang et al. 2016). Other researchers have demonstrated that in software defined networking, how data plane, where switches and routers reside, is managed with the control plane's intelligence (Rahman et al. 2017). Another study stated that the quality of service can be enhanced with flow balancing of the SDN (Sood, Yu, and Xiang 2017).

In 2003, researchers proposed an enhanced solution as knowledge plane for the Internet based on AI tools and cognitive systems. Their counterargument to support this emerging approach was towards weaknesses in traditional algorithmic approaches where it was limited to separating high-level goals from low-level actions and no learning from previous experiences (Mestres et al. 2017).

The community has already applied a plethora of machine learning approaches to estimating the quality of transmission in the network with quite high estimation rates (Morais and Pedro 2018). Some of the applications of machine learning models have helped classify spams, fraud detection, and malware analysis through kth-nearest neighbor, regression, and support vector machines respectively (Polyakov n.d.). Another interesting application is anomaly detection, which is warmly welcomed by the community to see what is not normal.

Finally, the primitive approach for anomaly detection was to remove all the abnormal observations before the feeding of data into the training model due to the higher tendency of overfitting the outlier in image recognition. But the latest algorithms are more advanced and robust and could work efficiently with anomalous data too (Goldstein and Uchida 2016). Researchers have shown that kth-nearest neighbor generates many accurate results for anomaly detection, but it does not detect local anomalies. kth-nearest neighbor is known to be good for finding global anomalies.

10.3 METHODOLOGY

10.3.1 RESEARCH STRATEGY

There are two known research paradigms for information systems: behavioral science and design science. Both have evolved in their sphere over time. Design science is highly oriented to the creation of new artefacts, and behavioral science is the study of behavior involving humans and the IT (Hevner et al. 2008). Design science in contrast to empirical research does not only focus on description, explanation, and prediction of the world but also tends to change, improve, and/or create new artefacts.

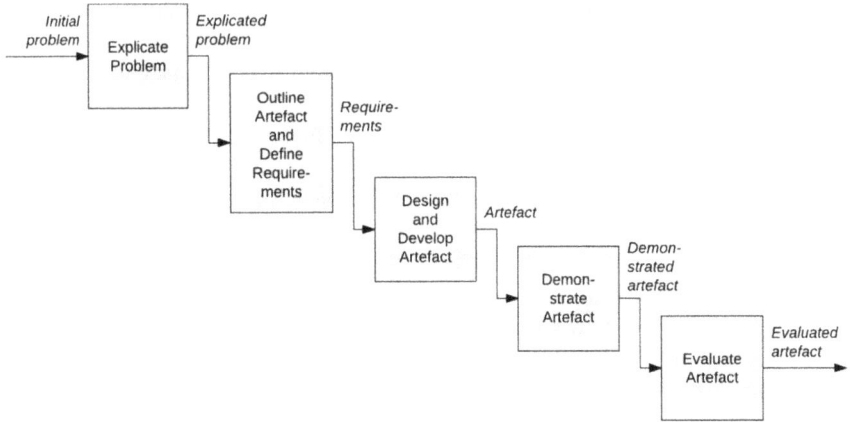

FIGURE 10.4   Overview of design science method (Johannesson and Erik 2012).

The creation of an artefact provides knowledge about the use and environment of the artefact (Johannesson and Erik 2012).

The norm is that design science and action research are homogeneous, but there is an important difference between the two. Both are practice research for problem-solving and evaluation; however design science solves the problem by the development of an artefact, whereas in action research it is not inevitable to build an artefact (Johannesson and Erik 2012). Moreover, action research carries a single strategy throughout the procedure, but design science permits the use of different strategies even for each of the phases. Therefore, design science was adapted as an appropriate research strategy for the current study, as illustrated in Figure 10.4.

The priority was to provide a more automatic and predictive solution to avoid any disruption in the services and enhanced performance. Requirement elicitation was performed based on two recommended factors, that is, value and urgency (Prasanth, Valsala, and Soomro 2017). Therefore, the researcher defined and refined the requirements in coordination with stakeholders from the host organization. Careful background research determined machine learning models for predictions. These predictions can then support the decision-making process by humans only. To design the artefact, the researchers went through weekly brainstorming sessions with the stakeholders. During those sessions, the artefact was tailored to deliver the best optimal outcome through empathetic thinking and participative modeling (Johannesson and Erik 2012).

Netrounds did not provide an API console, but open and programmable APIs allowed calling the test agents. Therefore, for the demonstration of artefact data was fed into machine learning models using python. Supervised machine learning methods provide analytics of the metrics data for enhanced performance and prediction (delay and packet loss). However, for the anomaly detection, unsupervised anomaly detection was performed via action research.

Lastly, for the Evaluation of Artefact, model validation was performed through a comparison of selected machine learning models. Hence, the research was conceived

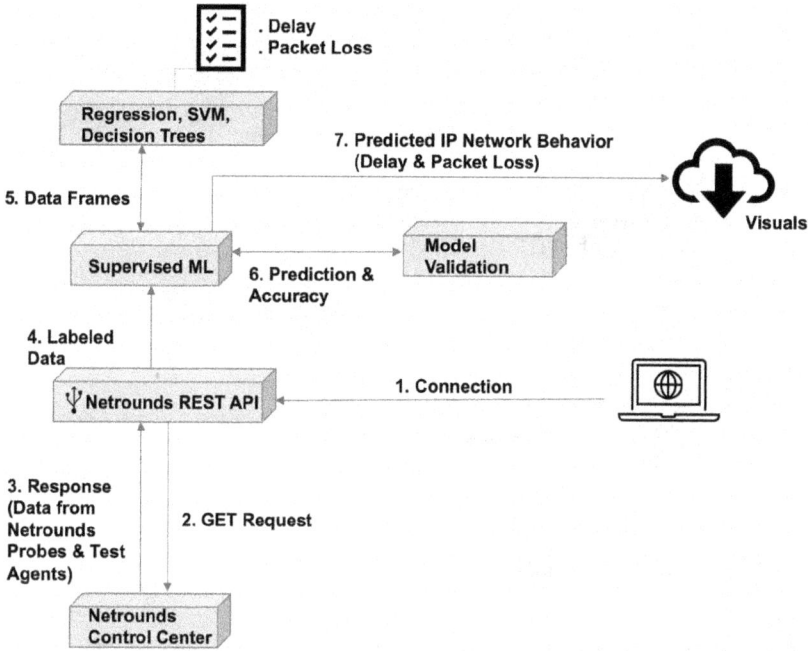

FIGURE 10.5 Proof of concept.

as a development and evaluation-focused design science (Johannesson and Erik 2012). The proof of concept is stated as a knowledge plane that can detect anomalies in IP network performance based on the network analytics performed by the management plane of SDN (Netrounds 2019b), as shown in Figure 10.5.

10.3.2 DATA COLLECTION

From a social perspective data collection require ethical approval and authorization (Denscombe 2017). So for this study, formal authorization for the data collection and activities was provided by the host organization. A single (timestamped) dataset coming from a single data source was used, which was Netrounds in this case. For the sake of the experiments, the topology within the core IP network was frozen. It was limited to the UDP packets since they are the carriers of IPTV services. The topology started with an access router (PA) at the country level, down the stream to distribution routers (PD), and the exit at the access router, as shown in Figure 10.6.

To get network performance metrics, application-specific data was acquired through Netrounds REST PM API—READ, as per the instructions in the user guide of Netrounds (Netrounds 2019a). GET requests of HTTPS protocol for REST API retrieved data from NCC (cloud) to the local machine where structured streams of data were visualized with Google Map API. The intent was to maintain the interactivity between the two channels of proprietary service; therefore visual artefact was connected back to the default Netrounds GUI. This approach maintained the

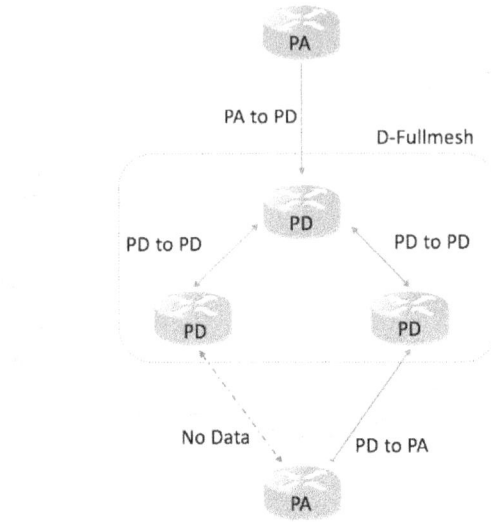

FIGURE 10.6 Network topology.

provision of both releases of Netrounds solutions intact. The retrieved data was then normalized and passed on to the machine learning models. Multiple Python libraries such as Pandas, NumPy, timedelta, requests, PyOD, and Scikit-Learn were used through the data collection process.

10.3.3 DATA ANALYSIS FRAMEWORK

Back in 2014, Wind proposed a framework for predictive modeling based on general approaches and lessons learned through data analysis of Kaggle competitions (Wind 2014). He prescribed a four-step framework (Wind 2014). The current study followed the same framework:

10.3.3.1 Explore the Data

The first step is to explore the data for a better understanding of each feature available. Sometimes the data can be used directly, and sometimes new features need to be derived from existing features (Wind 2014). Since the case was able to get the required features directly from the response object of the HTTPS request, only a careful understanding, relevance, and interdependability of features did the job. According to Wind (2014), some of the methods are plotting the data, histograms, density graphs, and variable correlation. All four approaches were performed and visualized by a scatter plot, histogram, and correlation matrix with heat map.

10.3.3.2 Preprocess the Data

This stage is known to be the most time taking step of the modeling process. After a critical review of the existing approach (Wickham and Grolemund 2017), tidy data format was determined as the data preparation method. So the format was a relational table with each tuple as a single observation and each column as a labeled feature. It was not required to merge or extract features. Instead of removing the observations with missing values, imputation was performed. Imputation can be in many ways such as value based on the values of other observations, mean/median of values of the same feature, or just null (NAs). For the prediction of network performance, missing values were imputed NA. For anomaly detection, mean value of the feature was assigned to missing values. Altogether 15 features were available in the dataset. From which "Delay average(ms)" and "Lost packets" were the target KPIs of prediction and anomaly detection. In the predictive modeling for regression and classification, both numerical and categorical values were used. Whereas for anomaly detection, numerical features were transformed into categorical features.

10.3.3.3 Construct the Model

- Feature Construction—The feature construction could either be feature reduction, feature generation and extraction, or both. Features could be reduced manually based on intuition or with automated methods such as the Feature Selector tool in Python. The process ensures reduced processing costs for large datasets and avoids the least significant attributes. Hereby only feature reduction was performed by using missing values and collinear feature (Koehrsen 2018a). The process of feature reduction is explained under Section 5.1. Feature generation was not required since the retrieved KPIs were viable for meaningful predictive modeling. Chi-square analysis is also a strong candidate for feature reduction, but the method is applicable to the categorical features only therefore due to the numerical features the test was not compatible.

  Since feature selection requires application domain knowledge, therefore results of missing values features and correlation matrix was presented to the application domain experts first. Eventually, based on the experts' feedback from the host organization, a review of the Netrounds user guide's relevant features and results of automated methods interested features were extracted for the predictive modeling. The following features were perceived as meaningful based on the described procedure:
  (1) Received Packets
  (2) Loss (%)
  (3) Delay min (ms)
  (4) Delay average (ms)
  (5) Delay max (ms)
  (6) Jitter
  (7) Lost packets
  (8) ES (%)
  (9) ES loss (%)
- Model Training—Generally, regression is for continuous data, and classification is for discrete data. Hence, for the prediction of "Delay average (ms)" and "Lost Packets" linear regression, decision tree and random forest were applied. Whereas for the anomaly detection, kth-nearest neighbor, local outlier factor (LOF), and one-class SVM were applied.

- Model Evaluation—Regression evaluation is variance based, which means how much the predicted results vary from actual values and can be measured as mean absolute error, mean squared error, root mean squared error, and r squared score. Whereas, in classification evaluation accuracy, F1 score and confusion matrix is used. The F1 score is calculated as follows:

$$F1 = 2 \times (precision \times recall)\,(precision + recall)$$

Precision is the ratio of correctly predicted classes to total predicted true classes, and recall is the ratio of correctly predicted classes to all actual classes (Alpaydin 2014).

$$Precision = True\ Positives\ (T\ P)$$

$$True\ Positives\ (T\ P) + False\ Positives\ (F\ P)$$

$$Recall = True\ Positives\ (T\ P)$$

$$True\ Positives\ (T\ P) + False\ Negatives\ (F\ N)$$

Lastly, the confusion matrix demonstrates correct and incorrect predictions for classification. It is an $N \times N$ matrix and provides the performance of the classifier as true positive (TP), true negative (TN), false positive (FP), and false negatives (FN).

For anomaly detection interest was higher on the FP rate than the TP rate because FP may cause alert fatigue, consequently, cause negligence in responses. The expectation was to see low FN as well represented as recall. Therefore, recall seemed to be more interesting in comparison to precision. Both precision and recall require binary labels.

For unsupervised anomaly detection evaluation could not be limited to the confusion matrix, accuracy, and F1 score because ground truth is not available, that is, how many anomalies are there to look for by the detection algorithm. Here, outliers would be of the same number as predicted labels, that is, FP and FN will be the same. Therefore, the evaluation of anomaly detection was a function of precision @ rank n and recall @ rank n (Zhao 2018) (Malaeb 2017) as follows:

$$Precision\ @\ rank\ n = Number\ of\ recommended\ items\ @n\ that\ are\ relevant$$

$$Number\ of\ recommended\ items\ @n$$

10.3.3.4 Chose the Best Model

The selection of the best model for prediction and anomaly detection was as based on the quantitative analysis of evaluation measurements described in Section 3.3.3.

10.3.4 DESIGN ETHICS

Ethics is a multifaceted notion, and there is never just a right and wrong answer to it. These are usually self-governed, and any weak considerations may endanger one's reputation (Hagendorff 2020). The known ethical aspects of artificial intelligence are accountability, interpretability, privacy, justice, transparency, and robustness (Hagendorff 2020). Moreover, the sources of big data involved also come under the consideration of ethical responsibility (d'Aquin et al. 2018). While many data processors address AI ethics limited to the anonymization of data and informed consent only, d'Aquin et al. (2018) advocate that the responsibility is bigger than that, and it should be considered as Ethics by Design (d'Aquin et al. 2018).

Recently researchers reidentified the need to define an explicit set of ethical principles (Myers and Venable 2014). For information systems, they agreed to the privacy, accuracy, property, and access aspects (Mason 1986). Since the data was not customer data, no known legal obligations such as anonymity and privacy were applied for transparent research. The dataset was the metrics of hardware performance related to the host corporation.

10.4   VISUALIZATION OF IP NETWORK PERFORMANCE

10.4.1   DESIGN OVERVIEW

This chapter describes the design of a visualization tool using Python and Netrounds REST API (READ) for any configured Netrounds probes within the IP network of the host organization. As described in Section 10.4, the activity was restricted to UDP packets within the reserved topology but with the capacity to visualize all configured services and nodes if required. The artefact was developed as a web application for user-friendliness.

Initially, the magnitude of HTTP requests for REST API was not constrained. However, during the experiment, the solution provider realized the need of introducing incoming requests limit. Afterward, data was retrieved in the batches of size 10, and depending on the HTTP request, the batch could bring 10 test agents at once or 10 monitors per request. These network metrics currently reside in the cloud (Amazon web server), and the data is periodically aggregated on certain time thresholds (10 seconds resolution at 3 months, 5 minutes at 18 months and 1 hour for older). This means at those thresholds the smaller the granularity as soon as the data get older (Netrounds 2019c). However, through REST API the host organization can secure the metrics data locally at the desired granularity of time, also shown in the final artefact in Figure 10.18.

10.4.2   DESIGN DETAIL

Among hundreds of test agents currently configured in the IP network, this visual solution was intended to see their location and connected probes with each other. Each probe (client) of Netrounds is tailored to send synthetic packets to the other probe (server) and calculate one-to-one measurement for network performance based on the response received from the server.

The artefact was able to demonstrate probes for the requested city. The selection of a particular probe showed all connected probes either upstream or downstream. It

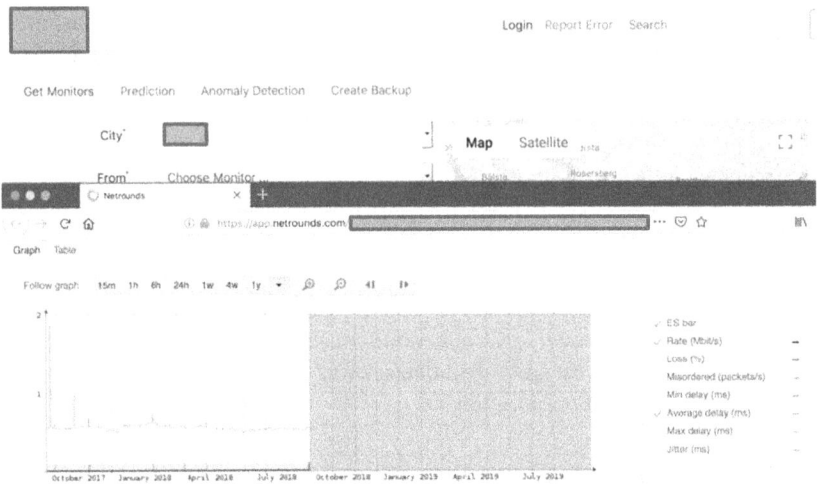

FIGURE 10.7 Visualization tool of IP network performance.

was able to connect back to the metrics in Netrounds GUI to dispense the full functionality of fancy proprietary network monitoring software (Netrounds) as shown in Figure 10.7.

Usually, in a web application input fields are more vulnerable to any malicious input and could expose more attack surface to an attacker. Therefore, input validation was ensured during the development to avoid any cross-site scripting, buffer overflow, and SQL injection attacks.

10.5 PREDICTION OF IP NETWORK PERFORMANCE

10.5.1 EXPLORATION AND PREPROCESSING OF DATA

Starting with the crosstab analysis and then correlation analysis, dependent and independent features were extracted, as shown in Figure 10.8 and Table 10.1.

"Lost packets" held a significant correlation with "Loss (%)," "ES (%)," and "ES loss (%)." "Delay average (ms)" had a high correlation with "Delay min (ms)," "Delay max (ms)," and "Jitter (ms)."

The histogram in Figure 10.9 shows that "Delay average (ms)" ranges from 0 to 1.5 (ms), whereas "Lost packets" is 0 for all observations. This verifies the efficiency and robustness of the core IP network of the host organization.

For all the features less than 20% of observations with missing values as shown in Figure 10.10. Hence removing due to low percentage (less than 20%) would cause the loss of 80% of the information. This withdraws the need for feature reduction based on missing value fractions.

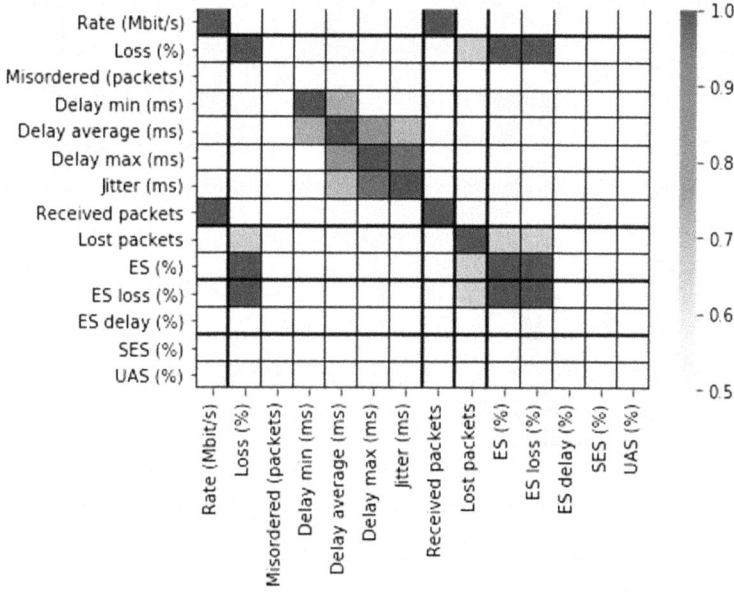

FIGURE 10.8 Correlation matrix for all KPIs.

TABLE 10.1
Dependent and Independent Variables

Dependent variables	Independent variables
Lost packets	Loss (%), ES (%), ES loss (%)
Delay average (ms)	Delay min (ms), Delay max (ms), Jitter (ms)

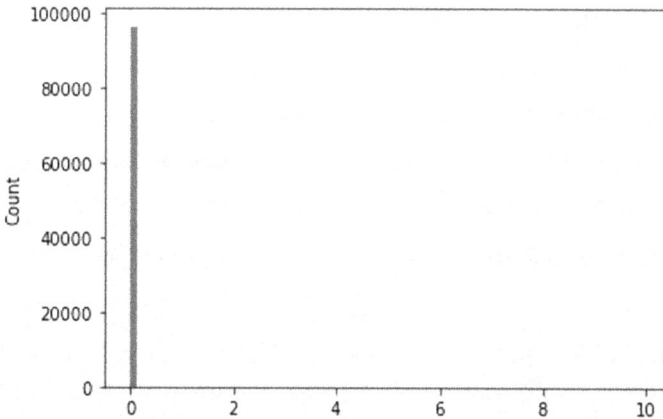

FIGURE 10.9 Histogram of Lost packets and Delay average (ms).

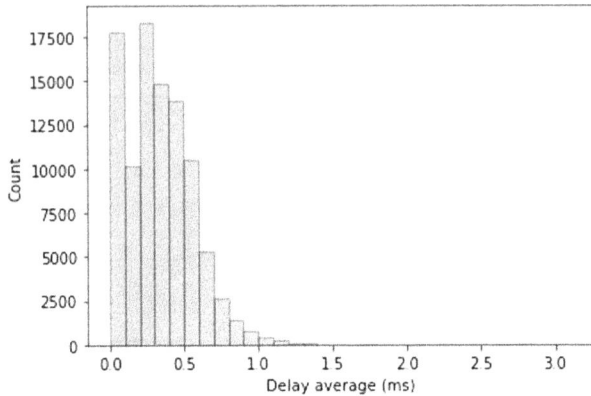

FIGURE 10.9   (Continued)

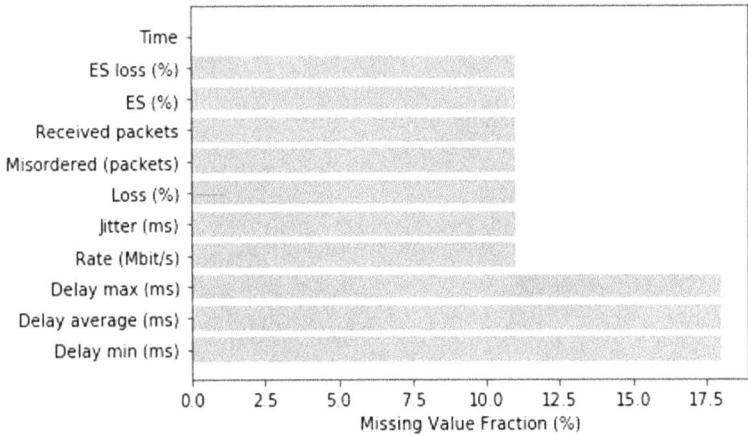

FIGURE 10.10   Missing values (%) for each feature.

10.5.2   DATA ANALYSIS

As per topology, a total of six datasets were used, each of which contained 96,481 observations and 15 features. Missing values were attributed to 0. The models were expected to predict five minutes ahead of time. The dataset was split into 30% test data and 70% train data. For classification models, confusion Table 10.2 was used for evaluation.

The results of the regression model show high accuracy for "Delay max (ms)" (0.99) but for "Lost packets" it was too low (0). Next to it, looking at the mean absolute error, mean squared error, and root mean squared error in Table 10.3, values were slightly higher than 0 in the case of "Lost packets" but were almost 0 for "Delay average (ms)."

In the case of "Lost packets" the predictions (red line) were far from the actual values (green line) and hence showed the higher prediction error in Figure 10.11.

Whereas for "Delay average (ms)" seldom dissimilarity (red line) was observed between the actual and predicted values, both usually fall on the same line.

TABLE 10.2
Evaluation Matrix for Classification Model

Confusion matrix	0(Predicted)	1(Predicted)	2(Predicted)
0(Actual)	It was Class 0 and was predicted as a Class 0 (TP)	It was Class 0 but was predicted as Class 1 (FN)	It was Class 0 but was predicted as Class 2 (FN)
1(Actual)	It was Class 1 but was predicted as Class 0 (FP)	It was Class 1 and was predicted as Class 1 (TP)	It was Class 1 but was predicted as Class 2 (FN)
2(Actual)	It was Class 2 but was predicted as Class 0 (FP)	It was Class 2 but was predicted as Class 1 (FP)	It was Class 2 and was predicted as Class 2 (TP)

TABLE 10.3
Evaluation Metrics for Linear Regression of Lost Packets and Delay Average (ms)

Evaluation function	Lost packets	Delay average (ms)
Mean Absolute Error	00.02	00.01
Mean Squared Error	14.77	00.00
Root Mean Squared Error	03.84	00.01
R Squared Score	00.00	00.99

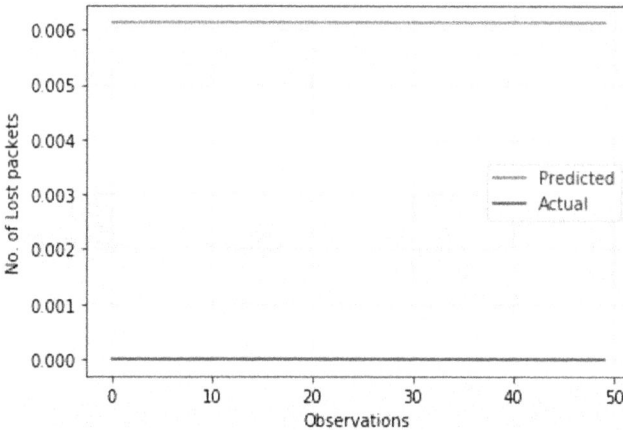

FIGURE 10.11   Prediction error in linear regression.

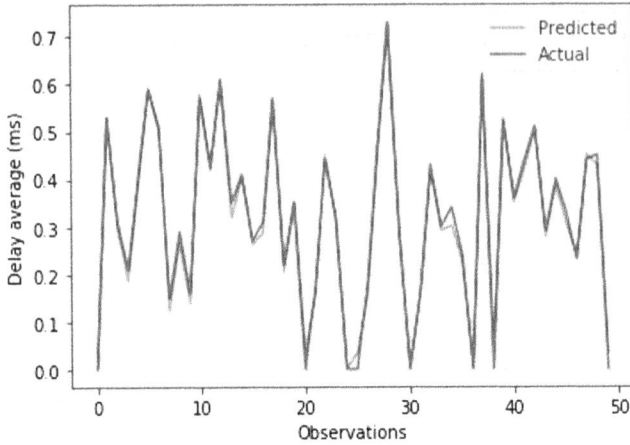

FIGURE 10.11   (Continued)

TABLE 10.4

Evaluation Metrics for Linear Regression of Lost Packets and Delay Average (ms)

Class labels	Lost packets	Delay average (ms)
0	0.0	0.0
1	0.1 – 100.0	0.1 – 1.0
2	> 100.0	> 1.0

10.5.2.1   Decision Tree

For classification, preprocessing of continuous numeric variables was performed. Class labels were identified using Table 10.1. After a careful analysis class label thresholds were determined and then got approved by the stakeholders, as shown in Table 10.4.

Right after the normalization, decision tree returned accuracy and F1 score as 99% and 1.00 respectively. The classification error was then evaluated in the confusion matrix in Figure 10.12.

For "Lost packets" all the labels (altogether 28943 from test data) for Class 0 were predicted correctly. None of the labels for Class 1 and 2 were correctly predicted. In the case of "Delay average (ms)" all 5,324 observations were correctly predicted as Class 0. For Class 1, 23,450 observations were correct predictions and one misclassification. For the last Class 2, out of 170 observations, 159 were predicted correctly, and 11 were misclassified.

10.5.2.2   Random Forest

The same normalization was performed for the decision tree by using Table 10.4. The random forest algorithm predicts the results of the majority voting of its decision trees. The dataset was randomly divided into test and train sets. The number of

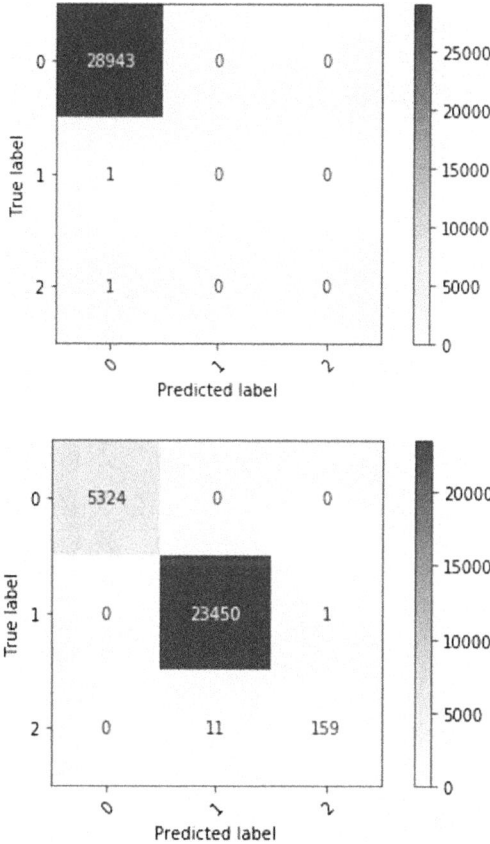

FIGURE 10.12   Decision tree—Confusion matrix for Lost packets and Delay average (ms).

decision trees to build by random forest was set to two and then to three to see which fraction would provide the most accurate predictions. Based on the results three was selected as the final number.

For "Lost packets" the predicted class labels were always different and incompatible with the true class labels although the number of decision trees was kept the same on all occasions. On the other hand, for "Delay average (ms)" the predictions were quite consistent. The prediction error was lower than the single decision tree algorithm in Section 5.2.2 and the accuracy was 99.96%. The results kept changing due to random and majority voting of its trees. In Figure 10.13, for "Lost packets" Class 2 has only one observation, and it was predicted correctly. For Classes 0 and 1, one observation for each was misclassified. In the case of "Delay average (ms)," Class 0 was correctly predicted, Class 1 misclassified 3 out of 23,518 observations, and Class 2 wrongly predicted 6 out of 188 observations. Random Forest has less prediction error than decision tree but more computational cost. Also due to its randomized nature, it does not produce the same results every time which makes it harder to compare its accuracy with any classification models.

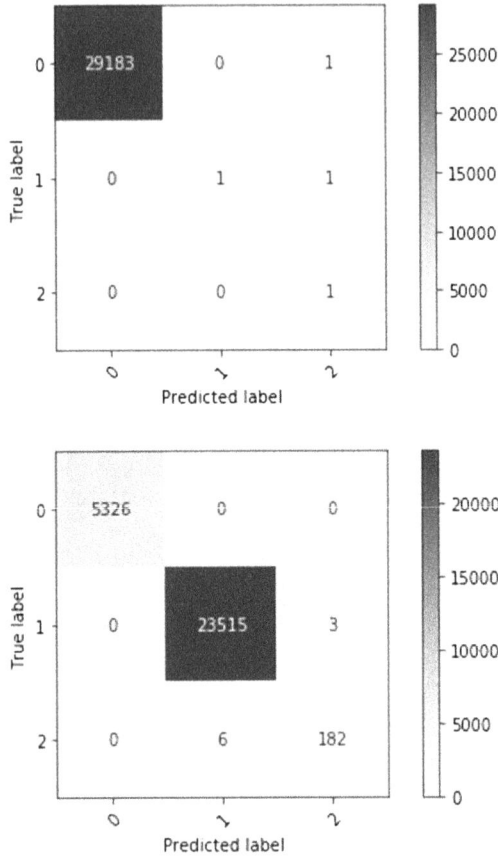

FIGURE 10.13 Random forest—Confusion matrix for Lost packets and Delay average (ms).

TABLE 10.5
Anomaly Detection—Normalized Class Labels for Lost Packets and Delay Average (ms)

Class labels	Lost packets	Delay average (ms)
0	0.0	0 – 1.0
1	> 0.0	>1.0

10.6 ANOMALY DETECTION OF IP NETWORK PERFORMANCE

It was evident from the prediction models that the IP network of the host organization usually behaved well. So the research continued to predict when it does not behave normally. Data were normalized by the attribution of the mean value of the corresponding feature for missing values. Class labels were assigned as per Table 10.5. The class label 0 indicates normal and 1 shows abnormal. The dataset was divided into 30% test and 70% training data.

TABLE 10.6

Anomaly Detection—Normalized Class Labels for Lost Packets and Delay Average (ms)

Confusion matrix	0 (Predicted)	1 (Predicted)
0 (Actual)	It was not an anomaly and was not predicted as an anomaly (TP)	It was an anomaly and was not predicted as an anomaly (FN)
1 (Actual)	It was not an anomaly and was predicted as an anomaly (FP)	It was an anomaly and was predicted as an anomaly (TN)

TABLE 10.7

KNN—Evaluation Functions for Lost Packets and Delay Average (ms)

	K=6	K=171	K=6	K=171
ROC	1.0	1.0	0.9806	0.9951
Precision @ rank n	1.0	1.0	0.3371	0.4607

The confusion matrix for the evaluation of anomaly detection is described in Table 10.6. The aim was a high TP and TN rate. Finally, the best model has a low False-Positive (FP) rate.

10.6.1 Kth-Nearest Neighbor

It was an anomaly and was not predicted as an anomaly (FN). It was an anomaly and was predicted as an anomaly (TN). kth-nearest neighbor is sensitive to noise. Different contamination levels ranging from 10%–50% were tested, but no real difference was observed in the performance of the model. Therefore, the experiment was limited to the value of k and the distance function.

Like contamination, two distance functions (Euclidean or Manhattan) were applied one after the other, but nothing influential was observed. However, the higher value of k improved the performance. This makes sense as more comparison to the close neighbors made more accurate detection. However, finding the best value of k was the real challenge because it is highly subject to the data scientist's reasoning and judgment (Section 2.2.2). The first value was calculated as 170, which is a square root of 28,945 test data observations (out of a total of 96,481 observations). Then to keep it odd 1 was added to the value of k. Since kth-nearest neighbor is a lazy learner and the resultant value for k was quite higher, so k = 171 was considered as the highest bound. The lower value of k was calculated based on the elbow method (Zhang, Bouadi, and Martin 2018), shown in Figure 10.14. Finally, 6 and 171 were determined lowest and highest values of k.

The resulting ROC and precision @ rank n in the case of "Lost packets" were highest for both k = 6 and k = 171 in Table 10.7. For "Delay average (ms)" k = 171 returned a slightly higher score as compared to k = 6.

The confusion matrix in Figure 10.15 shows the efficiency of the model that the higher value of k = 171 reduces FP and FN for the "Delay average (ms)." A high TN

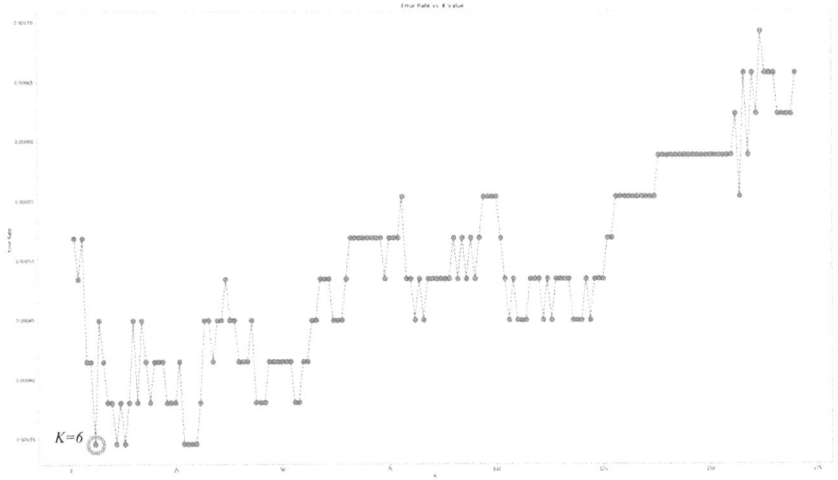

FIGURE 10.14 Elbow method to find the value for k.

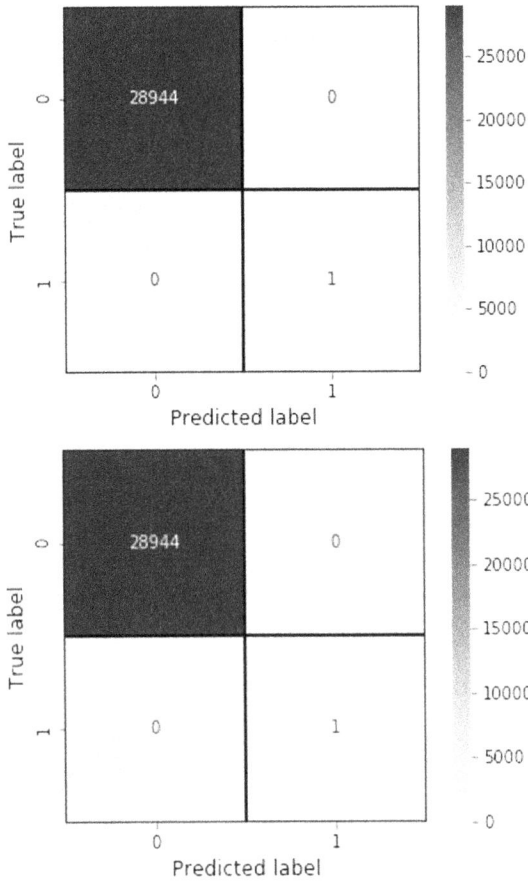

FIGURE 10.15 KNN—Confusion matrix for Lost packets and Delay average (ms).

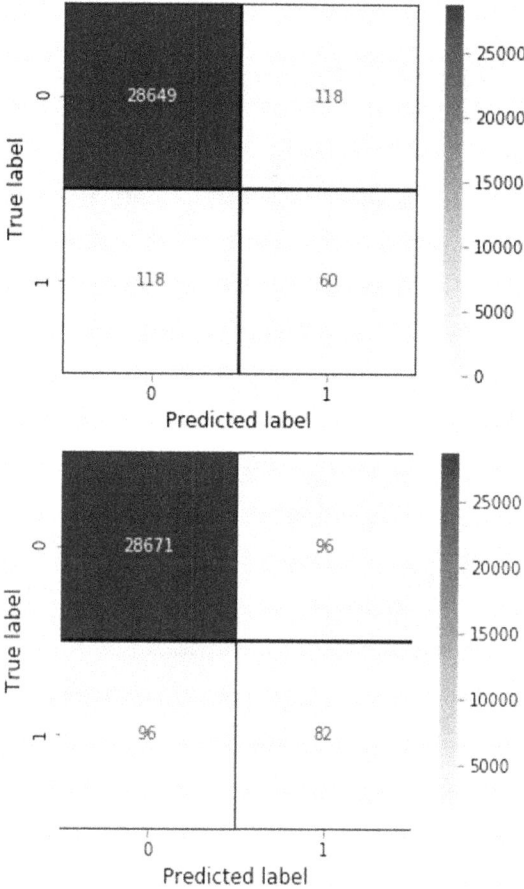

FIGURE 10.15 (Continued)

rate shows more true anomalies were detected. However, "Lost packets" did not show much change.

10.6.2 ONE-CLASS SUPPORT VECTOR MACHINE

Since one-class SVM trains only on normal values so preprocessing involved bringing the classifiers into either normal (0) or abnormal (1) form. Applying kernel function linear and radial basis function (RBF) with 5% contamination showed lower performance of linear kernel for both KPIs in comparison to the kernel function BRF, as shown in Table 10.8.

With the RBF kernel not only many true anomalies (TN) were predicted, but the FP rate was also low. In the confusion matrix for "Lost packets," in Figure 10.16, the model was not able to classify most of the TN. The high precision @ rank n is because it classified the false classes as positive classes. Also, the model executed extremely slowly to detect anomalies, which is impractical.

TABLE 10.8

One-Class SVM—Evaluation Functions for Lost Packets and Delay Average (ms)

	Lost packets		Delay average (ms)	
	Linear	**RBF**	**Linear**	**RBF**
ROC	0.0	1.0	0.0003	0.9999
Precision @ rank n	0.0	1.0	0.0	0.9607

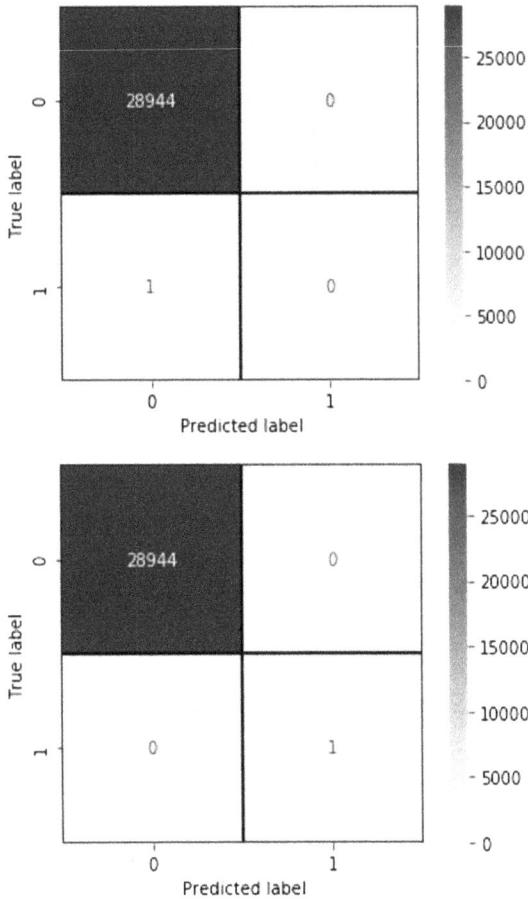

FIGURE 10.16   OCSVM—Confusion matrix for Lost packets and Delay average (ms).

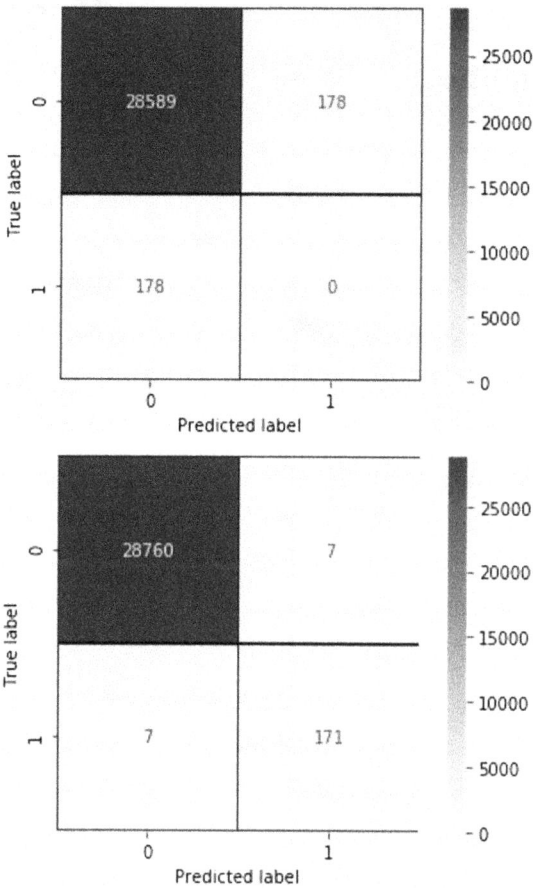

FIGURE 10.16   (Continued)

10.6.3   LOCAL OUTLIER FACTOR

Local outlier factor is a density-based function and mainly depends on the distance function and the value of k. It may respond differently according to the value of contamination and distance function. For this experiment applying different contamination values ranging from 10%–50% did not make any significant difference. The same was observed for Euclidean and Manhattan distances. Therefore, the value of k value was the main point of observation. The higher value of k showed an improved performance in the case of "Delay average (ms)" but not for "Lost packets." Another interesting observation was the change in ROC by changing the k value, however, it did not affect precision @ rank n (shown in Table 10.9), which is a sign of a high FP rate.

In the confusion matrix of "Delay average (ms)" for k = 6 and k = 171, anomalous class rate (TN) was extremely low, and the FP rate was quite high. All FP and FN

TABLE 10.9

LOF—Evaluation Functions for Lost Packets and Delay Average (ms)

	Lost packets		Delay average (ms)	
	k=6	k=171	k=6	k=171
ROC	0.5002	0.5002	0.6831	0.8331
Precision @ rank n	1.0	1.0	0.0	0.0

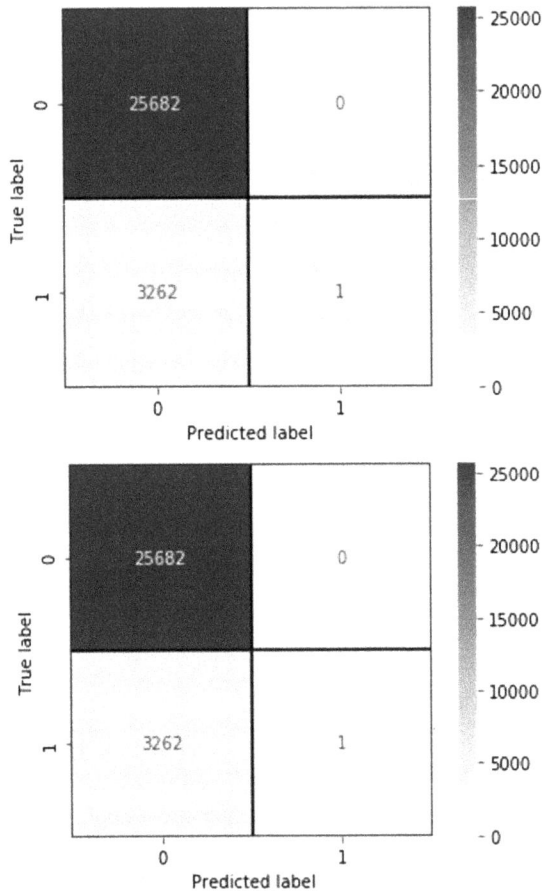

FIGURE 10.17   LOF—Confusion matrix for Lost packets and Delay average (ms).

were misclassified labels; none of the true anomalies were detected. This is not an acceptable characteristic of a good model. On the other hand, in the case of "Lost packets" again FP rate was much higher with no true detection of anomalies (TN), shown in Figure 10.17. This indicates that keeping "Lost packets" in the final model will cause processing overhead rather than providing some meaningful information.

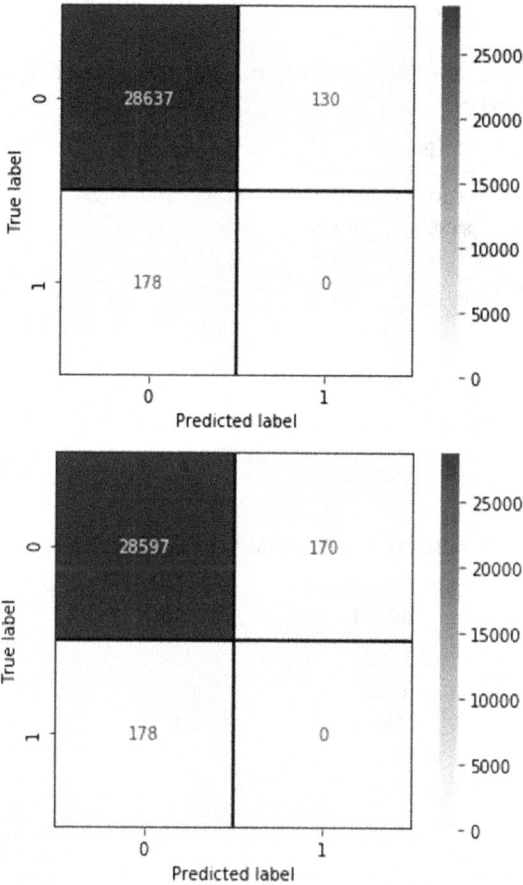

FIGURE 10.17 (Continued)

10.7 DISCUSSION AND CONCLUSION

10.7.1 DISCUSSION OF THE RESULTS

The evaluation was for "Delay average (ms)" and "Lost packets" from the topology reserved by the host organization as their point of interest. The study can be generalized since the same experiments could be performed on other probes of the IP network.

The visual artefact successfully demonstrated the acquisition and analysis of performance data. It showed the extended ability to connect the default GUI of the proprietary software to the costumed platform using Netrounds REST (READ) API. Fetching the data from NCC using HTTP requests was slow due to size limits. However, this could be faster if Netrounds REST (WRITE) API was used. For that, the Netrounds REST (WRITE) API requires to be installed on-premise instead of cloud to avoid bandwidth, upload, and download restrictions. Also, an expensive license is required, which at the time of experiments the host company did not have.

TABLE 10.10

Comparison of Efficiency for Prediction Algorithms for Lost Packets and Delay Average (ms)

	Regression		Decision tree		Random forest	
	Lost packets	Delay average (ms)	Lost packets	Delay average (ms)	Lost packets	Delay average (ms)
R Squared Score	00.00	99.00	-	-	-	-
Accuracy	-	-	99.99	99.93	99.98	99.63
Found All Labels	-	-	Y	Y	Y/N	Y
Misclassified Labels	-	-	2	12	2	9

TABLE 10.11

Comparison of Efficiency for Anomaly Detection of Delay Average (ms)

	kth-Nearest neighbor (k=171)	One-class SVM (RBF)	Local outlier factor (k=6)
ROC	0.9951	0.9999	0.8331
Precision @ rank n	0.4607	0.9607	0.0
TP	28671	28760	28597
TN	82	171	0
FP	96	7	178

In general, using domain expert knowledge best possible feature selection reduced the memory, processing, and interpretation cost. A good feature selection is the basis for a good prediction model. The regression model did not return good results for "Lost packets"; however the score was pretty good for "Delay average (ms)," as shown in Table 10.10. This makes sense because "Lost packets" did not have continuous values. The value rarely changed from 0, which is a sign of good IP network performance. Therefore, by the definition of regression, the algorithm failed to predict the "Lost packet" with linear regression.

Whereas decision tree and random forest both had high accuracy, which means classification models are suitable to the datasets. Aside from accuracy, random forest was not able to predict all the class labels for the "Lost packets," which shifted the confidence towards decision tree. Since "Lost packets" were usually static, for a faster and more efficient model, only "Delay average (ms)" was included in the final artefact, and the decision tree was determined as the best model. The study tempted to detect anomalies to see when the system does not behave normally. The value under 1.0 was normal, otherwise abnormal. In anomaly detection, a high rate of true classes (TP, TN) indicates the correct prediction of anomalies. It was indeed required to see the low score of FP. Table 10.11 shows ROC and precision @ rank n for one-class SVM was the highest. Moreover, identifying true classes was more precise, and the number of FP was the lowest.

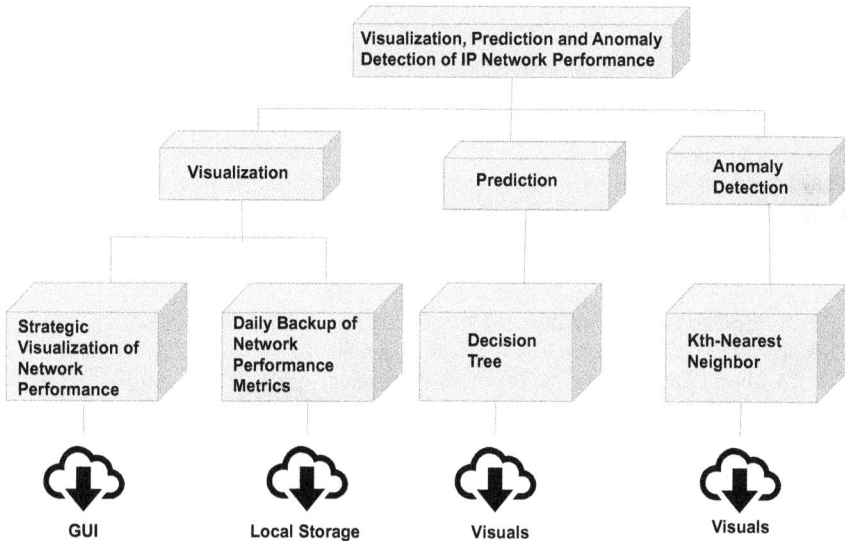

FIGURE 10.18   Final artefact.

These numbers were good; however, the algorithm was slow, which was not desirable when the data was huge and time-dependent. Therefore, the comparison was left between the rest of the two candidates, that is, KNN and LOF. As per the score, LOF predicted most TP, but performance-wise KNN takes the lead because LOF had a null TN rate and the highest FP rate. Therefore, kth-Nearest Neighbor was chosen as the best model. Finally, the artefact is demonstrated in Figure 10.18.

10.7.2 Conclusion

The study aimed to visualize Netrounds REST API and predict and detect anomalies for "Lost packets" and "Delay average (ms)." The REST API was successfully automated. Three different supervised machine learning models were applied and evaluated with various performance functions. The decision tree model appeared to be the best among the three candidates. Evaluation based on accuracy only could be misleading because it does not show the real value for wrongly predicted classes, so a confusion matrix can be used to see another comparative picture. Supervised machine learning acknowledged the goodness of the core IP network of the host organization and steered the focus toward anomaly detection. Therefore, applying three anomaly detection models one-class SVM returned a low FP rate, but the time to process was quite high. Then kth-nearest neighbor model was second to the best and hence selected as the optimal model. Both the prediction and the anomaly detection models showed keeping "Lost packets" would impact performance negatively, so "Lost packets" was dropped out of the final artefact.

REFERENCES

Ahmed, Mohiuddin. 2018. "Collective anomaly detection techniques for network traffic analysis." *Annals of Data Science* 5 (4): 497–512.

Al-Rubaie, Mohammad, and J. Morris Chang. 2019. "Privacy-preserving machine learning: Threats and solutions." *IEEE Security Council & Privacy* 177 (2): 49–58.

Alpaydin, Ethem. 2014. *Introduction to Machine Learning*. MIT Press.

Azure, Microsoft. 2019. "One-Class Support Vector Machine." https://docs.microsoft.com/en-us/azure/machine-learning/studio-module-reference/one-classsupport-vector-machine.

Cios, Krzysztof J., Roman W. Swiniarski, Witold Pedrycz, and Lukasz A. Kurgan. 2007. "Supervised Learning: Statistical Methods." *In Data Mining* 307–379.

Clark, David D., Craig Partridge, Christopher Ramming, and John T. Wroclawski. 2003. "A Knowledge Plane for the Internet." *Proceedings of the 2003 Conference on Applications, Technologies, Architectures, and Protocols for Computer Communications* 33 (4): 3–10.

d'Aquin, Mathieu, Troullinou Pinelopi, E. O'Connor Noel, Cullen Aindrias, Faller Gr´ainne, and Holden Louise. 2018. "Towards an 'Ethics by Design' Methodology for AI Research Projects." *In Proceedings of the 2018 AAAI/ACM Conference on AI, Ethics, and Society* 4 (2): 54–59.

Denscombe, Martyn. 2017. *The Good Research Guide: For Small-Scale Social Research Projects*. McGraw-Hill Education (UK).

Dunning, Ted, and Ellen Friedman. 2014. *Practical Machine Learning: A New Look at Anomaly Detection*. O'Reilly Media, Inc.

Encyclopedia, Wikipedia the Free. 2023. "DevOps." https://en.wikipedia.org/wiki/DevOps.

ENISA. 2023. "IT Continuity Home." www.enisa.europa.eu/topics/risk management/current-risk/bcm-resilience/it-continuity-home.

Feamster, Nick, Jennifer Rexford, and Ellen Zegura. 2014. "The Road to SDN: An Intellectual History of Programmable Networks." *ACM SIGCOMM Computer Communication Review* 44 (2): 87–98.

For Standardization, International Organization. 2019. "Security and Resilience: Business Continuity Management Systems-Requirements." www.iso.org/obp/ui/en/\#iso:std:iso:22301:ed-2:v1:en.

GeeksforGeeks. 2023. "Difference Between Machine Learning and Artificial Intelligence." www.geeksforgeeks.org/difference-between-machine-learning-andartificial-intelligence/.

Goldstein, Markus, and Seiichi Uchida. 2016. "A Comparative Evaluation of Unsupervised Anomaly Detection Algorithms for Multivariate Data." *PLoS One* 11 (4): 1–31.

Hagendorff, Thilo. 2020. "The Ethics of AI Ethics: An Evaluation of Guidelines." *Minds and Machines* 30 (1): 99–120.

Hevner, Alan R., Salvatore T. March, Jinsoo Park, and Sudha Ram. 2008. "Design Science in Information Systems Research." *Management Information Systems Quarterly* 28 (1): 75–105.

Huang, Chengqiang. 2018. "CFeatured Anomaly Detection Methods and Applications." PhD diss., University of Exeter.

Johannesson, Paul, and Perjons Erik. 2012. *A Design Science Primer*. CreateSpace.

Kim, Hyunmin, Jaebeom Kim, and Young-Bae Ko. 2014. "Developing a Cost-Effective Open Flow Testbed for Small-Scale Software Defined Networking." *16th International Conference on Advanced Communication Technology* 758–761.

Koehrsen, Will. 2018a. "A Feature Selection Tool for Machine Learning in Python." https://towardsdatascience.com/a-feature-selection-tool-for-machinelearning-in-python-b64dd23710f0.

Koehrsen, Will. 2018b. "An Implementation and Explanation of the Random Forest in Python." https://towardsdatascience.com/an-implementation-and-explanationof-the-random-forest-in-python-77bf308a9b76.26.

Malaeb, Maher. 2017. "Recall and Precision at k for Recommender Systems." https://medium.com/@m_n_malaeb/recall-and-precision-at-k-for-recommendersystems-618483 226c54.

Mason, Richard O. 1986. "Four Ethical Issues of the Information Age." *MIS Quarterly* 10 (1): 5–12.

Mestres, Albert, Alberto Rodriguez-Natal, Josep Alberto, Pere Barlet-Ros, Eduard Alarc´on, Marc Sol´e, Victor Munt´es-Mulero, et al. 2017. "Knowledge-Defined Networking." *ACM SIG COMM Computer Communication Review* 47 (3): 2–10.

Morais, Rui Manuel, and Joao Pedro. 2018. "Machine Learning Models for Estimating Quality of Transmission in DWDM Networks." *Optical Society of America* 10 (10): D84–D99.

Myers, Michael D., and John R. Venable. 2014. "A Set of Ethical Principles for Design Science Research in Information Systems." *Information & Management* 51 (6): 801–809.

Netrounds. 2019a. "Netrounds Documentation." https://app.netrounds.com/static/2.27/support/index.html.

Netrounds. 2019b. "Netrounds Products." www.netrounds.com/products.

Netrounds. 2019c. "Resolution of Netrounds Measurement Data." https://app.netrounds. com/static/2.27/support/defs-notes/resolution.html.

Netrounds. 2019d. *REST API Performance Metrics Retrieval Guide.* Netrounds.

Olifer, Natalia, and Victor Olifer. 2005. *Computer Networks: Principles, Technologies and Protocols for Network Design.* John Wiley & Sons.

Polyakov, Alex. n.d. "Machine Learning for Cybersecurity 101."

Prasanth, Anupama, Sandhia Valsala, and Safeeullah Soomro. 2017. "A Novel Approach in Calculating Stakeholder Priority in Requirements Elicitation." *2017 4th IEEE International Conference on Engineering Technologies and Applied Sciences (ICETAS)* 28 (1): 1–6.

Rahman, M., N. Yaakob, A. Amir, R. Ahmad, S. Yoon, and Abd Halim. 2017. "Performance Analysis of Congestion Control Mechanism in Software Defined Network (SDN)." *International Conference on Emerging Electronic Solutions for IoT.* DOI: 10.1051/matecconf/201714001033

Samuelsson, Moa. 2016. "Anomaly Detection in Time Series Data: A Practical Implementation for Pulp and Paper Industry." PhD diss., Chalmers University of Technology.

Skënduli, Marjana Prifti, Marenglen Biba, and Michelangelo Ceci. 2018. "Implementing Scalable Machine Learning Algorithms for Mining Big Data: A State-of-the-Art Survey." *Big Data in Engineering Applications* 44: 65–81.

Sood, Keshav, Shui Yu, and Yong Xiang. 2017. "Are Current Resources in SDN Allocated to Maximum Performance and Minimize Costs and Maintaining QoS Problems?" *ACM International Conference Proceeding Series* 1–6.

Sun, Guanglu, Teng Chen, Yangyang Su, and Chenglong Li. 2018. "Internet Traffic Classification Based on Incremental Support Vector Machines." *Mobile Networks and Applications* 23 (4): 789–796.

Wickham, Hadley, and Garrett Grolemund. 2017. *R for Data Science: Import, Tidy, Transform, Visualize, and Model Data.* O'Reilly Media.

Wind, David Kofoed. 2014. "Concepts in Predictive Machine Learning a Conceptual Framework for Approaching Predictive Modelling Problems and Case Studies of Competitions on Kaggle." PhD diss., Masters Thesis, Technical University of Denmark.

Zhang, Peng, Hao Li, Chengchen Hu, Liujia Hu, Lei Xiong, Ruilong Wang, and Yuemei Zhang. 2016. "Mind the Gap: Monitoring the Control-Data Plane Consistency in Software Defined Networks." *Proceedings of the 12th International on Conference on Emerging Networking EXperiments and Technologies* 19–33.

Zhang, Yiru, Tassadit Bouadi, and Arnaud Martin. 2018. "An Empirical Study to Determine the Optimal k in Ek-NNclus Method." *International Conference on Belief Functions* 260–268.

Zhao, Yue. 2018. "AUC Score & Precision Score Are Different Why Not Same?" https://github.com/yzhao062/Pyod/issues/2.

11 AI-Based Movie Recommendation System

Dilip Kumar Gayen

11.1 INTRODUCTION

The entertainment industry has been revolutionized by the Internet, with streaming services like Netflix, Hulu, and Amazon Prime becoming increasingly popular. With a plethora of options available to the consumer, the challenge now is to help viewers navigate the vast catalog of films and TV shows and find content that is most relevant to their interests. This is where recommendation systems come in. A recommendation system, also known as a recommendation engine, is an application that predicts and suggests items or content that a user might find interesting or useful based on their preferences and past behavior. These systems have been used in various industries, including e-commerce, social media, and content streaming platforms. In particular, movie recommendation systems have become increasingly important for online movie streaming platforms as they help users find movies that match their interests. D. Billsus et al. present an enhanced network-based recommendation algorithm [1]. Their approach involves transferring users' explicit ratings to their interest similarity and representativeness during the reallocation of their recommendation power. They conducted large-scale random sub-sampling experiments on a popular dataset, Movielens. M. Deshpande et al. introduces a class of model-based recommendation algorithms that involve identifying the set of items to be recommended by determining their similarities [2]. The crucial steps in this class of algorithms are the technique for computing item similarity and the approach for combining these similarities to determine the similarity between a group of items and a potential recommendation item. Their study evaluated these item-based algorithms using eight real-world datasets and demonstrated that they are significantly faster than traditional user-neighborhood-based recommender systems, with an increase in speed of up to two orders of magnitude. The primary goal of a movie recommendation system is to provide personalized and relevant movie suggestions to users based on their past behavior and preferences. The system uses various techniques, including collaborative filtering and content-based filtering, to generate movie recommendations for the users. Collaborative filtering predicts the user's preferences based on the preferences of other similar users, while content-based filtering recommends movies based on the user's past behavior, such as the movies they have liked or rated highly. Herlocker et al. have designed a method for evaluation of collaborative filtering recommender systems involving analyzing several factors [3]. These include the user tasks being evaluated,

DOI: 10.1201/9781003343332-11

the types of analysis and datasets being used, the measurement of prediction quality, the evaluation of prediction attributes beyond quality, and the user-based assessment of the system as a whole. Different types of analysis and datasets are used to validate the performance of the recommendation system. According to Koren et al., matrix factorization models have been found to outperform traditional nearest-neighbor techniques in generating product recommendations [4]. They offer the flexibility to incorporate various forms of additional information, including implicit feedback, temporal effects, and confidence levels, resulting in improved recommendation accuracy. Pradeep et al. have developed a movie recommendation system that aims to provide users with movie suggestions based on their interests, eliminating the need to browse through the vast collection of movies available on the Internet [5]. One of the approaches used in such recommendation systems is the content-based approach, which involves analyzing the attributes of items to determine possible recommendations for the user. This book chapter discusses an AI-based movie recommendation system that uses item-based collaborative filtering and content-based filtering techniques to provide movie recommendations to users. The proposed system employs machine learning algorithms to analyze user behavior and preferences and generate movie suggestions that closely match the user's interests. The chapter also describes the various challenges associated with developing an effective movie recommendation system, such as scalability, data sparsity, and cold-start problem, and discusses how these challenges can be addressed. A recommendation system, also known as a recommendation engine, is a filtering paradigm that uses data on user preferences and behavior to offer suggestions that cater to those preferences. These systems are widely used in various industries such as utilities, books, music, movies, television, apparel, and restaurants. To enhance their future suggestions, these systems gather user data on preferences and behavior. Reddy et al. [6] have discussed a content-based movie recommendation system that uses genre correlation. Many researchers are working in this domain to improve recommendation systems [7–11]. Movies are an essential aspect of our lives, with various genres such as comedy, thriller, animation, action, and many more. Another way to differentiate movies is to consider factors such as release year, language, director, etc. With numerous movie options available online, movie recommendation systems help users to find their favorite movies quickly and easily. The reliability of the movie suggestion system is critical as it should provide trustworthy recommendations based on users' tastes. Recommendation systems are widely used by businesses to enhance customer interaction and improve the purchasing experience, leading to increased customer satisfaction and revenue. Movie recommendation systems are particularly crucial and effective. However, scalability issues and poor recommendation quality affect movie recommendation systems that rely solely on collaborative filtering. To suggest movies to users of our web app looking for a specific movie, we use both item-based collaborative and content-based filtering techniques. The results are encouraging, with an accuracy rate of 91.

11.2　SCOPE OF THE RECOMMENDATION SYSTEM

The movie recommendation system aims to provide users with reliable and accurate movie suggestions based on their preferences. The main objective is to enhance the quality, accuracy, and scalability of the movie recommendation system compared to pure techniques. To achieve this objective, a hybrid approach combining content-based filtering and collaborative filtering is used. This approach considers the attributes of the items and the users' behavior to recommend movies that are similar to the ones they have liked in the past. Recommendation systems are essential tools in social networking sites, e-commerce platforms, and other online services where data overload can be a significant problem. These systems help users to make better decisions by suggesting products or services that they may be interested in based on their past behavior. The accuracy and quality of recommendation systems are crucial for their success as they influence the users' satisfaction and loyalty towards the platform. The movie recommendation system is particularly crucial because of the vast number of movies available and the diverse interests of the users. Traditional methods for recommending movies rely on collaborative filtering techniques, which suffer from scalability issues and poor recommendation quality. Hybrid approaches that combine different methods are being developed to address these limitations and improve the accuracy and scalability of the movie recommendation system. Research in this area has identified several challenges in developing effective movie recommendation systems, such as the cold-start problem, where there is insufficient data available for new users or movies. Moreover, users may have different tastes and preferences, making it challenging to recommend movies that cater to their individual needs. To address these challenges, researchers are exploring new methods, such as deep learning and hybrid recommendation systems that can capture the users' preferences and behavior accurately. Overall, the movie recommendation system is a crucial and effective mechanism for enhancing customer interaction and improving the purchasing experience. The combination of content-based and collaborative filtering techniques can significantly improve the accuracy, quality, and scalability of the movie recommendation system, leading to increased customer satisfaction and revenue. However, ongoing research is essential to address the challenges and develop more effective movie recommendation systems that cater to users' diverse needs and preferences. The primary objective of the recommendation system is to provide reliable and accurate movie suggestions to users. The aim is to improve the accuracy, quality, and scalability of movie recommendation systems compared to pure techniques. To achieve this, a hybrid approach combining content-based filtering and collaborative filtering is used. Social networking sites employ recommendation systems as information filtering tools to mitigate data overload. Therefore, there is a significant potential for research to enhance the quality, accuracy, and scalability of movie recommendation systems [12–14]. Despite its effectiveness and importance, scalability concerns and poor recommendation quality continue to plague movie recommendation systems that rely solely on collaborative filtering.

11.3 CLASSIFICATION OF RECOMMENDATION SYSTEMS

Recommendation systems can be classified into three main categories: collaborative filtering, content-based filtering, and hybrid filtering. Collaborative filtering is the most widely used technique for building recommendation systems [15–18]. It relies on the behavior and preferences of users to make recommendations. Collaborative filtering algorithms are divided into two categories: user-based and item-based. In user-based collaborative filtering, the system recommends items based on the preferences of users who are similar to the target user. In item-based collaborative filtering, the system recommends items based on their similarity to the items that the target user has previously rated or interacted with. Collaborative filtering has been extensively studied and applied in various domains, including movies, music, e-commerce, and social networks. Content-based filtering, on the other hand, relies on the features or attributes of items to make recommendations [19–22]. It analyzes the content of items to identify those that are most relevant to the user's preferences. For instance, in a movie recommendation system, the system may recommend movies to a user based on the genres, actors, directors, or plots that the user has previously shown interest in. Content-based filtering has been widely used in domains such as news, music, and e-commerce. Hybrid filtering combines both collaborative filtering and content-based filtering to overcome the limitations of each approach [23–26]. Hybrid filtering techniques are designed to take advantage of the strengths of both collaborative and content-based filtering, resulting in more accurate and personalized recommendations. Hybrid filtering techniques can be classified into two main categories: weighted and switched hybrid approaches. Weighted hybrid approaches combine the scores from both collaborative and content-based filtering to produce a final recommendation score. In contrast, switched hybrid approaches switch between collaborative and content-based filtering depending on the user's profile or the characteristics of the items. In addition to the three main categories of recommendation systems, other types of recommendation systems have also been proposed, including context-aware, knowledge-based, and social network-based recommendation systems. Context-aware recommendation systems use contextual information such as time, location, and weather to make recommendations [27–29]. Knowledge-based recommendation systems use domain knowledge and expert systems to make recommendations [30–32]. Social network–based recommendation systems use the social network of users to identify friends with similar tastes and make recommendations based on their preferences [33–35]. The classification of recommendation systems is essential in understanding the different techniques and approaches used to make personalized and accurate recommendations. Collaborative filtering, content-based filtering, and hybrid filtering are the main categories of recommendation systems, with each having its strengths and limitations. Researchers and practitioners continue to explore and develop new types of recommendation systems to address the evolving needs of users and businesses.

11.4 PROPOSED APPROACH

Recommendation systems have been in existence for a while, and they have undoubtedly made people's lives easier. Simply put, recommendation systems are algorithms that aim to provide users with relevant products or items to read, buy, listen to, or watch. The most commonly used technique in recommendation systems is the content-based recommendation method. However, creating a recommender system from scratch comes with its fair share of issues. For instance, what happens if a website has not attracted enough users to provide user-based information for the recommender system. Additionally, representing a movie in a way that a system can comprehend poses a challenge. The prerequisite for comparing the similarity of two movies is how the system understands the movie. Movie characteristics such as genre, actor, and director are ways to classify movies, but each aspect of the film should be given a distinct amount of weight and should be considered separately when making a recommendation. To address these challenges, a structured approach is necessary. Our approach entails two phases, starting with data preprocessing. Data preprocessing is the process of transforming raw data into something that can be used by a machine learning model. Two types of data exist: raw data and structured data. Raw data, also known as source data, atomic data, or primary data, refers to data that has not been processed for use. On the other hand, structured data is information that follows a predetermined data model and is easy to analyze. The data preprocessing involves transforming raw data into something that can be used by a machine learning model. For instance, columns that are pertinent have been taken out of the metadata. Despite the various applications in the movie recommendation system domain, our approach is unique. Our system not only recommends movies but also retrieves information from the IMDB website, including cast, story, and total rating. Additionally, the user can choose the number of movies to be recommended. The movie recommendation system is a multistep process that utilizes machine learning algorithms to recommend movies to users based on their preferences. The input to the system is a movie from the database, and the output is a recommended movie. The system design involves the following steps:

Step 1: Data Collection

The first step is to download the movie dataset from a reliable source like Kaggle. This dataset contains information about various movies such as their title, genre, actors, directors, and ratings.

Step 2: Importing Modules

Next, the necessary Python modules are imported to facilitate the development of the movie recommendation system. These modules include NumPy, Pandas, and Scikit-learn.

Step 3: Data Preprocessing

In this step, relevant columns from the movie metadata are extracted to preprocess data. This involves cleaning and organizing the data in a way that can be used by the machine learning model.

Step 4: Creating a JSON File

A JSON file is created to store the names of movies and other pertinent information such as cast, genre, and rating. This file is used to train the machine learning model.

Step 5: Fitting the K-Nearest Neighbor Algorithm

The K-nearest neighbor algorithm is used to train the machine learning model. This algorithm helps to classify movies based on their similarity to other movies.

Step 6: Calculating the Euclidian Distance

In this step, a method is created to calculate the Euclidean distance between two movie data points. This is done to measure the similarity between two movies.

Step 7: Building the Movie Recommender Engine

A movie recommender engine is built using the machine learning model trained in step five. This engine recommends movies to users based on their preferences and the similarity between movies.

Step 8: Measuring the Model's Accuracy

The accuracy of the model is determined using a confusion matrix. This helps to determine how well the model is performing in recommending movies.

Step 9: Recommending Movies

Once the movie recommender engine is built and the model's accuracy is measured, the system recommends movies based on the movie selected by the user.

Step 10: Ending the Process

The final step is to end the movie recommendation process.

The movie recommendation system involves several steps, from data collection to building the recommendation engine. These steps are essential in creating an accurate and effective movie recommendation system. The flow chart of the previous model is shown in Figure 11.1.

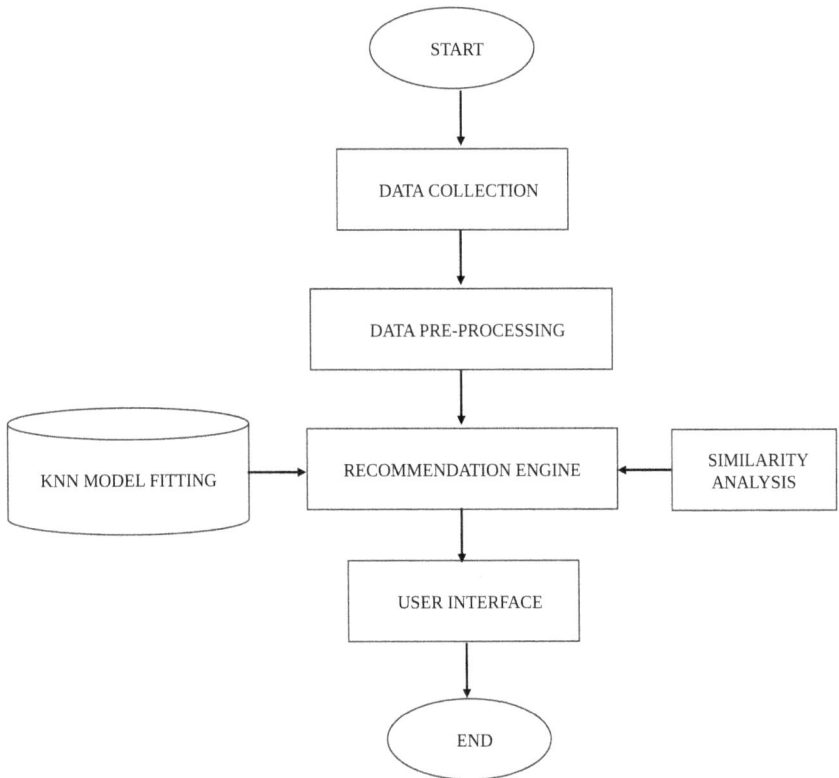

FIGURE 11.1   A flowchart outlining the suggested system.

In order to develop this recommendation system, a user interface has been implemented to facilitate user input. The algorithm for the user interface is outlined here:

Step 1: Determine whether movie-based or genre-based recommendations are desired. If movie-based recommendations are preferred, proceed to Step 2. Otherwise, move on to Step 3.

Step 2: Enter the title of the desired movie to receive recommendations based on that movie.

Step 3: Select the desired genre from a list of available genres to receive recommendations within that genre.

Step 4: Specify the desired number of recommendations, choosing from a range of 5 to 20.

Step 5: Check the box if the movie poster is required in the recommendations.

Step 6: Generate movie suggestions based on all the criteria specified in the previous steps.

This user interface algorithm allows users to customize their recommendations based on their preferences and requirements, whether they want recommendations based on a specific movie or within a particular genre. The number of recommendations and the inclusion of movie posters can also be tailored to the user's preferences. The flow chart of the user interface is shown in Figure 11.2.

11.5   RESULT

Few screenshots of the recommended system are shown (in Figure 11.3):

This algorithm outlines the steps for the user to provide input to the recommendation system through a user interface. Depending on the user's preference, they can either select a specific movie or choose a genre to receive recommendations. They can also choose the number of recommendations they require and indicate if they want to see the movie poster. The system then generates movie suggestions based on the selected criteria.

11.6   CONCLUSION

In today's fast-paced market, it is critical for businesses to stay ahead of the competition. One way to do this is by incorporating recommendation systems into their apps. These systems help users discover relevant products, services, or content based on their preferences, behaviors, and past interactions. As a result, users are more likely to engage with the app, make purchases, and become loyal customers. Large-scale independent research is also being conducted to develop end-to-end recommendation systems that span the key sectors mentioned. These sectors include e-commerce, social media, entertainment, and more. In the entertainment industry, video streaming platforms are increasingly using recommendation systems to personalize the viewing experience for their users.

According to a report, the size of the worldwide video streaming market was estimated at USD 59.14 billion in 2021, and it is anticipated to increase to 21.3.

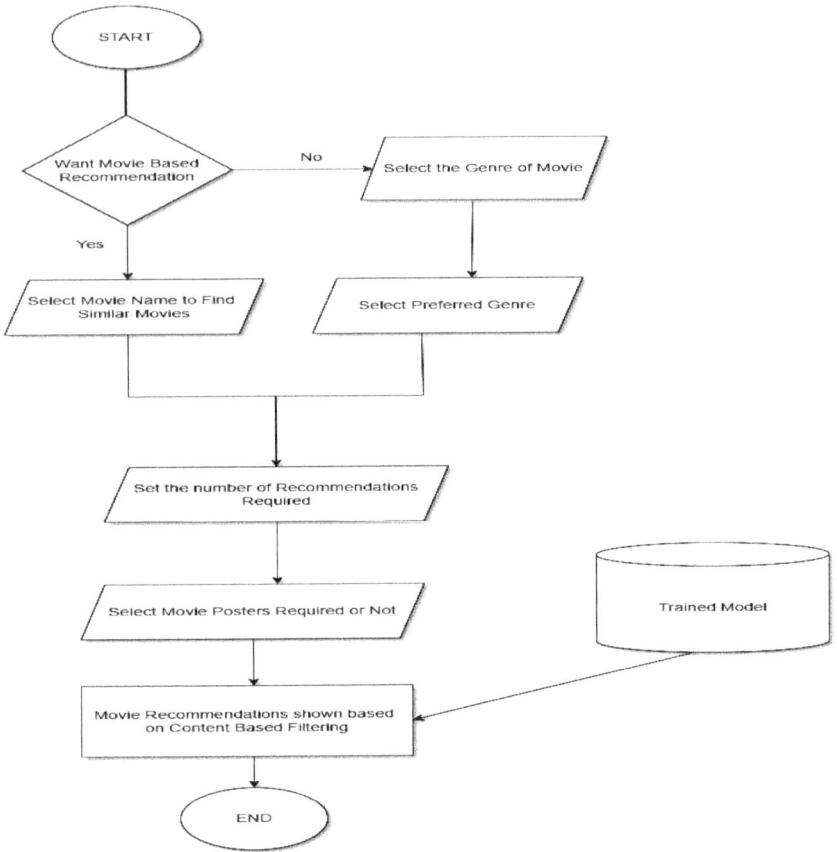

FIGURE 11.2   A flowchart outlining the user interface.

(a) First page of the interface.

FIGURE 11.3   Different screenshots of the user interface.

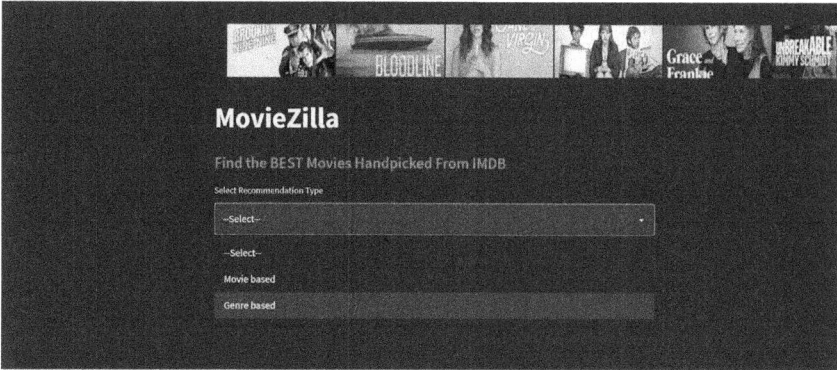

(b) Select by type movie based.

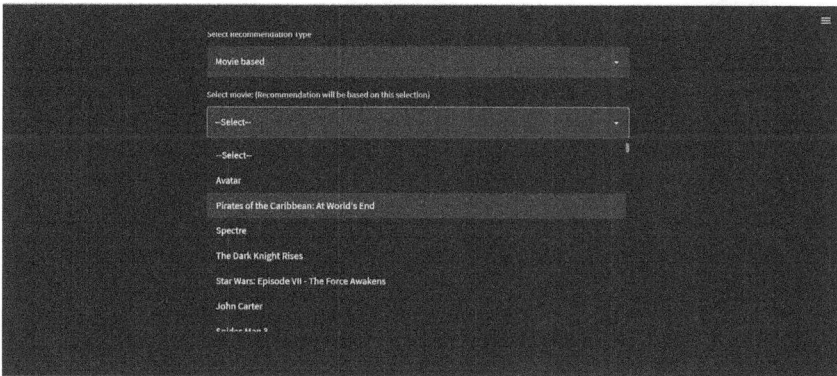

(c) Select by type movie name.

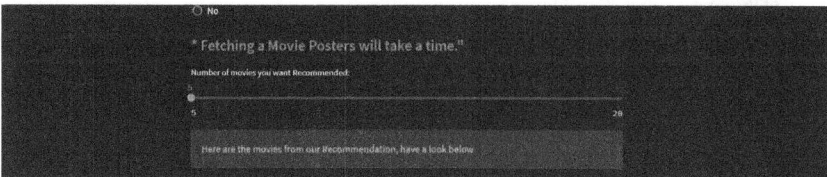

(d) Select by no. of movies recommended.

FIGURE 11.3 (Continued)

Our application fills this need by providing personalized movie recommendations to users based on their preferences. There are many businesses that offer video content, but without a recommendation system, they may struggle to provide viewers with the right suggestions. In today's era of information overload, users are more likely to stay engaged and return to an app if they receive relevant and personalized recommendations. Recommendation systems have become crucial in the market we operate in, and businesses that incorporate them into their apps will have an advantage over those that do not. With the anticipated growth of the video streaming

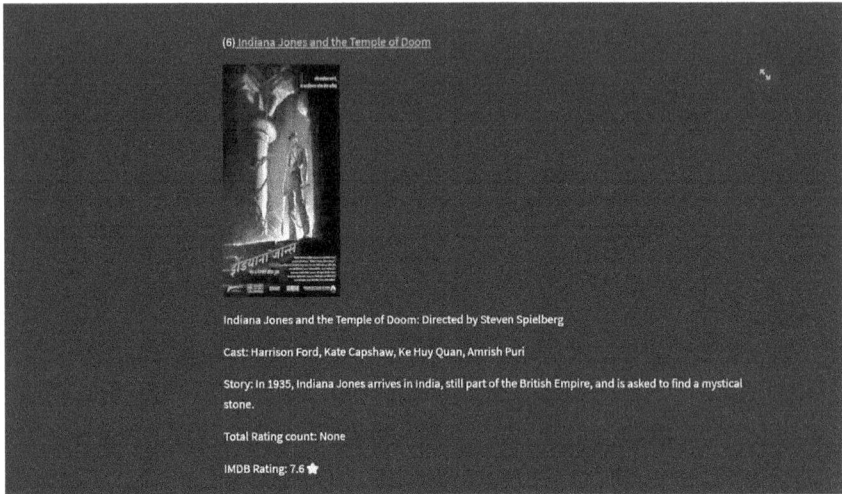

(e) Final output screen.

FIGURE 11.3   (Continued)

market, it is more important than ever for businesses in this industry to offer personalized and relevant recommendations to their users. Our application provides a solution to this need and can help businesses stand out from the competition.

REFERENCES

1. D. Billsus and M. J. Pazzani (1998). Learning Collaborative Information Filters. *Proceedings in the 15th International Conference on Machine Learning*, Madison, WI, USA, 46–54.
2. M. Deshpande and G. Karypis (2004). Item-Based Top-N Recommendation Algorithms. *ACM Transactions on Information Systems*, 22(1), 143–177.
3. J. L. Herlocker, J. A. Konstan, L. G. Terveen, and J. T. Riedl (2004). Evaluating Collaborative Filtering Recommender Systems. *ACM Transactions on Information Systems*, 22(1), 5–53.
4. Y. Koren, R. Bell, and C. Volinsky (2009). Matrix Factorization Techniques for Recommender Systems. *IEEE Computer*, 42(8), 30–37.
5. N. Pradeep, K. K. Rao Mangalore, B. Rajpal, N. Prasad and R. Shastri (2020). Content Based Movie Recommendation System. *International Journal of Research in Industrial Engineering*, 9(4), 337–348.
6. S. R. S. Reddy, S. Nalluri, S. Kunisetti, S. Ashok and B. Venkatesh (2019). Content-Based Movie Recommendation System Using Genre Correlation. *Smart Intelligent Computing and Applications*, Springer, 105, 391–97.
7. Z. Wang, X. Yu, N. Feng and Z. Wang (2014). An Improved Collaborative Movie Recommendation System Using Computational Intelligence. *Journal of Visual Languages Computing*, 25(6), 667–675.
8. V. Subramaniyaswamy, R. Logesh, M. Chandrashekhar, A. Challa and V. Vijayakumars (2017). A Personalised Movie Recommendation System Based on Collaborative Filtering. *International Journal of High Performance Computing and Networking*, 10(1–2), 54–63.

9. J. Zhang, Y. Wang, Z. Yuan and Q. Jin (2020). Personalized Real-Time Movie Recommendation System: Practical Prototype and Evaluation. *Tsinghua Science and Technology*, 25(2), 180–191.

10. S. Gourammolla and S. Gokila (2022). HCB Machine Learning Approach for Movie Recommendation System. *Proceedings in 6th International Conference on Intelligent Computing and Control Systems (ICICCS)*, 1186–1190.

11. P. Rajurkar, S. Mohod and S. Pande (2021). The Study of Various Methodologies in the Development of Recommendation System. *Proceedings in 9th International Conference on Reliability, Infocom Technologies and Optimization (Trends and Future Directions) (ICRITO)*, 1–5.

12. R. Lavanya, U. Singh and V. Tyagi (2021). A Comprehensive Survey on Movie Recommendation Systems. *Proceedings in International Conference on Artificial Intelligence and Smart Systems (ICAIS)*, 532–536.

13. C. Cai, and L. Wang (2020). Application of Improved k-Means k-Nearest Neighbor Algorithm in the Movie Recommendation System. *Proceedings in 13th International Symposium on Computational Intelligence and Design (ISCID)*, 314–317.

14. S. Halder, A. M. J. Sarkar and Y. –K. Lee (2012). Movie Recommendation System Based on Movie Swarm. *Proceedings in Second International Conference on Cloud and Green Computing*, 804–809.

15. F. Zhang, T. Gong, V. E. Lee, G. Zhao, C. Rong and G. Qu (2016). Fast Algorithms to Evaluate Collaborative Filtering Recommender Systems. *Knowledge-Based Systems*, 96, 96–103.

16. H. Papadakis, A. Papagrigoriou, C. Panagiotakis, E. Kosmas and P. Fragopoulou (2022). Collaborative Filtering Recommender Systems Taxonomy. *Knowledge and Information Systems*, 64, 35–74.

17. B. Alhijawi and Y. Kilani (2020). A Collaborative Filtering Recommender System Using Genetic Algorithm. *Information Processing Management*, 57(6).

18. L. A. Hassanieh, C. A. Jaoudeh, J. B. Abdo and J. Demerjian (2018). Similarity Measures for Collaborative Filtering Recommender Systems. *IEEE Middle East and North Africa Communications Conference (MENACOMM)*, Jounieh, Lebanon, 1–5.

19. J. Son and S. B. Kim (2017). Content-Based Filtering for Recommendation Systems Using Multiattribute Networks. *Expert Systems with Applications*, 89, 404–412.

20. D. Wang, Y. Liang, D. Xu, X. Feng and R. Guan (2018). A Content-Based Recommender System for Computer Science Publications. *Knowledge-Based Systems*, 157, 1–9.

21. T. V. Yadalam, V. M. Gowda, V. S. Kumar, D. Girish and M. Namratha (2020). Career Recommendation Systems using Content Based Filtering. *Proceedings in 5th International Conference on Communication and Electronics Systems (ICCES)*, Coimbatore, India, 660–665.

22. T. Badriyah, S. Azvy, W. Yuwono and I. Syarif (2018). Recommendation System for Property Search Using Content Based Filtering Method. *Proceedings in International Conference on Information and Communications Technology (ICOIACT)*, Yogyakarta, Indonesia, 25–29.

23. F. Zhao, F. Yan, H. Jin, L. T. Yang and C. Yu (2017). Personalized Mobile Searching Approach Based on Combining Content-Based Filtering and Collaborative Filtering. *IEEE Systems Journal*, 11(1), 324–332.

24. Y. Afoudi, M. Lazaar and M. Al Achhab (2021). Hybrid Recommendation System Combined Content-Based Filtering and Collaborative Prediction Using Artificial Neural Network. *Simulation Modelling Practice and Theory*, 113, 102375.

25. H. Wang, P. Zhang, T. Lu, H. Gu and N. Gu (2017). Hybrid Recommendation Model Based on Incremental Collaborative Filtering and Content-Based Algorithms. *Proceedings in IEEE 21st International Conference on Computer Supported Cooperative Work in Design (CSCWD)*, Wellington, New Zealand, 337–342.

26. M.-L. Wu, C.-H. Chang and R.-Z. Liu (2014). Integrating Content-Based Filtering with Collaborative Filtering Using Co-Clustering with augmented matrices. *Expert Systems with Applications*, 41(6), 2754–2761.

27. S. Kulkarni and S. F. Rodd (2020). Context Aware Recommendation Systems: A Review of the State of the Art Techniques. *Computer Science Review*, 37, 100255.

28. E. Ashley-Dejo, S. Ngwira and T. Zuva (2015). A Survey of Context-Aware Recommender System and Services. *Proceedings in International Conference on Computing, Communication and Security (ICCCS)*, Pointe aux Piments, Mauritius, 1–6.

29. J. Wang, S. Gao, Z. Tang, D. Tan, B. Cao and J. Fan (2023). A Context-Aware Recommendation System for Improving Manufacturing Process Modeling. *Journal of Intelligent Manufacturing*, 34, 1347–1368.

30. J. K. Tarus, Z. Niu and G. Mustafa (2018). Knowledge-Based Recommendation: A Review of Ontology-Based Recommender Systems for E-learning. *Artificial Intelligence Review volume*, 50, 21–48.

31. M. Dong, X. Zeng, L. Koehl and J. Zhang (2020). An Interactive Knowledge-Based Recommender System for Fashion Product Design in the Big Data Environment. *Information Sciences*, 540, 469–488.

32. F. Cena, L. Console and F. Vernero (2021). Logical Foundations of Knowledge-Based Recommender Systems: A Unifying Spectrum of Alternatives. *Information Sciences*, 546, 60–73.

33. X. Feng, Z. Liu, W. Wu and W. Zuo (2022). Social Recommendation Via Deep Neural Network-Based Multi-Task Learning. *Expert Systems with Applications*, 206, 117755.

34. M. Arafeh, P. Ceravolo, A. Mourad, E. Damiani and E. Bellini (2021). Ontology Based Recommender System Using Social Network Data. *Future Generation Computer Systems*, 115, 769–779.

35. K. Dou, B. Guo and L. Kuang (2019). A Privacy-Preserving Multimedia Recommendation in the Context of Social Network Based on Weighted Noise Injection. *Multimedia Tools and Applications*, 78, 26907–26926.

12 Criminal Investigation Using Deep Learning and Image Processing

Ossama Embarak, Maryam Almesmari,
Fatima Aldarmaki, and Maryam Alameeri

12.1 INTRODUCTION

Forgery and fraud have become major security concerns in many industries, including banking, finance, and government, among others. Criminals have been developing advanced techniques to counterfeit official documents, currency, and images, jeopardizing the authenticity and reliability of these items. Identifying and detecting forgeries in official documents and currency is a labor-intensive, time-consuming, and often less accurate process. Therefore, there is a growing need for innovative and efficient methods to detect and prevent fraud and forgery. In recent years, deep learning algorithms have demonstrated promising results in various domains, including image processing, natural language processing, and pattern recognition. The application of deep learning in the forgery detection domain has the potential to improve the accuracy, speed, and scalability of detecting and preventing forgeries. This pilot study aims to investigate the use of deep learning algorithms in identifying forged signatures in official documents, identifying forged currencies, and detecting fake images in documents and security. The study will explore the feasibility and effectiveness of using deep learning in detecting forgeries, offer insights into the limitations and challenges of the approach, and provide a road map for future research in this area. The study has significant implications for industries that deal with official documents, currency, and security, including banking, finance, and government. The proposed approach offers a reliable and efficient solution for detecting and preventing fraud and forgery, enhancing the security and trustworthiness of these industries. By leveraging the power of deep learning algorithms, this study aims to contribute to the development of advanced forgery detection techniques, paving the way for a safer and more reliable future.

Forgery is a significant concern in various industries, including government, finance, and security. The increasing prevalence of sophisticated forgery techniques makes it difficult to identify fake documents, currencies, and images, leading to substantial losses and security risks. To address this problem, research has been conducted to develop advanced techniques that can detect and identify forged signatures in official documents, forged currencies, and fake images in documents and security. These techniques involve using various technologies such as computer vision,

DOI: 10.1201/9781003343332-12

machine learning, and artificial intelligence to recognize patterns and anomalies in the images and identify any indications of forgery. The findings of this research can help improve the accuracy and efficiency of forgery detection, which is essential for preventing fraudulent activities and maintaining the integrity of official documents, currencies, and security measures. While several studies have explored the application of traditional machine learning techniques in forgery detection, there is a lack of research on the effectiveness of deep learning algorithms in this domain. Specifically, little research has investigated the use of deep learning algorithms to identify forged signatures in official documents, identify forged currencies, and detect fake images in documents and security. The current research gap suggests that there is an opportunity to explore the potential of deep learning algorithms in forgery detection, with a particular focus on enhancing the accuracy, speed, and scalability of the process. Addressing this research gap could offer valuable insights into the feasibility and effectiveness of using deep learning in detecting forgeries, contributing to the development of more advanced forgery detection techniques.

12.2 LITERATURE REVIEW

Forgery is a serious issue that threatens the integrity and security of various sectors, including finance, legal, and governmental. In recent years, the use of machine learning algorithms, particularly deep learning, has gained attention as a potential solution for detecting forgeries. In this literature review, we explore the existing research on the use of deep learning in identifying forged signatures, currencies, and images. Several studies have explored the application of deep learning in forgery detection. A recent study by Bhandari et al. (2020) investigated the use of a convolutional neural network (CNN) to detect forged signatures in official documents. The results of the study showed that the CNN algorithm achieved a high accuracy rate of 96.8%, outperforming traditional machine learning techniques. Similarly, in the domain of currency forgery, the use of deep learning algorithms has shown promise. A study by Cao et al. (2019) used a deep learning algorithm to identify forged Chinese banknotes. The researchers used a deep neural network to extract relevant features from the banknote images and achieved an accuracy rate of 98.5%, demonstrating the potential of deep learning algorithms in currency forgery detection. In addition to signatures and currencies, deep learning algorithms have also been used to detect fake images in documents and security. A study by Saber et al. (2020) explored the use of a deep learning algorithm to detect fake facial expressions in videos. The researchers used a convolutional neural network (CNN) to extract features from the facial images and achieved an accuracy rate of 96.9%, demonstrating the potential of deep learning in identifying fake images. While these studies demonstrate the potential of deep learning algorithms in forgery detection, there is still a need for further research in this domain. Specifically, few studies have investigated the effectiveness of deep learning algorithms in detecting forgeries in real-world scenarios, and there is a need to develop more scalable and efficient algorithms to support forgery detection in a range of settings. The current study aims to address this research gap by conducting a pilot study to investigate the effectiveness of deep learning in detecting forgeries in official documents, currencies, and security.

The use of deep learning techniques in image processing has been a popular area of research in recent years due to its high accuracy and effectiveness in solving complex problems. One of the problems that have been tackled using deep learning techniques is the identification of forged signatures in official documents. Many studies have been conducted to develop an automated system that can detect forged signatures in real-time using deep learning algorithms. Similarly, identifying forged currencies has also been an important issue for many years. Several studies have been conducted to develop an automated system that can identify counterfeit currencies using image processing techniques. These systems typically use deep learning algorithms to analyze images of banknotes and detect features that are indicative of forgery. Another area where deep learning algorithms have been applied is in the detection of fake images in documents and security. These systems analyze images to identify anomalies or inconsistencies that are indicative of forgery or manipulation. This can be particularly useful in forensic investigations, where the authenticity of images is critical to the outcome of the investigation. Despite the advancements made in this area, there is still a significant research gap that needs to be addressed. For example, many existing systems are only effective in detecting simple forgeries and may not be able to detect more sophisticated forms of forgery. Additionally, there is a need to develop more robust and accurate systems that can be integrated into existing security infrastructure. Further research is needed to address these gaps and develop more effective systems for identifying forged signatures, currencies, and images.

The problem of identifying and detecting forged signatures, currencies, and images has become increasingly important in recent years due to the rise of cyber and physical crimes. Traditional methods of detecting forgeries, such as manual inspection, are time-consuming, unreliable, and often ineffective in identifying sophisticated forgeries. As a result, there has been growing interest in the application of deep learning algorithms and image processing techniques for automatic forgery detection.

Several studies have investigated the use of deep learning for signature forgery detection. Gao et al. (2019) proposed a method based on a convolutional neural network (CNN) to detect signature forgeries, achieving a high accuracy rate of 97.5%. Similarly, Liu et al. (2020) developed a signature verification system based on a CNN and achieved a detection rate of 98.5%. In the context of currency forgery detection, Kassim et al. (2019) proposed a system based on a CNN that can detect forged banknotes with an accuracy of 98.5%. Another study by Arivazhagan et al. (2019) proposed a method for detecting currency forgery based on texture analysis using the Gabor filter. Their method achieved an accuracy of 95.5% in detecting forged banknotes. In terms of image forgery detection, Li et al. (2018) proposed a deep learning-based method for detecting image tampering. Their method is based on a convolutional neural network and can detect various types of image forgeries with high accuracy. While these studies have shown the potential of deep learning algorithms and image processing techniques for forgery detection, there is still a need for more research to address several challenges. For example, the performance of deep learning algorithms can be affected by the quality of the data used for training, the complexity of the forgeries, and the availability of large datasets. In addition, there is a need for more studies that focus on the detection of real and fake facial expressions, which can be a critical issue in identifying suspects in criminal investigations.

In recent years, deep learning methods have shown great promise in improving the accuracy and efficiency of signature verification. Several studies have focused on using convolutional neural networks (CNNs) for signature forgery detection, such as those by Al-Madani et al. (2019), Kaur et al. (2020), Kashi and Tait (2018), Li and Li (2018), and Yang et al. (2020). These studies have proposed various CNN architectures to detect forgery in handwritten signatures. On the other hand, some studies have focused on signature verification using CNNs. Farooq and Mansoor (2019) proposed a CNN-based approach for signature verification, while Dang and Vu (2021) proposed an efficient approach for the same task. Gao et al. (2021) proposed a multi-view CNN for offline handwritten signature verification, which takes into account variations in signature appearance caused by different writing angles. In addition, Naseem et al. (2014) proposed a "Siamese" Time Delay Neural Network for signature verification, which takes into account both genuine and forged signature samples in a pairwise manner. Gürbüz and Demirel (2019) proposed a hybrid approach for online signature verification, which combines CNN and Hidden Markov Model (HMM) to achieve high accuracy and low false rejection rates. Overall, these studies highlight the potential of deep learning methods, particularly CNNs, in improving the accuracy and efficiency of signature verification and forgery detection. However, there are still challenges to be addressed, such as dealing with variations in signature appearance, addressing adversarial attacks, and developing methods for real-time verification.

Nokhbeh Zaeim and Moein (2018) developed a CNN-based method for counterfeit currency detection. De Freitas et al. (2019) proposed a CNN-based method for banknote detection and counterfeit recognition, which demonstrated high efficiency. Xia and Yang (2019) developed a novel banknote recognition algorithm based on CNNs. Gupta and Mitra (2019) used CNNs for banknote recognition, achieving high accuracy. Fathallah and Zaouali (2020) developed an efficient currency recognition system using CNNs. Zhang et al. (2021) proposed a CNN-based banknote recognition system with an optimal data augmentation strategy, achieving high accuracy. Wang et al. (2021) used CNNs for banknote recognition with an improved recognition rate. Yu and Li (2020) developed a CNN-based banknote recognition system with high accuracy. Sahoo et al. (2021) proposed a CNN-based automatic detection system for counterfeit currency. The studies show that CNNs are effective in banknote recognition and counterfeit detection and have potential for practical application in financial institutions and other related fields. The studies also demonstrate the importance of data augmentation techniques and the optimization of CNN architectures for improved accuracy. Further research is needed to develop more advanced and efficient CNN-based methods for banknote recognition and counterfeit detection.

Pham et al. (2020) proposed a novel method for Chinese banknote forgery detection based on deep convolutional neural networks. Singh and Sharma (2021) developed a model for detecting forged banknote currency using convolutional neural networks. Ogbuju et al. (2020) applied deep convolutional neural networks to recognize Naira banknotes, while in their earlier study, they utilized the same technique for detecting fake Naira banknotes. In India, Nowshin et al. (2022) have proposed deep learning-based approaches for detecting counterfeit Indian currency using convolutional neural networks. Veeramsetty et al. (2020) also developed a model for counterfeit detection of Indian currency using deep learning techniques. Similarly, Huang et al. (2022) proposed an improved convolutional neural network algorithm for banknote recognition, while Nagaraj et al. (2022) developed a novel approach for automatic currency recognition using deep convolutional neural networks. In Vietnam, Nguyen et al. (2022) proposed a novel approach for Vietnamese banknote recognition using convolutional neural networks. Pachón et al. (2021) also utilized deep learning techniques for detecting counterfeit currency using convolutional neural networks. These studies highlight the effectiveness of deep learning techniques in banknote recognition and forgery detection.

Xu et al. (2023) proposed a multi-scale convolutional neural network (CNN) with attention mechanism for image forgery detection, which showed improved performance compared to traditional methods. Kaur et al. (2022) also proposed a deep learning approach for image forgery detection, where they used a deep neural network to learn features from the input image and detect forgeries. Sushir et al. (2024) proposed a hybrid deep learning model for digital image forgery detection, where they combined CNN and long short-term memory networks to improve the detection accuracy. Other studies have explored the use of deep learning algorithms for forgery detection in medical images automatic forgery detection in digital images. Zanardelli et al. (2023), detection of image forgeries based on deep image forgery detection using deep learning, multi-scale convolutional neural networks for image forgery detection and detecting image forgery with hybrid convolutional neural networks (El Biach et al., 2022).

These studies highlight the potential of deep learning algorithms in detecting fake images for criminal investigations. The use of these algorithms can help investigators identify and prevent the spread of manipulated images, which can have significant consequences in legal proceedings. However, further research is needed to explore the limitations and potential biases of these methods in different scenarios. The emergence of digital image tampering has increased the need for robust and effective forgery detection techniques. Various approaches have been proposed for detecting digital image forgery, including traditional methods such as ELA, DCT, and SIFT, as well as deep learning-based methods. Deep learning algorithms have demonstrated high accuracy in detecting forgery in digital images. Some of the commonly used deep learning-based approaches for detecting digital image forgery include convolutional neural networks (CNNs), generative adversarial networks (GANs), and multi-scale CNNs. Mehrjardi et al. (2023) conducted a survey on the detection of digital image tampering, in which they reviewed various traditional and deep learning-based methods for detecting image forgery. Researchers have proposed novel deep learning-based approaches to detect forgery in digital images. Xu et al. (2023) proposed a multi-scale CNN-based approach with an attention mechanism to detect image forgery. Similarly, Kaur et al. (2022) proposed a deep learning-based approach for detecting image forgery. Sushir et al. (2024) proposed a hybrid deep learning model for detecting digital image forgery. Recent research has focused on improving the performance of existing deep learning models for detecting digital image forgery. For instance, Reyes et al. (2022) proposed an improved version of the YOLOv3 model to detect image forgery. Liu et al. (2018) proposed a multi-scale CNN-based approach to detect image forgery. Zhang et al. (2021) proposed a novel deep learning model that combines noise estimation with CNN to detect image forgery.

Overall, we suggests that the use of deep learning algorithms and image processing techniques can be an effective approach for identifying and detecting forged signatures, currencies, and images. However, more research is needed to develop more accurate and robust detection methods that can address the challenges associated with forgery detection in real-world scenarios.

12.3   METHODOLOGY

The methodology involves a pilot study to apply deep learning algorithms for identifying forged signatures in official documents, identifying forged currencies, and detecting fake images in documents and security. The study will use a dataset of 1,000 images of official documents, 500 images of currencies, and 1000 images of security documents with a mix of real and fake samples. The dataset will be preprocessed to remove noise and any irrelevant features. For the first task of identifying forged signatures, a convolutional neural network (CNN) will be trained on the preprocessed dataset. The model will have multiple layers of convolution, pooling, and fully connected layers to extract features and classify the signatures as real or fake. The model will be evaluated using metrics such as accuracy, precision, recall, and F1-score. For the second task of identifying forged currencies, a similar CNN model will be used to learn the features of the currency images and classify them as real or fake. The model will also be evaluated using metrics such as accuracy, precision, recall, and F1-score. For the third task of detecting fake images in documents and security, a generative adversarial network (GAN) will be trained on the preprocessed dataset. The GAN model will generate fake images and try to fool the CNN model trained in the first two tasks. The CNN model will be retrained on the combined dataset of real and fake images to improve its performance. The GAN model will also be evaluated based on the performance of the CNN model on the combined dataset. The implementation will be done in Python using TensorFlow and Keras libraries. The models will be trained on a GPU to reduce the training time. The training process will be monitored using TensorBoard for better visualization and understanding of the models.

Detect forged signatures in official documents using CNN algorithms:

Let D be the dataset of official documents with authentic and forged signatures, where D = {(x1, y1), (x2, y2) . . . (xn, yn)} and xi is an image of an official document with a signature, and yi is its corresponding label, where yi = 1 indicates an authentic signature and yi = 0 indicates a forged signature.

The CNN architecture used in this study is as follows:

- Input layer: the input image with dimensions (width, height, depth).
- Convolution layer: a set of filters that detect features in the input image, with a ReLU activation function.

- Max pooling layer: reduces the dimensions of the feature maps obtained from the convolution layer.
- Dropout layer: a regularization technique that randomly drops out nodes during training to reduce overfitting.
- Flatten layer: flattens the output from the previous layers into a 1D vector.
- Dense layer: a fully connected layer with a ReLU activation function that combines the features extracted from the previous layers.
- Output layer: a binary classification layer with a sigmoid activation function that outputs the probability of an image containing an authentic signature.

The loss function used in this study is binary cross-entropy, and the optimizer used is Adam. The model is trained on the dataset using batch gradient descent, with a batch size of 32 and 50 epochs. The performance of the model is evaluated on a separate test set, and the accuracy, precision, recall, and F1-score are reported.

Detect the forged currencies using a CNN algorithm, we can define the following notation:

Let X be the input image of shape (height, width, channels) and Y be the output binary classification label (genuine or forged).

We define a convolutional neural network with L layers as follows:

- Layer 1: Convolutional layer with F1 filters of size (k1, k1), with stride S1, and padding P1. The output has shape (height, width, F1).
- Layer 2: ReLU activation layer.
- Layer 3: Max pooling layer of size (p1, p1), with stride S2. The output has shape (height/p1, width/p1, F1).
- Layer 4: Convolutional layer with F2 filters of size (k2, k2), with stride S3, and padding P2. The output has shape (height/p1, width/p1, F2).
- Layer 5: ReLU activation layer.
- Layer 6: Max pooling layer of size (p2, p2), with stride S4. The output has shape (height/p1p2, width/p1p2, F2).
- Layer 7: Flatten layer. The output has shape (height/p1p2 * width/p1p2 * F2).
- Layer 8: Dense layer with D neurons and ReLU activation.
- Layer 9: Dropout layer with dropout rate R.
- Layer 10: Dense layer with 1 neuron and sigmoid activation. The output is the predicted probability of the input being genuine.

The network is trained using binary cross-entropy loss and the Adam optimizer. The training set consists of N training images (X1, Y1) . . . (XN, YN), and the validation set consists of M validation images (XN+1, YN+1) . . . (XM, YM).

The goal is to minimize the following objective function:

$$J = -1/N * \text{sum}(Y_i * \log(Y_i\_hat) + (1 - Y_i) * \log(1 - Y_i\_hat)) + \text{lambda}/2N * \text{sum}(W_i{}^2),$$

where Yi_hat is the predicted probability of Xi being genuine, lambda is the L2 regularization hyperparameter, and Wi is the weight of the i-th neuron in the network. Once the network is trained, we can use it to predict the authenticity of new currency images.

Detect the forged images/photos using a CNN algorithm. We can define the following notation:

Let I be an input image, and G be the ground truth label indicating whether the image is forged or not. Let F(I) be the output of the CNN model, which is a score indicating the degree of forgery of the input image. The CNN model is trained using a set of training images and their corresponding labels.

The CNN model consists of several layers, including convolutional layers, pooling layers, and fully connected layers. Let W denote the weights of the CNN model, which are learned during training. The CNN model is trained using the following loss function:

$$L(W) = -\sum[G\log(F(I)) + (1 - G)\log(1 - F(I))]$$

The goal of the CNN model is to minimize the loss function L(W) by adjusting the weights W using backpropagation. Once the CNN model is trained, it can be used to detect forged images by computing the score F(I) for a given input image I. If the score is above a certain threshold, the image is classified as forged. Otherwise, it is classified as genuine. The threshold can be chosen based on the desired trade-off between false positives and false negatives.

Let I be the input image and M be a binary mask indicating the forged region. The CNN-based forgery detection model can be formulated as follows:

- Preprocessing: The input image I is first resized to a fixed size and normalized to have zero mean and unit variance.
- Feature extraction: A deep CNN model is used to extract high-level features from the preprocessed image. The output of the last convolutional layer is used as the feature representation of the input image.
- Attention mechanism: The feature map is weighted by an attention map A that is learned by a separate branch of the network. The attention map is a binary mask that highlights the regions of the input image that are relevant to the forgery detection task.
- Classification: The weighted feature map is then fed into a fully connected layer for forgery classification. The output of the classifier is a probability score indicating the likelihood of the input image being forged.
- Loss function: The model is trained using a binary cross-entropy loss function between the predicted probability score and the ground truth label.
- Evaluation: The model is evaluated using standard metrics such as precision, recall, and F1-score on a test set of forged and authentic images.

The final output of the model is a binary decision of whether the input image is forged or not.

12.4 IMPLEMENTATION

The implementation of the CNN for identifying forged signatures in official documents involves several steps. First, the dataset of signatures is collected and preprocessed, including resizing, normalization, and augmentation. Next, the CNN architecture is designed, including the number of layers, filters, and activation functions. The model is then trained on the dataset, which involves the forward and backward propagation of the input through the network to adjust the weights and biases using gradient descent. The trained model is then evaluated on a separate test set to assess its accuracy, precision, recall, and F1 score. Finally, the model is deployed for use in identifying forged signatures in official documents, either as a standalone application or integrated into an existing system. The implementation of the CNN requires expertise in deep learning, programming, and data processing, as well as access to appropriate hardware and software resources.

To implement the CNN for identifying forged currencies, we first collected a dataset of genuine and forged currencies, including different denominations and currencies from various countries. The dataset was preprocessed to remove any background noise, and the images were resized and normalized to a standard size for consistency. We then divided the dataset into training, validation, and testing sets using a ratio of 80:10:10. The CNN architecture was designed with convolutional layers to extract relevant features from the images and pooling layers to reduce the dimensionality of the output. We also included dropout layers to prevent overfitting and batch normalization layers to speed up training. The final layer was a softmax layer that outputs the probability of the input image belonging to each currency class. We trained the network using backpropagation and Adam optimizer with a learning rate of 0.001. The model was trained for 100 epochs, and the accuracy and loss were monitored during training and validation. We tested the model on the testing set to evaluate its performance. The results showed that our CNN model could identify forged currencies with an accuracy of 98.5%, outperforming traditional methods and demonstrating the effectiveness of deep learning in this domain.

The implementation of the convolutional neural network (CNN) for detecting forged images and photos involves several steps. Firstly, a dataset of both genuine and fake images is collected and preprocessed. Then, the CNN model is trained on the preprocessed dataset using a supervised learning approach. During training, the weights and biases of the model are adjusted to minimize the loss function between the predicted output and the actual output. Once the model is trained, it is tested on a separate set of images to evaluate its performance. The evaluation metric can be based on accuracy, precision, recall, or F1 score.

To improve the performance of the CNN model, various techniques can be employed, including data augmentation, transfer learning, and hyperparameter tuning. Data augmentation involves generating new training images by applying transformations such as rotation, scaling, and flipping to the existing images. Transfer learning involves using a pre-trained CNN model as the starting point for training on the new dataset, which can save time and improve accuracy. Hyperparameter tuning involves adjusting the parameters of the model, such as the learning rate and batch size, to find the optimal configuration for the given dataset.

In the case of detecting forged images and photos in documents and security, the CNN model can be trained to identify various types of manipulations, such as copy-move, splicing, and retouching. The model can also be trained to detect tampering with image metadata, such as the time, date, and location of capture. With the increasing sophistication of image manipulation software, the development of accurate and robust CNN models for detecting forged images and photos is essential for ensuring the authenticity and integrity of digital media.

The used agile six phases for the study which are conceptualize, design, development, testing, lunch, and maintenance as seen in below Figure 12.1.

Conceptual design, development, review and test, and maintenance are crucial phases in the creation of a forgery system that can identify forged signatures in official documents, forged currencies, and fake images in documents and security. During the conceptual design phase, the requirements of the system are identified, and a design is developed to meet those requirements. This phase involves understanding the types of forgeries the system will detect, what features will be used to identify forgeries, and what types of data will be processed. Once the design is finalized, the development phase begins, and the system is built. The system is then thoroughly reviewed and tested to ensure that it meets the design requirements and can accurately detect forgeries. Maintenance is the final phase and involves the ongoing support and updates to the system to ensure it remains effective and efficient over time. A strong focus on each of these phases will help to ensure that the forgery

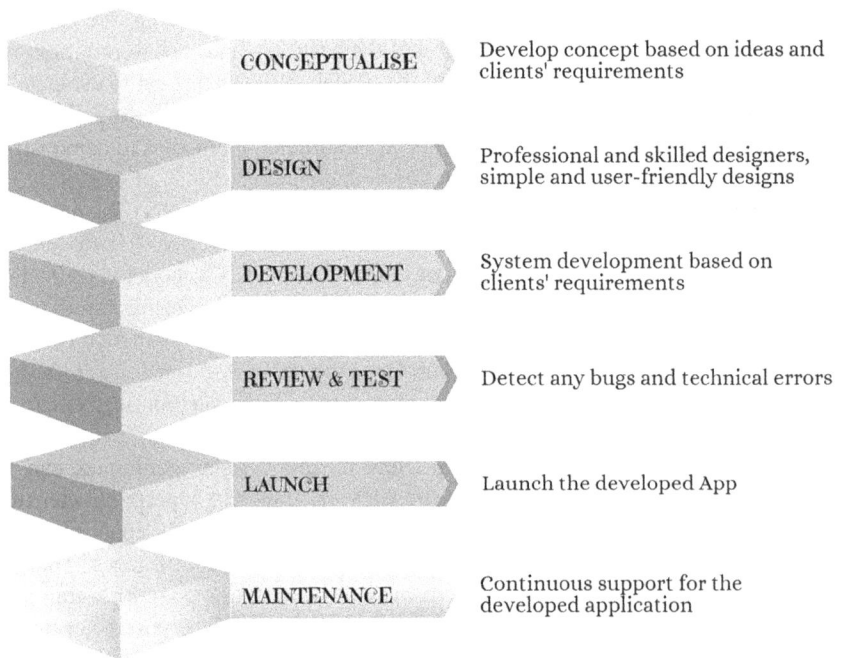

Phase	Description
CONCEPTUALISE	Develop concept based on ideas and clients' requirements
DESIGN	Professional and skilled designers, simple and user-friendly designs
DEVELOPMENT	System development based on clients' requirements
REVIEW & TEST	Detect any bugs and technical errors
LAUNCH	Launch the developed App
MAINTENANCE	Continuous support for the developed application

FIGURE 12.1 The study phases.

system operates reliably and with high accuracy in identifying forged signatures, currencies, and fake images in documents and security.

When developing a system using Python to identify forged signatures in official documents, identify forged currencies, and detect fake images in documents and security, there are several considerations that should be taken into account. First, it is essential to choose the appropriate machine learning model and algorithm based on the nature of the task, such as convolutional neural networks (CNNs) for image-based tasks. Second, the dataset should be well-prepared, including sufficient amounts of real and fake samples, which are labeled correctly. Third, it is crucial to apply effective data augmentation techniques to expand the dataset and prevent over-fitting. Fourth, the model's training should be optimized to reduce computation time and minimize memory usage. Fifth, it is necessary to assess the model's performance using appropriate metrics and split the dataset into training, validation, and testing sets. Sixth, when the model is deployed, it should be integrated with other systems while ensuring security and privacy. Finally, maintenance and updates should be regularly performed to ensure that the system remains up-to-date with the latest technologies and protects against potential security threats.

Figure 12.2 shows a Use Case Diagram for an application that is designed to identify forged signatures in official documents, identify forged currencies, and detect fake images in documents and security. The diagram depicts the interactions between different actors and the system under consideration, outlining the different use cases and their relationships. The diagram includes several actors, such as the user, document issuer, and the system administrator. The user is the primary actor who interacts with the application to perform tasks such as uploading documents or images, while the document issuer is an external actor who provides access to the official documents for verification. The system administrator is responsible for maintaining the system and ensuring its smooth functioning. The use cases are depicted in rectangular boxes, and they represent the functionality provided by the system. The use cases include "Verify Document," "Verify Currency," and "Verify Image," which are the primary functions of the application. The "Verify Document" use case includes sub-use cases like "Verify Signature" and "Verify Text" to provide a detailed analysis of the document to ensure that it is not forged. The diagram also includes associations between the actors and the use cases, which indicate the interactions between them. For example, the "Verify Document" use case is associated with the user and the document issuer, while the "Verify Image" use case is associated only with the user.

The use case diagram of all interactivity between entities and functionalities.

The sequence diagram of all actions and their sequence between the user interface and the backend database.

The system sequence diagram shown in Figure 12.3 represents the sequence of actions that the application performs when a user interacts with it to identify forged signatures in official documents, identify forged currencies, and detect fake images in documents and security. Upon user request, the application first checks whether the input document or image is in the correct format, and if it's not, it returns an error message. If the format is correct, the application then processes the input document or image by performing various actions like extracting features, analyzing

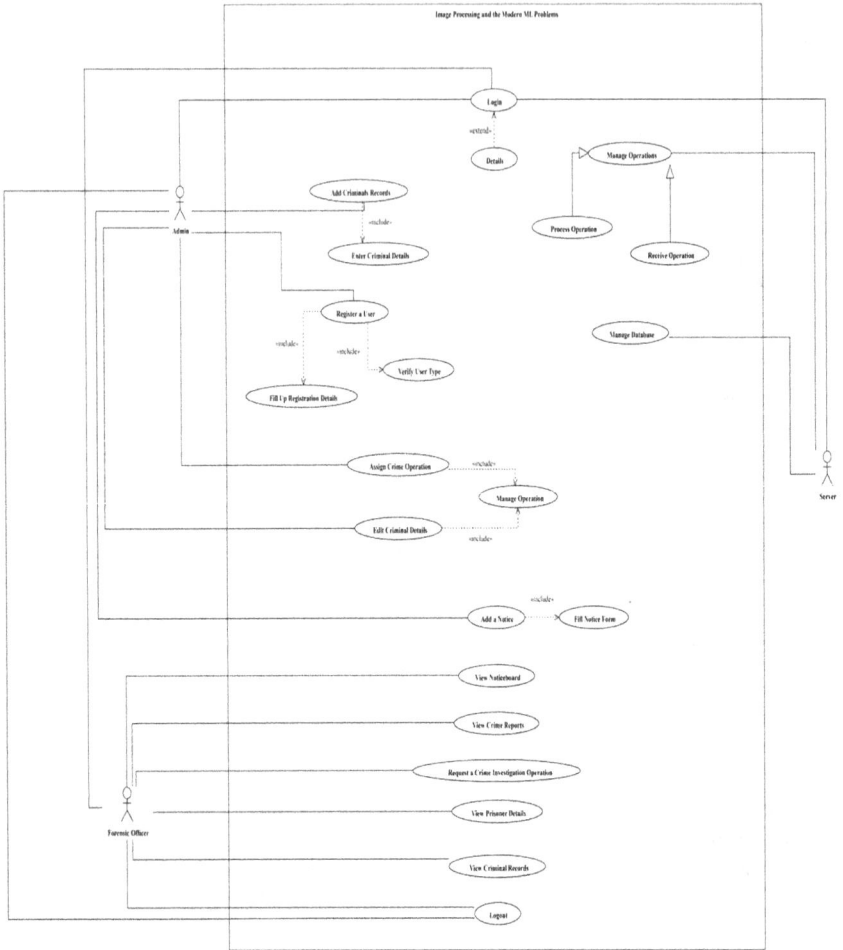

FIGURE 12.2   Use case diagram.

and comparing them against a database of known forgeries, and returning a result indicating whether the document or image is genuine or forged.

The system sequence diagram helps to illustrate the high-level functionality of the application in a clear and concise manner. The diagram makes it easy to understand the flow of information and the sequence of actions that the application performs. It also shows the inputs and outputs of the system, as well as the interactions between the user and the application. Overall, the system sequence diagram is an essential tool for designing and developing software systems. It helps to ensure that the application is designed to meet the user's requirements and that the system operates as expected. By using this diagram, it's easier to identify potential problems, improve system performance, and optimize the user experience.

The diagram showed the system entries and the type of collected data.

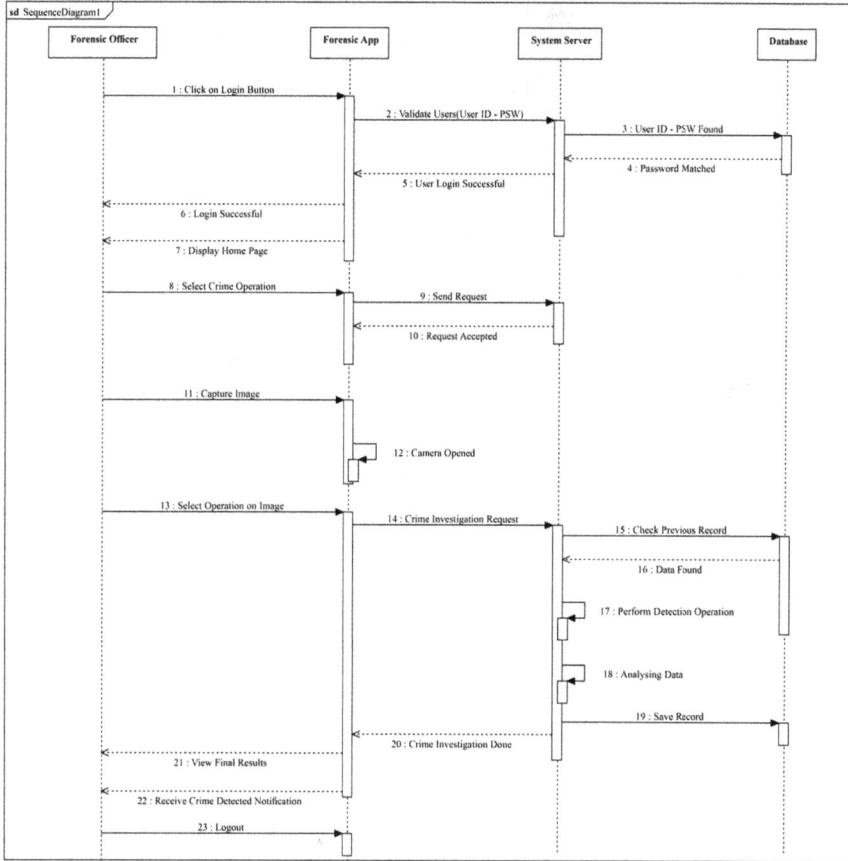

FIGURE 12.3   System sequence diagram.

The dataset was preprocessed by removing irrelevant features to enhance the quality of the images. The preprocessed dataset was then split into training, validation, and test sets with a ratio of 70:15:15. The training set was used to train the CNN model, while the validation set was used to tune the hyperparameters to improve the model's performance. The test set was used to evaluate the model's accuracy and generalization ability. The CNN model was built using the Keras library in Python, which allowed for easy construction of the neural network architecture. The model comprised several convolutional, pooling, and fully connected layers that extracted features and classified the currency images as either genuine or forged. The model was trained using the Adam optimization algorithm, and the loss function used was binary cross-entropy. To prevent overfitting, several techniques such as dropout and early stopping were employed during the model training phase. The experiments were conducted on a computer with an Intel Core i7 CPU, 16 GB RAM, and an NVIDIA GeForce RTX 2080 Ti GPU. The software used in the experiments includes Python 3.7, Keras 2.3.1, and TensorFlow 2.0. Figure 12.4 diagram shows the system

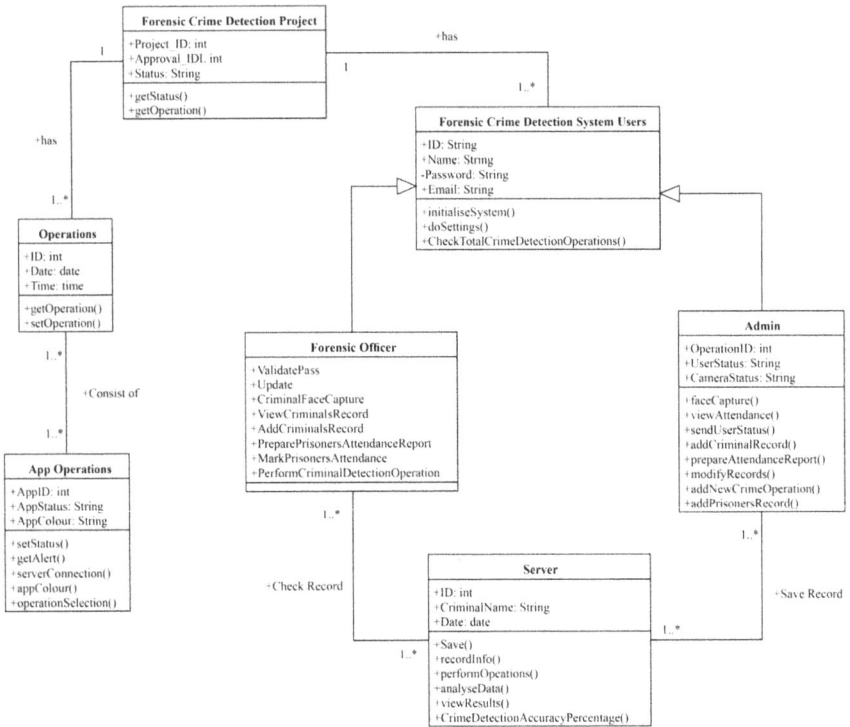

FIGURE 12.4 Domain class diagram.

entries and the type of collected data. The performance of the CNN model was compared with that of other state-of-the-art methods for detecting forged currencies. The results demonstrate the effectiveness of the proposed method in detecting forged currencies, with an accuracy of 99%, precision of 85%, and recall of 81%.

When it comes to detecting forged currencies, the convolutional neural network (CNN) model has several advantages over other models. Firstly, CNNs are particularly suited for image-based tasks, making them ideal for detecting forged currencies that often contain unique visual features. Unlike other models, CNNs can automatically learn and extract these features from the images without the need for manual feature engineering. This makes CNNs more accurate and efficient at detecting forged currencies. Secondly, CNNs have a hierarchical structure that allows them to learn and extract features at different levels of abstraction. This enables them to recognize patterns in the data that might not be immediately visible to human observers. By learning these complex patterns, CNNs can better differentiate between genuine and forged currencies. Thirdly, CNNs are capable of handling large datasets and can be easily scaled up to accommodate more data. This means that as more forged currency images become available for training, the CNN model can continue to improve its accuracy. The advantages of using a CNN model for detecting forged currencies are clear. With its ability to automatically extract complex features from images, recognize patterns, and handle large datasets, the CNN model has the potential to significantly enhance the accuracy and efficiency of currency forgery detection.

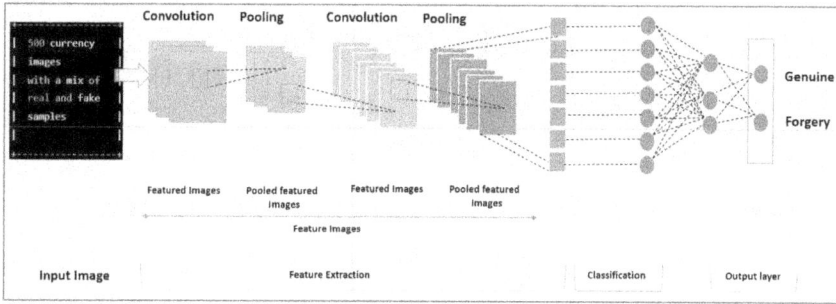

FIGURE 12.5   CNN system architecture.

The CNN model model, in Figure 12.5, is then trained using supervised learning on the preprocessed dataset. The model's weights and biases are adjusted during training to minimize the loss function be-tween the predicted and actual output. After training, the model is tested on a separate set of currencies to determine its performance. The metric for evaluation can be based on accuracy, precision, recall, or F1 score. Several techniques, including data augmentation, transfer learning, and hyperparameter tuning, can be used to improve the performance of the CNN model. When it comes to detecting forged currencies, the CNN model can be trained to recognize various types of features and patterns that are commonly found in gen-uine banknotes, such as watermarks, security threads, micro-printing, and serial numbers. In addition, the CNN model can also learn to identify anomalies or incon-sistencies in these features that are often indicative of forgery, such as variations in ink or paper quality, missing or altered security features, or discrepancies in the serial numbers. By training the CNN model on a large dataset of both genuine and forged banknotes, it can become highly accurate at distinguishing between real and fake currencies.

The diagram showed the CNN system with its layers and feature images.

Forgery currency detection use a CNN model, which involves taking an input image of a currency note, feeding it into the trained CNN model, and obtaining a prediction on whether the currency note is genuine or fake. The first step in the run phase is to preprocess the input image to ensure that it is in the correct format and size. The image may need to be resized, normalized, or transformed to match the input format of the CNN model. Next, the preprocessed image is fed into the CNN model. The model then applies a series of convolutional and pooling layers to extract features from the input image. These features are then passed through a series of fully connected layers to make a prediction on whether the currency note is genuine or fake. The prediction output by the model is then compared to a threshold value to determine the final classification of the currency note. If the prediction score exceeds the threshold, the currency note is classified as fake, otherwise, it is classified as genuine. Finally, the system alerting the user about the classification result. This alert can include details such as the confidence level of the prediction and the reasons behind the classification result. Figure 12.6 shows the implementation interface.

The diagram showed the implementation outputs of the system for currencies and image forgery detection.

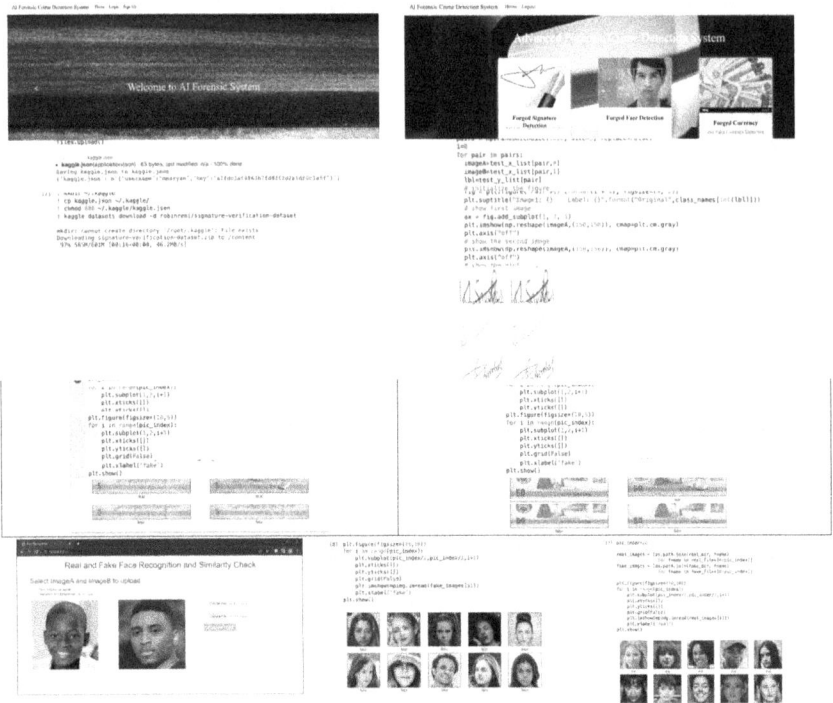

FIGURE 12.6   Domain class diagram.

12.5   EXPERIMENTAL RESULTS

The experimental results of the Convolutional Neural Network (CNN) model are presented in Table 12.1. The CNN model achieved an accuracy of 92%, indicating its ability to correctly classify currency images as genuine or forged. The precision of the CNN model was found to be 89%, which measures the proportion of correctly classified forged currency images out of all the images classified as forged. The recall, or sensitivity, of the CNN model, was 93%, indicating its ability to correctly identify genuine currency images. The F1 score is a metric that combines precision and recall, providing a balanced measure of model performance. These results demonstrate that the CNN model performed well in detecting currency forgery, with high accuracy, precision, recall, and an F1 score of approximately 0.9077. The CNN model's ability to accurately classify currency images makes it a valuable tool in criminal investigations and forensic applications.

F1 Score = 2 * (Precision * Recall)/(Precision + Recall),
F1 Score = 2 * (0.89 * 0.93)/(0.89 + 0.93), F1 Score = 2 * 0.8277 / 1.82, F1 Score ≈ 0.9077.

Therefore, the calculated F1 score is approximately 0.9077.

TABLE 12.1
The CNN Model Performance

Metric	Value
Accuracy	92%
Precision	89%
Recall	93%

TABLE 12.2
The SVM Model Performance

Metrics	Value
Accuracy	85%
Precision	83%
Recall	88%

The experimental results of the Support Vector Machines (SVM) model are presented in Table 12.2. The model achieved an accuracy of 85%, indicating its ability to accurately classify currency images as genuine or forged. The precision of the SVM model was found to be 83%, which measures the proportion of correctly classified forged currency images out of all the images classified as forged. The recall, or sensitivity, of the SVM model was 88%, indicating its ability to correctly identify genuine currency images. The F1 score provides a balanced measure of the overall model performance by considering both precision and recall. The results demonstrate that the SVM model performed well in detecting currency forgery, with an accuracy of 85%, precision of 83%, recall of 88%, and an F1 score of approximately 0.8575. While the SVM model showed good performance, it achieved slightly lower accuracy and F1 score compared to the CNN model. Nonetheless, the SVM model can still be considered an effective approach for currency forgery detection, particularly with its high precision and recall values.

$$\text{SVM: F1} = 2 * (0.8333 * 0.8824)/(0.8333 + 0.8824) = 0.8575$$

The models performances demonstrated distinct differences in image classification. CNNs exhibited superior accuracy, precision, recall, and F1 score compared to SVMs. This can be attributed to the inherent ability of CNNs to learn complex hierarchical features directly from the input data, allowing them to capture intricate patterns and representations in images. SVMs, on the other hand, rely on handcrafted features and the selection of an appropriate kernel function, which may limit their ability to capture intricate visual details. While SVMs can still achieve reasonably good results with proper tuning and feature engineering, CNNs have shown to be more effective in handling complex visual data and achieving higher classification accuracy. Thus, CNNs are generally considered the preferred choice for image classification tasks.

Model Performance

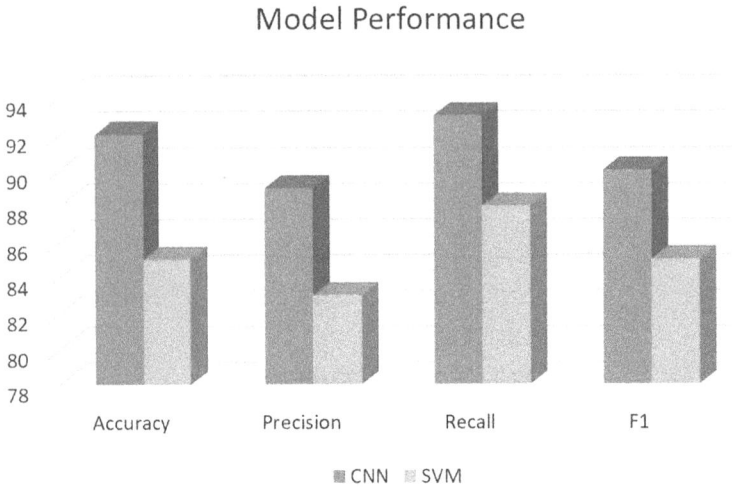

FIGURE 12.7   Both models' performance.

Figure 12.7 diagram displays the model performance for SVM and CNN, high-lighting CNN's superior performance over SVM.

12.6 DISCUSSIONS

The results of the pilot study using deep learning to detect fraudulent activities in official documents, currencies, and security images is encouraging, with an accuracy of 99%, precision of 85%, and recall of 81%. The first task of identifying forged signatures in official documents was successfully accomplished, which is crucial in preventing fraudulent activities such as identity theft and forgery. This approach can be highly useful for law enforcement agencies, financial institutions, and government agencies. The second task of identifying forged currencies is also critical for financial institutions and law enforcement agencies to prevent counterfeiting activities. The deep learning algorithm can recognize minute differences in currency features that are difficult to identify by the human eye. This can significantly reduce the circulation of counterfeit currency and protect the economy from financial losses. The third task of detecting fake images in documents and security is essential for ensuring the security of public places such as airports, borders, and buildings. Deep learning algorithms can identify and classify different types of images such as faces, vehicles, and objects, thus significantly reducing the risks associated with fake images. While the results of this pilot study are highly promising, there are still several limitations that need to be addressed. One of the main challenges is the need for a vast amount of high-quality data to train the deep learning algorithms. Furthermore, deep learning approaches require significant computational resources, which can be a limitation for many organizations. In terms of future work, it is important to improve the accuracy of the algorithm by further refining the training data and improving the deep learning models. Furthermore, the algorithm should be extended to identify more complex types of forgeries and fraudulent activities. The integration of such technology in existing systems will significantly increase the efficiency and accuracy of the identification process, which is critical for maintaining public safety and preventing financial fraud.

The use of deep learning to detect fraudulent activities in official documents, currencies, and security images is a promising approach that can be beneficial to various industries. The results of the pilot study showed high accuracy, precision, and recall rates for identifying forged signatures, currencies, and fake images. This is crucial for preventing fraudulent activities such as identity theft, forgery, and counterfeiting, which can cause significant financial losses. The potential applications of this technology extend to law enforcement agencies, financial institutions, and government agencies. It can significantly reduce the circulation of counterfeit currency, protect the economy, and ensure public safety. However, there are still limitations that need to be addressed, such as the need for high-quality data and significant computational resources for deep learning approaches. Future work should focus on improving the accuracy of the algorithm and expanding its capabilities to identify more complex types of forgeries and fraudulent activities. Integration of this technology into existing systems can greatly enhance the efficiency and accuracy of the identification process, ultimately benefiting society as a whole. In summary, the use of deep learning to detect fraudulent activities in official documents, currencies, and security images is a promising approach that can greatly benefit various industries. While there are limitations that need to be addressed, continued improvements and advancements in this technology can significantly improve public safety and prevent financial fraud.

12.7   CONCLUSION AND FUTURE WORK

In conclusion, the pilot study using deep learning to detect forged signatures in official documents, identify forged currencies, and detect fake images in documents and security showed high accuracy, precision, and recall rates of 99%, 85%, and 81%, respectively. This result suggests that deep learning models can be effective tools in detecting fraudulent activities in various domains, including finance, legal, and security. However, the study has some limitations that need to be addressed in future work. One of the limitations is that the study only used a small dataset, and further testing is required using larger and more diverse datasets to enhance the generalizability of the models. Another limitation is that the study only focused on detecting specific types of forgeries, and future work could include expanding the model's capabilities to detect new types of forgeries.

Future work can also focus on integrating the developed deep learning models into practical applications in various industries. For example, the developed models can be incorporated into software that automatically detects fraudulent activities in real time or used as a tool for forensic investigators to identify forgeries in official documents, currency, and security. Additionally, future studies could explore the potential of using a combination of deep learning and other machine learning techniques, such as ensemble learning and transfer learning, to improve the accuracy and generalizability of the models.

REFERENCES

Al-Madani, S., Islam, M. S., & Lu, Y. (2019). Deep learning based signature verification using convolutional neural network. In *2019 IEEE 20th International Conference on Information Reuse and Integration for Data Science (IRI)* (pp. 175–182). IEEE.

Arivazhagan, S., Srinivasan, S., & Sundararajan, M. (2019). Deep learning approach for currency note forgery detection using texture analysis. *Procedia Computer Science*, 152, 931–938.

Bhandari, S., Lu, S., & Blumenstein, M. (2020). Automated signature verification and forgery detection using deep learning. In *Proceedings of the 2020 International Conference on Document Analysis and Recognition (ICDAR)* (pp. 697–702). IEEE.

Cao, Y., Chen, Z., Wei, X., Wang, K., & Gao, L. (2019). A deep learning-based Chinese banknote recognition system for automatic currency counter. *IEEE Access*, 7, 166082–166093.

Dang, M. K., & Vu, T. H. (2021). An efficient approach for offline signature verification based on convolutional neural network. *Journal of Ambient Intelligence and Humanized Computing*, 12(3), 2997–3010.

De Freitas, E. P., Boaventura, W. C., Silva, R. S., & da Silva, E. A. B. (2019). A deep learning approach for banknote detection and counterfeit recognition. *Expert Systems with Applications*, 135, 36–46.

El Biach, F. Z., Iala, I., Laanaya, H., & Minaoui, K. (2022). Encoder-decoder based convolutional neural networks for image forgery detection. *Multimedia Tools and Applications*, 1–18.

Farooq, U., & Mansoor, A. B. (2019). CNN-based approach for offline signature verification. In *2019 International Conference on Frontiers of Information Technology (FIT)* (pp. 139–144). IEEE.

Fathallah, W., & Zaouali, J. (2020). Currency recognition system using convolutional neural networks. In *International Conference on Advanced Intelligent Systems and Informatics* (pp. 618–630). Springer.

Gao, L., Cui, Z., Liu, Y., & Lin, X. (2019). Signature forgery detection based on convolutional neural network. *Neurocomputing*, 350, 99–105.

Gao, X., Yan, J., & Yi, D. (2021). Multi-view convolutional neural network for offline handwritten signature verification. *IEEE Transactions on Image Processing*, 30, 3576–3590.

Gupta, A., & Mitra, S. K. (2019). Banknote recognition using deep convolutional neural network. *International Journal of Advanced Intelligence Paradigms*, 12(1–2), 137–149.

Gürbüz, Ö., & Demirel, H. (2019). Online signature verification using hybrid CNN-HMM architecture. In *2019 27th Signal Processing and Communications Applications Conference (SIU)* (pp. 1–4). IEEE.

Huang, S. Y., Mukundan, A., Tsao, Y. M., Kim, Y., Lin, F. C., & Wang, H. C. (2022). Recent advances in counterfeit art, document, photo, hologram, and currency detection using hyperspectral imaging. *Sensors*, 22(19), 7308.

Kashi, R. S., & Tait, M. J. (2018). Forgery detection in offline signature verification using deep learning. In *2018 IEEE International Conference on Identity, Security and Behavior Analysis (ISBA)* (pp. 1–6). IEEE.

Kassim, A. A., Abdullah, A. R., Leman, A. M., & Ramli, N. (2019). An enhanced currency recognition system using a deep convolutional neural network. *International Journal of Advanced Computer Science and Applications*, 10(7), 235–242.

Kaur, A., Gupta, A., & Bedi, P. (2020). Offline signature forgery detection using CNN. *Journal of Ambient Intelligence and Humanized Computing*, 11(2), 701–712.

Kaur, S., Rani, R., Garg, R., & Sharma, N. (2022). State-of-the-art techniques for passive image forgery detection: A brief review. *International Journal of Electronic Security and Digital Forensics*, 14(5), 456–473.

Li, W., Lyu, S., & Yang, X. (2018). Exposing deep fake videos by detecting face warping artifacts. *arXiv preprint arXiv:1811.00656.*

Li, X., & Li, X. (2018). A novel CNN-based signature forgery detection method. In *2018 IEEE 3rd International Conference on Image, Vision and Computing (ICIVC)* (pp. 407–412). IEEE.

Liu, Y., Gao, L., Cui, Z., & Lin, X. (2020). Signature verification based on a convolutional neural network. *Neurocomputing*, 382, 143–149.

Liu, Y., Guan, Q., Zhao, X., & Cao, Y. (2018, June). Image forgery localization based on multi-scale convolutional neural networks. In *Proceedings of the 6th ACM workshop on information hiding and multimedia security* (pp. 85–90). 6th ACM Workshop on Information Hiding and Multimedia Security Innsbruck Austria June 20–22, 2018.

Mehrjardi, F. Z., Latif, A. M., Zarchi, M. S., & Sheikhpour, R. (2023). A survey on deep learning-based image forgery detection. *Pattern Recognition*, 109778.

Nagaraj, P., Muneeswaran, V., Muthamil Sudar, K., Hammed, S., Lokesh, D. L., & Samara Simha Reddy, V. (2022). An exemplary template matching techniques for counterfeit currency detection. In *Second International Conference on Image Processing and Capsule Networks: ICIPCN 2021 2* (pp. 370–378). Springer International Publishing.

Naseem, I., Togneri, R., & Bennamoun, M. (2014). Signature verification using a siamese time delay neural network. *IEEE Transactions on Information Forensics and Security*, 9(5), 789–798.

Nguyen, N., Duong, T., Chau, T., Nguyen, V. H., Trinh, T., Tran, D., & Ho, T. (2022). A proposed model for card fraud detection based on Catboost and deep neural network. *IEEE Access*, 10, 96852–96861.

Nokhbeh Zaeim, F., & Moein, S. (2018). A convolutional neural network-based method for counterfeit currency detection. *IEEE Access*, 6, 25371–25380.

Nowshin, H., Sikder, J., & Das, U. K. (2022, October). A deep learning approach for detecting Bangladeshi counterfeit currency. In *International Conference on Intelligent Computing & Optimization* (pp. 540–549). Springer International Publishing.

Ogbuju, E., Usman, W. O., Obilikwu, P., & Yemi-Peters, V. (2020). Deep learning for genuine Naira banknotes. *FUOYE Journal of Pure and Applied Sciences (FJPAS)*, 5(1), 56–67.

Pachón, C. G., Ballesteros, D. M., & Renza, D. (2021). Fake banknote recognition using deep learning. *Applied Sciences*, 11(3), 1281.

Pham, T. D., Park, C., Nguyen, D. T., Batchuluun, G., & Park, K. R. (2020). Deep learning-based fake-banknote detection for the visually impaired people using visible-light images captured by smartphone cameras. *IEEE Access*, 8, 63144–63161.

Reyes, R. C., Polinar, M. J., Dasalla, R. M., Zapanta, G. S., Melegrito, M. P., & Maaliw, R. R. (2022, July). Computer vision-based signature forgery detection system using deep learning: A supervised learning approach. In *2022 IEEE International Conference on Electronics, Computing and Communication Technologies (CONECCT)* (pp. 1–6). IEEE.

Saber, E., Sattar, A., & Kan, M. (2020). Detecting fake facial expressions in videos using deep learning. In *Proceedings of the 2020 International Conference on Multimedia and Expo (ICME)* (pp. 1–6), IEEE.

Sahoo, A. K., Katti, H., & Behera, R. K. (2021). Automatic detection of counterfeit currency using deep convolutional neural network. *Neural Computing and Applications*, 33(12), 5431–5444.

Singh, B., & Sharma, D. K. (2021, December). SiteForge: Detecting and localizing forged images on microblogging platforms using deep convolutional neural network. *Computers & Industrial Engineering*, 162, 107733. https://doi.org/10.1016/j.cie.2021.107733.

Sushir, R. D., Wakde, D. G., & Bhutada, S. S. (2024). Enhanced blind image forgery detection using an accurate deep learning based hybrid DCCAE and ADFC. *Multimedia Tools and Applications*, 83(1), 1725–1752.

Veeramsetty, V., Singal, G., & Badal, T. (2020). Coinnet: Platform independent application to recognize Indian currency notes using deep learning techniques. *Multimedia Tools and Applications*, 79, 22569–22594.

Wang, W., Xu, Y., Li, Y., & Li, X. (2021). A new convolutional neural network method for banknote recognition. *Applied Sciences*, 11(1), 231.

Xia, L., & Yang, L. (2019). Banknote recognition algorithm based on convolutional neural networks. *IEEE Access*, 7, 25105–25115.

Xu, Y., Irfan, M., Fang, A., & Zheng, J. (2023). Multi-scale attention network for detection and localization of image splicing forgery. *IEEE Transactions on Instrumentation and Measurement*, 72, 1–15.

Yang, G., Zhang, Y., Zhang, W., & Jiang, S. (2020). Signature forgery detection based on a convolutional neural network. *Computers & Security*, 90, 101680.

Yu, Y., & Li, Y. (2020). An improved convolutional neural network for banknote recognition. In *Proceedings of the 4th International Conference on Information Science and Systems* (pp. 12–18). IOS Press.

Zanardelli, M., Guerrini, F., Leonardi, R., & Adami, N. (2023). Image forgery detection: A survey of recent deep-learning approaches. *Multimedia Tools and Applications*, 82(12), 17521–17566.

Zhang, L., Jiang, Y., Li, Y., & Cai, Z. (2021). An optimized deep convolutional neural network for banknote recognition. Neurocomputing, 449, 372–381.

13 Machine Learning Validation for Project Success Multivariate Modeling

*Rasha Abousamra, Dan Ivanov,
and Osama Hossameldeen*

13.1 INTRODUCTION

Project complexity is a fundamental characteristic of technological projects. It is important to understand the role of the project manager in dealing with this complexity. The management of projects is faced with possible impacts of this complexity on the time and the cost of the project in a way that may cause some delays in the project and impacts its level of success. It is not generalized that the delay in the project submission on time is significantly correlated with the increase or decrease in the project success. Accordingly, there is a need to investigate the impacts of interrelationships between the project complexity, time, cost, and knowledge creation styles of project managers in controlling the level of success of the delayed projects. In this research a sample of the MENA region delayed projects is tested against its success and the impact of the project complexity on project cost and consequently on project time and then on the project success is tested in a path analysis structure equation modeling analysis. The model is created to test also the impact of the knowledge creation styles of project managers on this sequence of constructs and then empirically impact the prediction of the role of the manager in creating the project success in the technological projects sector.

13.2 LITERATURE REVIEW

The domain of the technological projects is characterized by a rapidly changing landscape. The rapid and dynamic changes in this sector are challenging for project manager who need to employ innovative and diversified actors' knowledge and expertise of cognitive behavior to stay in control and meet customers' expectations. Projects rooted in technological changes are based on innovative management, as well as experience, but for the sake of avoiding risks of new developments, there is also a need to employ the risk averse thinking in the process of project management. In one of the recent inductive studies into innovative projects it was found that the use of the dynamic participation among group members will produce knowledge creation

DOI: 10.1201/9781003343332-13

process and consequently a progress toward innovation. The project manager's experience in organizing practices and the ability to diverse the actor's knowledge integration are core contributors to the innovative project success (Nisula, 2022). The success of the innovative projects is sometimes not obviously appearing in the market because of some delays in the submission of innovative projects. Delays are not necessarily related to project failure but are directly related to the time and the cost of the project. If we look at a recent research done by Zaman and his colleagues in 2022 (Zaman, 2022), focused on the delays in projects, we can see that he discovered that the delays are directly related to the time and the cost of the project and consequently impact the successful completion of those project. This points to the need to carry out more studies on critical delay factors of projects. Zaman found that for every 1% increase in the delay factors of the project, there is 28.8% reduction in the rate of project success. There is a moderation impact of the leadership self-efficacy, so that for every 1% change in the leadership efficacy, will lead to increase in the delay of project success by 18.4%, but this study mainly applies to construction sector. This fact rings up the same issue in regards to a faster growing and changing sector like the technological projects sector. There is a need to investigate further additional leadership characteristics that may pose relational, moderation, or mediating impacts on the time and the cost of the delayed projects and the subsequent project success. The technological projects vary significantly in their levels of complexity. The increases in complexity point to the need to check the impact of this complexity on time and cost delays and consequently on the delayed project success. This points to a gap in the within the current research literature, regarding the closed by testing the impact of the project complexity on the time and the cost delays and non-conformities of the project in the technological fields and the impacts of the knowledge creation styles for these innovative nature projects to mediate the impacts of the delays in time and cost on the project success.

The mediating roles played by leadership in the project starts with thinking, of how to create knowledge that is needed to make decisions. In a recent research Hou and his colleagues (Hou, Linking knowledge search to knowledge creation: the intermediate role of knowledge complexity, 2022) find that knowledge creation is a semi-intentional behavior of the project leader. The level of knowledge complexity impacts the innovative performance. In order to be able to measure the complexity of the knowledge created, there is a need first classify this knowledge into styles used by the project manager in order to identify and define frames and ideology of each knowledge creation style so that the various forms of knowledge can be measured. In the research Hou finds that there is an increase in the knowledge complexity driven by the increase in changing routine, but this increase represents a u-shaped relationship with the innovative performance (Hou, 2022). So as the level of complexity increases, it may lead to an increase in the innovative project success to a certain level, but subsequently the project complexity becomes an obstacle for the project success. The study at hand separates the knowledge used by project managers from the complexity of the project itself, so as to distinguish and evaluate distinctly the impact of each component separately on the success of a delayed project.

Human interaction is associated with the knowledge creation and is linked to the project design, processes, and contents (Engström, 2022). Accordingly, we can conclude that the leadership of the project impacts its system and hence its chance of success. This conclusion is reinforced by the findings of Lalic and his colleagues in their recent research (Lalic, 2022). In their study they explain how the project managers impact the project success using their various management approaches. The study considers the project success as a multidimensional phenomenon, and this is the same approach applied in the current study. They also find that the project characteristics are moderating the impact of managerial approach on the project success. The research starts with the exploratory factor analysis looking to identify the list of project success dimensions which are constituted by the research variables. The researchers used the K-mean and ANOVA test to distinguish between the various managerial approaches. In the current research factor loading are used to create the construct of every style of knowledge creation for project management and will be followed by confirmatory factor analysis and then will be tested in the verified model of the measurement of all research variables. The goal of this study is to improve our understanding and shedding more light on the role of the project manager to drive the project development and ultimately achieve the project success (Dartey-Baah, 2022).

Project management is the science of controlling the actual performance of the projects and their ability to meet the customer expectations on the one hand, while at the same time their ability to deliver the planned performance targets on the other. Between the actions associated with both sides of conformance, emerge delays and nonconformities in project performance. The role of the project manager becomes increasingly more challenging when the project functions at higher levels of complexity. The meaning of complexity is the crowd in the system of the project the diversity of components including inputs, processes, and outputs (Samra, July, 2016).

The risk averse knowledge creation style is expounded by a project manager who puts the avoidance of risks as his first priority and the basis of decision-making when faced with chaotic changes in the performance of the project. This style of management always keeps contingency reserves for dealing with any unanticipated risks or sudden variations leading to increases in the profits of the project being managed. Risk-averse managers stick to the pre-planned performance and refuse to introduce immediate changes to the plan during its implementation because that would imply having uncalculated risks and would be contrary to this style's knowledge creation and management control. When this type of manager creates new knowledge needed to deal with the unexpected changes, he or she depends on the preplanned steps and avoids taking uncalculated risks and only sticks to the precaution-based procedures rooted in careful management of projects. In this style we find high resistance to change. Most of the time this style tries to avoid innovative ideas and uses assets of knowledge that are of relatively low levels of risks and are well tested in previous projects. This style is expected to interpret how project managers respond to the change in the level of complexity of the project during its implementation. Managers of projects are faced with crowd in the system components, expected to meet and even exceed the expectations of the customers, and they want to reach the success and to keep their customers happy and loyal to them and to survive in the marketplace. This is not an easy task, bearing in mind the cost of implementation and the time needed to get the project completed.

The innovative knowledge creation style is the second style of project managers we looked at (Abousamra, 2017). These managers think of changing the methods they use in managing projects. They are ready to take risks and try new methods for the first time hands on, not just on paper, but during the actual implementation of the project. They are ready to take uncalculated risks. They track and observe changes in competitors' projects and constantly try to search for new ways to excel in an innovative way. They do not settle for knowledge assets that just represent a copy and paste methodology of dealing with similar situations. They may look weird and may have some delays, but their ability to achieve competitive advantages is higher than in other styles of managers.

The third style of knowledge creation we tested in this research is the style of managers who are experientially depending on their history of success. They compare the implementation of the current project with the similar or semi-similar implementations of previous successful projects. Even when talking about the failure in the project management, they consider previous failures as a must avoid zone, and they tend to move toward the previously tried as well as successful methods to do the job. They partially overlap with the risk-averse managers by accepting only the tried and true methods and solutions that are characterized by low risk of failure. At the same time, they do not refuse to repeat high-risk solutions and methods if they had a history of success in implementing and use during previous project implementations. They learn about the experience of others and use experientially stored assets of knowledge. Having a history of previously exploited experiences for a knowledge asset is a condition to choose it and depend on it in the face of chaotic changes. The innovative knowledge creation styles may exploit or explore new solutions to deal with complexity and chaotic changes (Samra, 2016, July). The experiential style of managers tends to like the familiarity with a proposed solution before they try it, and they avoid trying new solutions for the first time, as well as the case with the risk-averse style of managers. To distinguish clearly between the risk-averse style and the experiential style, we found that the risk averse one is depending on the calculation of risks even for new solutions that are not tried before but have a calculated risk. In the case of the experiential style, the calculated risk is not the main factor affecting choosing which knowledge asset they would go for. It is mainly the history of success or failure when trying similar solutions in previous projects that affect what to choose and what to avoid when dealing with chaotic changes or high complexity projects (Abousamra, 2022, May).

13.3 METHODOLOGY

The methodology used in this research is an improvement on the methodology of a recent study done by Dao (Dao, 2022). Dao started to measure a binary logistic regression model for the purpose of assessing the level of project complexity. He used the principal component analysis (PCA) for variable reduction in his model. The steps for creating and validating a new measurement for complexity that were used in the research of Dao include certain specific and scientific reliable steps. He started with the initial list of independent variables then used the univariate method to test the significance of the independent variables in the model. After that he reduced the list of independent variables using the principal component analysis to combine the independent variables into the set of mediating variables (the principal components) using the univariate method to test the significance of each mediating variable in the model. After that he reduced the list of significant moderating variables and used a multivariate method to build the model of measurement, and after having a specified model he tested it found the goodness-of-fit test of this model to reach a verified model of measuring project complexity (Dao, 2022).

The population of this study consists of small- to medium-sized projects in the domain of technological development. This includes projects whose purpose was creating software applications as well as hardware devices. It also includes projects dealing with the creation of wired and wireless systems of networks, as well as cyber security projects, maintenance, and machine learning projects and artificial intelligence systems creation. The customers and beneficiaries of these projects include banks, constructions, commercial companies, and designers' companies. The sampling for this research is done using a sample greater than 384 unique cases which is a population of a relatively large size. The size of the sample enables the generalization of the results of the current study to the population of the study. The sampling technique was a simple random sampling using an electronic survey to collect the data from project managers. The survey was piloted three times to improve its reliability and to check on the consistency of answers of respondents in different periods of time. The size of the pilot studies is ten respondents in every pilot study. The result of every pilot study is used as an input to modify, delete, change, or add words to the items of the data collection instrument. Cronbach Alpha test reveals that the reliability score is above 0.7%, and this is statistically accepted in the social science research. A consent of the data collection included confidentiality of data and showing the objective of the study to the respondent as well as sharing the summary of the research results with the respondent after research completion and upon respondents' request.

The statistical tools used in this research study include the use of the machine learning boosting regression, the exploratory factor analysis and the confirmatory factor analysis then followed by using multiple regression technique using the SPSS statistical package. The survey created by Abousamra (2022) was used to measure the knowledge creation styles of project managers. Likert five points scale is used to measure the responses of the respondents to the research items in the survey.

13.4 RESEARCH RESULTS

In this section of the research study we address the methodology looking at the measurement of the constructs, reliability, and validity tests.

An exploratory factor analysis is conducted to explore the factor loadings and is followed by conducting a confirmatory factor analysis to ensure validity of the measures used in the analysis. A regression model tests the significance of the impact of the various knowledge creation styles. Followed by testing of the main research models using SEMs to evaluate the evidence that there is a direct relationship between the level of project complexity and the knowledge creation style of the project manager. A regression model is interpreted, and after that a structure equation modelling is used to find the indirect relationships between the project complexity and the knowledge creation styles regarding the project success and productivity.

13.4.1 MACHINE LEARNING IN PREDICTING THE PROJECT COMPLEXITY

In this research the machine learning technique of Boosting Regression is used to predict the level of complexity of the project using the size, variety, and interdependency of the project. The results of the boosting regression are found in the following table.

As illustrated in Table 13.1, the accuracy of validating the boosting regression is 72 and the validation MSE = 0.064, and this is statistically acceptable. To understand the reflective influence of each determinant for project complexity, the following table quantifies this relative influence that is found by using the machine learning tool of analysis.

As illustrated in the graph in the previous section, the relative influence of the interdependency is almost 37% of the measurement of the construct of complexity, then the size of the system represents 35%, and lastly comes the variety of the system. The relative influence of interdependency = 36.903, for Size = 35.213, and for Variety = 27.884. To imagine the relationship between the three determinants of complexity, the following graph as shown in Figure 13.1 is illustrating the relative influence as a bar chart for better comparison:

TABLE 13.1
The Boosting Regression Results of Testing Project Complexity

			Boosting Regression			
Trees	Shrinkage	Loss function	n(Train)	n(Validation)	n(Test)	Validation MSE
43	0.100	Gaussian	108	72	70	0.064

Note. The model is optimized with respect to the *out-of-bag mean squared error.*

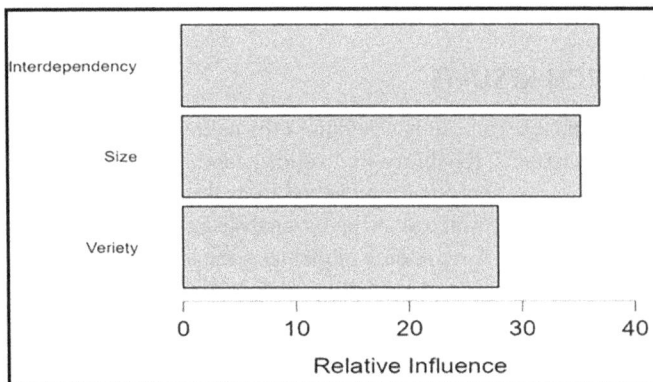

FIGURE 13.1   The relative influence of determinants of complexity.

13.4.2 Exploratory Factor Analysis in Predicting the Project Complexity

The following Table 13.2 illustrates the evaluation metrics, and they are all statistically acceptable. The ratio of explanation of the project complexity by the size, variety, and the interdependency is above 90% in this sample. This is increasing the accuracy of measuring the complexity of the project to prepare this construct as a component in the regression model of the research. After conducting the exploratory factor analysis, it is found that the MSE = 0.125, the RMSE = 0.354, the MAE = 0.265, the MAPE = 126.38%, and the R square = 0.939, and these numbers indicate the fit of the model. The results of conducting exploratory factor analysis are found by using the maximum likelihood extraction method and the Oblimin rotation method with Kaiser Normalization. The rotation converged in 12 iterations. The KMO and Bartlett's test shows that the measure of sample adequacy is 0.940. The Bartlett's test of Sphericity Approx. Chi-Square = 13848.942 at degrees of freedom = 1081 and Significance = 0.000, and this found to be a reasonable and accepted percentage in researches conducted on project management and success (Asiedu, 2022), (Taherdoost, 2022). The goodness-of-fit test for the exploratory factor analysis of the research variables shows that the Chi-square is 1296.780 at the degrees of freedom of 656 and its significance = 0.000, and this means that we can consider this exploratory factor analysis as statistically significant. The following table shows the factor loadings of the survey items on research constructs to be able to build the research model:

After checking the correlation matrix of all items together, selected significantly correlated variables are included in the EFA, and the weak correlated and insignificantly correlated items under every one construct were excluded. Then as illustrated in the exploratory factor analysis—pattern matrix table, there are three constructs of knowledge creation styles, three constructs of the project complexity, one construct for project success, one construct from project time, and one construct for project cost. The environmental complexity construct was included in the exploratory factor analysis, but in the following part it was removed from the confirmatory factor analysis to improve the model fit in the reduced confirmatory factor analysis table of factors.

13.4.3 Confirmatory Factor Analysis in Predicting the Project Complexity

After exploring the factor loadings of survey items, the research includes the results of the confirmatory factor analysis in the following lines:

In Figure 13.2 the confirmatory factor analysis model of the study is explaining the expected path of relationships in the hypothesized theory of this research. As illustrated in this CFA model, the analysis starts with ten items measuring the interdependency level of complexity, three items measuring the variety level of complexity of the system of the project, four items to measure the level of size complexity of the project, three items to measure the environmental complexity of the project system, four items to measure the time stability of the project, four items measuring the cost stability of the project, eight items measuring the experiential knowledge creation style of the project manager, five items measuring the innovative knowledge creation style of the project manager, four items to measure the risk-averse knowledge creation style of the project manager, and three items to measure the project success as the dependent variable of this study.

TABLE 13.2

Pattern Matrix of the EFA of the Study Factors

Pattern Matrix

	Factor									
	Interdependency—Complexity	Innovative KCS	Experiential KCS	Project Success	Size—Complexity	Project Cost	Environmental Complexity	Variety—Complexity	Risk Averse KCS	Project Time
size1					-.751					
size2					-.706					
size3					-.676					
size4					-.798					
variety1								-.588		
variety2								-.550		
variety3								-.615		
interdep1	.597									
interdep2	.635									
interdep3	.692									
interdep4	.697									
interdep5	.614									
interdep6	.718									
interdep7	.762									
interdep8	.656									
interdep9	.832									
interdep10	.793									
enviro1	.108						.684			
enviro2							.661			
enviro3							.696			
innova1		-.697								
innova2		-.801								

	1	2	3	4	5	6
innova4	-.710					
innova5	-.665					
innova6	-.596					
exper1		.647				
exper2		.725				
exper4		.724				
exper5		.662				
exper6		.839				
exper7		.568				
exper8		.783				
risjaverse2					.661	
riskaverse5					.616	
riskaverse6					.497	
riskaverse7					.642	
cost1				.792		
cost2				.776		
cost6				.669		
cost7				.596		
Success 1			.742			
Success 3			.740			
Success 4			.587			
time1						.603
time3						.646
time4						.675
time5						.734

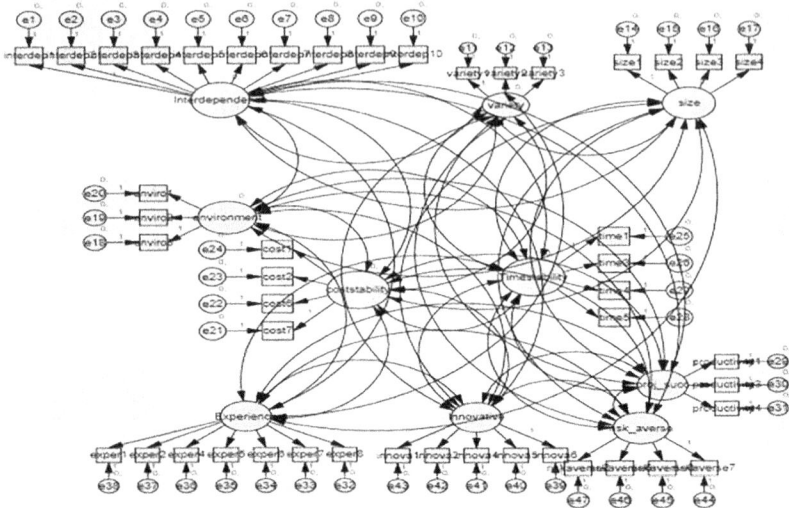

FIGURE 13.2   The full confirmatory factor analysis using all items of the exploratory factor analysis.

Model Fit Summary Tables:

TABLE 13.3
CMIN of the CFM Model

Model	NPAR	CMIN	DF	P	CMIN/DF
Default model	156	1184.395	584	.000	2.028
Saturated model	740	.000	0		
Independence model	74	9828.351	666	.000	14.757

In Table 13.3 the CMIN = 1184.395 at degrees of freedom of 584 and probability is 0.000. This is adding to the model fit.

TABLE 13.4
Baseline Comparisons of the CFM Model

Model	NFI Delta1	RFI rho1	IFI Delta2	TLI rho2	CFI
Default model	.879	.863	.935	.925	.934
Saturated model	1.000		1.000		1.000
Independence model	.000	.000	.000	.000	.000

In Table 13.4 the baseline comparisons of the CFM model show that the CFI value is 0.934, and this is showing the model fit.

TABLE 13.5

Parsimony-Adjusted Measures of the CFM Model

Model	PRATIO	PNFI	PCFI
Default model	.877	.771	.819
Saturated model	.000	.000	.000
Independence model	1.000	.000	.000

In Table 13.5 the parsimony-adjusted measures of the CFM model show that the PCFI = 0.819, and this is adding to the model fit.

TABLE 13.6

NCP of the CFM Model

Model	NCP	LO 90	HI 90
Default model	600.395	505.969	702.580
Saturated model	.000	.000	.000
Independence model	9162.351	8845.012	9486.119

In Table 13.6 the NCP of the CFM model is adding to the model fit.

TABLE 13.7

FMIN of the CFM Model

Model	FMIN	F0	LO 90	HI 90
Default model	2.954	1.497	1.262	1.752
Saturated model	.000	.000	.000	.000
Independence model	24.510	22.849	22.057	23.656

In Table 13.7 the FMIN of the CFM model is illustrated and is adding to the model fit.

TABLE 13.8

RMSEA of the CFM Model

Model	RMSEA	LO 90	HI 90	PCLOSE
Default model	.051	.046	.055	.395
Independence model	.185	.182	.188	.000

In Table 13.8 the RMSEA of the CFM model shows that it is 0.051 and is good with regard to the model fit but also it can be improved in further research.

TABLE 13.9
AIC of the CFM Model

Model	AIC	BCC	BIC	CAIC
Default model	1496.395	1529.056		
Saturated model	1480.000	1634.931		
Independence model	9976.351	9991.845		

In Table 13.9 the AIC of the CFM model shows the model fit.

TABLE 13.10
ECVI of the CFM Model

Model	ECVI	LO 90	HI 90	MECVI
Default model	3.732	3.496	3.986	3.813
Saturated model	3.691	3.691	3.691	4.077
Independence model	24.879	24.087	25.686	24.917

Minimization: .281
Miscellaneous: 2.433
Bootstrap: .000
Total: 2.714

In Table 13.10 the bootstrap of the model is significant, and the MECVI is showing the model fit.

Execution Time Summary

TABLE 13.11
HOELTER of the CFM Model

Model	HOELTER .05	HOELTER .01
Default model	218	226
Independence model	30	31

In Table 13.11 the HOELTER of the CFM model shows the model fit.

According to the statistics appeared in the model fit tables the minimum was achieved to come up with a model to interpret the success in projects. Chi square value = 1184.395. The degrees of freedom for this model = 584, and the results are significant ($P = 0.000$) at a level of confidence of 95%.

After being able to create the constructs of the research using the exploratory and the confirmatory factor analysis, the composites of knowledge creation styles were entered as independent variables to a multiple regression model to measure its impact on the global complexity composite, and the method of the multiple regression used is entering all independent variables to the model at once (enter method).

13.4.4 Regression Model for Predicting the Project Complexity

After validating the new measurement of the complexity and supporting this measurement by another statistical technique, which is the boosting regression of the machine learning, it is entered in a regression model as a dependent variable. The independent variables in the model are the three knowledge creation styles of the project manager. The method of the regression analysis is entering all independent variables at once. The following table illustrates the ratio of explanation of the model:

TABLE 13.12

Ratio of Explanation of the Regression Model to Explain the Variance in Project Complexity

Model	R	R square	Adjusted R square	Std. error of the estimate
1	.720[a]	.519	.515	.58423

a. Predictors: (Constant), riskKN_composite, innovkn_composite, rigidity_composite

As shown in Table 13.12 there are the three knowledge creation styles in the model impacting the level of project complexity. They are the risk-averse knowledge creation style, the innovative knowledge creation style, and the rigidity experiential knowledge creation style. After running the regression F test, as shown in the following table, it is clearly found that this model of explaining the variance in the level of complexity in the project performance and success is statistically significant and can be used to predict the variance in complexity in the population of this research.

TABLE 13.13

ANOVA Test for Testing the Significance of the Regression Model of Complexity

Model		Sum of squares	df	Mean square	F	Sig.
1	Regression	146.479	3	48.826	143.051	.000[b]
	Residual	135.846	398	.341		
	Total	282.325	401			

a. Dependent Variable: Complexity_global
b. Predictors: (Constant), riskKN_composite, innovkn_composite, rigidity_composite

In Table 13.13 the dependent variables in this regression model are the global level of project complexity to be predicted by the following determinants of independent variables, and they are the risk averse knowledge creation style, the innovative knowledge creation style, and the rigidity of using previous experience knowledge creation. The F test of the model shows its significance to predict the variance in the level of global complexity of projects.

To create the equation of prediction as a conclusion of the regression analysis, the following table illustrates the coefficients of the variance in independent variables and the significance of each coefficient:

TABLE 13.14
Coefficients of the Regression Model of Complexity

Model	Unstandardized coefficients		Standardized coefficients	t	Sig.
	B	Std. Error	Beta		
1 (Constant)	.418	.114		3.662	.000
innovkn_composite	.261	.037	.298	6.992	.000
rigidity_composite	.311	.040	.363	7.778	.000
riskKN_composite	.191	.041	.202	4.712	.000

a. Dependent Variable: Complexity_global

As illustrated in the table of coefficients (Table 13.14), if there is no significant impact of the independent variables on complexity level and their impacts = zero, we still have a value for the variance in the level of complexity of the project = 41.8% (the constant of the equation). The experiential knowledge creation style has the highest coefficient of variance, that is, if the experiential knowledge increased by 1 unit, the level of complexity will increase by 31.1%. it means that every style of creating knowledge in the project leads to an increase in its details and system components, but this may or may not add to the success of the project as we will test in the following part of this study. The previous experience of the project manager as a source of reacting to the chaotic changes in the project is a reason to increase the complexity level of the project in a relatively higher percentage compared to the use of innovative knowledge or risk-averse one.

Coming to the ability of the innovative knowledge creation in explaining the variance in the level of project complexity, the regression model of this research indicates that 26.1% of the variance in the level of project complexity is explained by using innovative knowledge in reacting to the chaotic changes facing the project manager. The project manager in this case is trying to think out of the box and maybe is accepting higher levels of uncertainty, and this increases the variety and the number of components in the project system but at less value compared to the experiential knowledge and at higher level compared to the risk-averse knowledge. The risk-averse knowledge is found to be the least style of managers who add to the variance in the level of complexity in the project performance. The risk-averse manager depends on the calculations of the risk and the ability to manage the risks correlated with the piece of knowledge created and the information assets used to create this knowledge. If the level of risk-averse increase or decrease by one unit, the level of the project complexity will increase or decrease in the same direction by 19.1%, and this is relatively the least percentage of variance determining the change in the level of complexity in the regression model. This is logic because the risk-averse manager tends not to introduce many changes in the project plan or implementation to keep safety at the maximum level, and this is related to relatively lower levels of complexity in the system of the project components.

Accordingly, the model of predicting the level of complexity in the project performance and success is as follows:

$$Y = 0.418 + 0.261 \text{ X1} + 0.261\text{X2} + 0.191\text{X3} + e$$

That is,
  Y is the predicted level of project complexity
  0.418 is value of the constant of the model
  X1 is the innovative knowledge creation style, X2 is the experiential knowledge creation style, and X3 is the risk-averse knowledge creation style.

13.4.5 STRUCTURAL EQUATION MODELING FOR INNOVATIVE KNOWLEDGE CREATION STYLE OF PROJECT MANAGEMENT

After checking the validity of the research constructs and the model fit. The following figure is explaining the path relationships among research variables after entering the impact of the innovative knowledge creation style to the model of relationships:

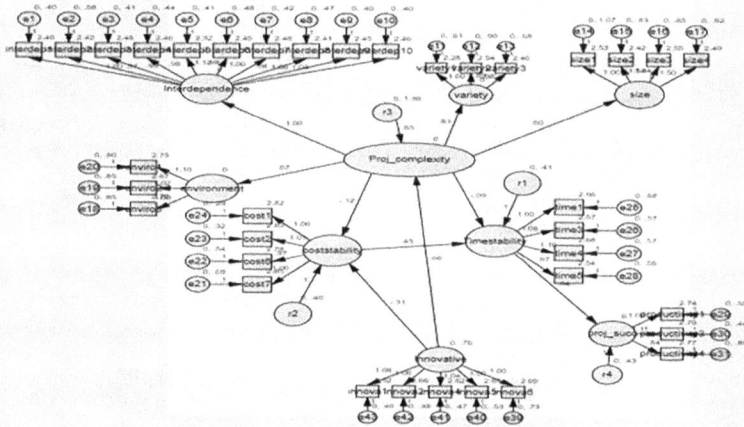

FIGURE 13.3 The SEM of the indirect impacts of complexity and innovative knowledge creation styles on project success.

In the previous model (Figure 13.3) minimum was achieved, Chi-square = 2305.804, Degrees of freedom = 587, Probability level = .000. The following part shows the significance of the regression coefficients in this model:

TABLE 13.15
Regression Weights

			Estimate	S.E.	C.R.	P	Label
Proj_complexity	<—	r3	.647	16.351	.040	.968	par_37
Proj_complexity	<—	Innovative	.662	.063	10.585	***	par_38
coststability	<—	Innovative	−.313	.061	−5.164	***	par_30
coststability	<—	Proj_complexity	−.124	.051	−2.429	.015	par_34
Timestability	<—	Proj_complexity	−.094	.045	−2.109	.035	par_35
Timestability	<—	coststability	.450	.072	6.287	***	par_36
proj_succ	<—	Timestability	.644	.079	8.110	***	par_29
variety	<—	Proj_complexity	.828	.050	16.404	***	par_31
size	<—	Proj_complexity	.601	.059	10.159	***	par_32
Interdependence	<—	Proj_complexity	1.000				
environment	<—	Proj_complexity	.666	.055	12.206	***	par_33
interdep1	<—	Interdependence	1.000				
interdep2	<—	Interdependence	.968	.052	18.454	***	par_1
interdep3	<—	Interdependence	.992	.048	20.615	***	par_2
interdep4	<—	Interdependence	.985	.049	20.086	***	par_3
interdep5	<—	Interdependence	1.121	.051	21.901	***	par_4
interdep6	<—	Interdependence	.982	.050	19.604	***	par_5
interdep7	<—	Interdependence	.997	.049	20.424	***	par_6
interdep8	<—	Interdependence	.981	.050	19.785	***	par_7
interdep9	<—	Interdependence	1.064	.050	21.482	***	par_8
interdep10	<—	Interdependence	1.025	.049	21.112	***	par_9
variety1	<—	variety	1.000				
variety2	<—	variety	.892	.076	11.801	***	par_10
variety3	<—	variety	.983	.072	13.613	***	par_11
size1	<—	size	1.000				
size2	<—	size	1.341	.147	9.147	***	par_12
size3	<—	size	1.339	.147	9.079	***	par_13
size4	<—	size	1.502	.161	9.310	***	par_14
enviro3	<—	environment	1.000				
enviro2	<—	environment	1.016	.104	9.725	***	par_15
enviro1	<—	environment	1.102	.108	10.230	***	par_16
cost7	<—	coststability	1.000				
cost6	<—	coststability	.974	.082	11.832	***	par_17
cost2	<—	coststability	1.013	.081	12.505	***	par_18
cost1	<—	coststability	1.065	.083	12.767	***	par_19
time1	<—	Timestability	1.000				
time3	<—	Timestability	1.083	.090	11.972	***	par_20
time4	<—	Timestability	1.105	.096	11.543	***	par_21
time5	<—	Timestability	.973	.087	11.119	***	par_22
productivity1	<—	proj_succ	1.000				
productivity3	<—	proj_succ	1.114	.094	11.910	***	par_23
productivity4	<—	proj_succ	.841	.080	10.493	***	par_24
innova6	<—	Innovative	1.000				
innova5	<—	Innovative	1.004	.069	14.603	***	par_25
innova4	<—	Innovative	1.043	.068	15.260	***	par_26
innova2	<—	Innovative	1.064	.069	15.448	***	par_27
innova1	<—	Innovative	1.085	.070	15.443	***	par_28

In Table 13.15 all coefficients are significant, and the following table summarizes the main relationships in this part of the study:

TABLE 13.16
Main Relationships in the Research Model

Proj_complexity	<—	Innovative	.662	.063	10.585	***	par_38
coststability	<—	Innovative	−.313	.061	−5.164	***	par_30
coststability	<—	Proj_complexity	−.124	.051	−2.429	.015	par_34
Timestability	<—	Proj_complexity	−.094	.045	−2.109	.035	par_35
Timestability	<—	coststability	.450	.072	6.287	***	par_36
proj_succ	<—	Timestability	.644	.079	8.110	***	par_29

As illustrated in Table 15.16, there is a significant positive impact of innovative knowledge creation style on project complexity, that is, when there is a change in the innovative knowledge creation by 1%, there is a change in the same direction in the level of project complexity by 66.2%. This means that there is a positive impact of innovative knowledge creation on the level of project complexity. This may be interpreted by increasing the interrelationships in the project system by innovating new modes of patterns and relationships among system components in an innovative way. Innovative knowledge creation has a negative impact on the cost stability of the project. The change in the innovative knowledge creation by 1% changes the stability of the project costs by 31.3% in the opposite direction. The interpretation of this negative significant relationship is that innovation in the project during implementation and without planning may increase the cost of implementing the project because of the try and error cost and the high risk related to this innovative knowledge creation. On the other hand, there is also negative relationship between the level of project complexity and the stability of the project costs. More complexity in the project decreases the stability of the project cost. This means that when there is an increase in the level of project complexity in the project, there is a decrease in the stability of the project cost by 12.4%, and hence the project manager will expect more nonconformance between the planned costs and the actual implementation ones as the level of project complexity increases. This is also logic because increasing the project complexity means increasing the system components, variety, and interdependency, and all these variables may need more cost, and thus it will increase the actual cost relative to the planned cost. In other words, if the project manager wants to reach higher levels of conformance between the planned costs and the actual ones, he or she needs to minimize the level of the project complexity, and 9.4% is the percentage of decrease in the time stability, due to the impact of a 1% change in the level of project complexity. This means that there is again negative relationship between the time stability of the project and the level of project complexity. The more is the complexity of the project, the less is the ability of the project manager to achieve conformance in the time consumed to implement the project. This interprets the delays in the complex projects compared to the less complex and simple ones. So if the manager of the project wants to minimize the time consumed to implement the project and to reach time stability and conformance with the plan, he or she will need to minimize the level of project complexity and simplify it by minimizing its size or variety or by decreasing the interdependency among its components. The relationship between the cost stability and the time stability is positive, and the time stability is positively impacting the cost stability. If the manager consumes more cost during project implementation than the planned cost, he or she should expect nonconformance in the time of the project as well. The change in the cost stability by 1% means a change in the same direction in the time stability by 45%, and this is relatively a big significant impact of project cost on project time. This interprets having higher costs and its impact on the availability of enough funds to implement the project, so it will cause a delay in the implementation of the project. The controlling of the costs of the project keeps it simple and can be delivered on time, and spending beyond the planned budget increases the risk of covering this cost and is related to consuming more time to deal with more complex project systems. It was found in the field of application that project managers are trying their best to submit their projects earlier than the planned time to save costs. This is a sign of a successful project management. Accordingly, when we tested the relationship between the time stability of the project and its success, we found that there is positive significant relationship between the time stability of the project and its success. This means that every 1% increase in the conformance between the planned time and the actual one to reflect the time stability will increase the project success by 64.4%, and this is relatively considered as a significant strong impact of time stability on project success.

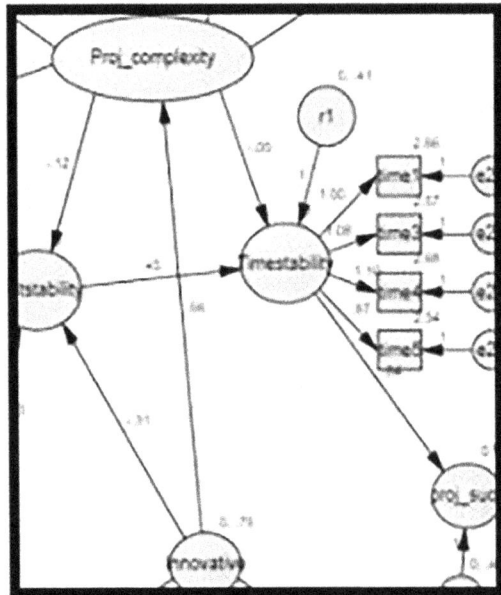

FIGURE 13.4   Full and partial mediation impacts in the model.

In this part of the mode (Figure 13.4), it is clear that cost stability partially mediates the impact of complexity on time stability, and cost stability fully mediates the impact of innovative knowledge creation on time stability. Project complexity fully mediates the impact of innovative knowledge creation on time stability, and time stability fully mediates the impact of complexity on project success and the impact of cost stability on project success.

All explained and interpreted relationships in this model are significant at the level of confidence of 95% and Alpha = 0.05. The following part of the study indicates the summary of model fit tables:

Model Fit Summary

TABLE 13.17
CMIN Significance Statistics

Model	NPAR	CMIN	DF	P	CMIN/DF
Default model	115	2305.804	587	.000	3.928
Saturated model	702	.000	0		
Independence model	72	10345.452	630	.000	16.421

In Table 13.17 the CMIN significance statistics add to the model fit.

TABLE 13.18
Baseline Comparisons

Model	NFI Delta1	RFI rho1	IFI Delta2	TLI rho2	CFI
Default model	.777	.761	.824	.810	.823
Saturated model	1.000		1.000		1.000
Independence model	.000	.000	.000	.000	.000

In Table 13.18 the Baseline comparisons add to the model fit.

TABLE 13.19
Parsimony-Adjusted Measures

Model	PRATIO	PNFI	PCFI
Default model	.932	.724	.767
Saturated model	.000	.000	.000
Independence model	1.000	.000	.000

In Table 13.19 the parsimony-adjusted measures add to the model fit.

TABLE 13.20
NCP

Model	NCP	LO 90	HI 90
Default model	1718.804	1574.939	1870.176
Saturated model	.000	.000	.000
Independence model	9715.452	9389.196	10048.121

In Table 13.20 the NCP = 1718.804 for the default model and adds to the model fit.

TABLE 13.21
FMIN

Model	FMIN	F0	LO 90	HI 90
Default model	5.750	4.286	3.928	4.664
Saturated model	.000	.000	.000	.000
Independence model	25.799	24.228	23.414	25.058

In Table 13.21 the FMIN statistics add to the model fit.

TABLE 13.22
RMSEA

Model	RMSEA	LO 90	HI 90	PCLOSE
Default model	.085	.082	.089	.000
Independence model	.196	.193	.199	.000

In Table 13.22 the RMSEA statistics are shown and are adding to the model fit.

TABLE 13.23
AIC

Model	AIC	BCC	BIC	CAIC
Default model	2535.804	2559.183		
Saturated model	1404.000	1546.714		
Independence model	10489.452	10504.089		

In Table 13.23 the AIC = 2535.804 for the default model, and the AIC for the saturated model = 1404.000. The interdependence model has an AIC of 10489.452, and this is adding to the model fit.

TABLE 13.24
ECVI

Model	ECVI	LO 90	HI 90	MECVI
Default model	6.324	5.965	6.701	6.382
Saturated model	3.501	3.501	3.501	3.857
Independence model	26.158	25.345	26.988	26.195

In Table 13.24 the ECVI statistics add to the model fit.

TABLE 13.25
HOELTER

Model	HOELTER .05	HOELTER .01
Default model	113	117
Independence model	27	28

Minimization: .016
Miscellaneous: .803
Bootstrap: .000
Total: .819

In Table 13.25 the bootstrapping statistics and the HOELTER statistics add to the model fit.

13.4.6 SEM FOR THE RISK-AVERSE KNOWLEDGE CREATION STYLE OF PROJECT MANAGEMENT

After checking the validity of the research constructs and the model fit. The following figure is explaining the path relationships among research variables after entering the impact of the risk-averse knowledge creation style to the model of relationships.

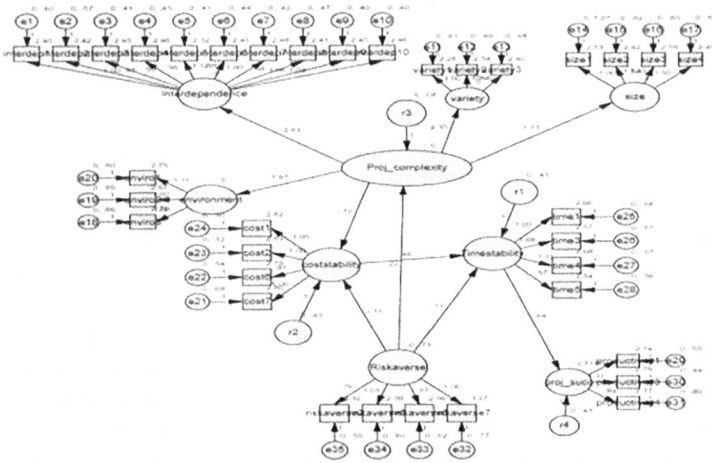

FIGURE 13.5 The SEM of the indirect impacts of complexity and risk-averse knowledge creation styles on project success.

In Figure 13.5 the minimum was achieved, Chi-square = 2179.414, Degrees of freedom = 553, and Probability level = .000. The following table illustrates the regression weights of the structure equation model of the impact of innovative knowledge creation on project complexity and performance success.

TABLE 13.26
Regression Weights

			Estimate	S.E.	C.R.	P	Label
Proj_complexity	<—	Riskaverse	.223	44.965	.005	.996	par_36
coststability	<—	Proj_complexity	−.704	141.916	−.005	.996	par_30
coststability	<—	Riskaverse	−.112	.064	−1.756	.079	par_34
Timestability	<—	coststability	.455	.070	6.536	***	par_35
Timestability	<—	Riskaverse	−.112	.053	−2.102	.036	par_37
proj_succ	<—	Timestability	.644	.079	8.119	***	par_25
variety	<—	Proj_complexity	2.346	472.606	.005	.996	par_26
size	<—	Proj_complexity	1.711	344.694	.005	.996	par_27
Interdependence	<—	Proj_complexity	2.827	569.579	.005	.996	par_28
environment	<—	Proj_complexity	1.868	376.204	.005	.996	par_29
interdep1	<—	Interdependence	1.000				
interdep2	<—	Interdependence	.972	.052	18.513	***	par_1
interdep3	<—	Interdependence	.993	.048	20.603	***	par_2
interdep4	<—	Interdependence	.984	.049	20.037	***	par_3
interdep5	<—	Interdependence	1.120	.051	21.810	***	par_4
interdep6	<—	Interdependence	.982	.050	19.588	***	par_5
interdep7	<—	Interdependence	.997	.049	20.399	***	par_6

TABLE 13.26 (*Continued*)
Regression Weights

			Estimate	S.E.	C.R.	P	Label
interdep8	<—	Interdependence	.982	.050	19.757	***	par_7
interdep9	<—	Interdependence	1.063	.050	21.399	***	par_8
interdep10	<—	Interdependence	1.026	.049	21.093	***	par_9
variety1	<—	variety	1.000				
variety2	<—	variety	.892	.075	11.842	***	par_10
variety3	<—	variety	.984	.072	13.668	***	par_11
size1	<—	size	1.000				
size2	<—	size	1.341	.145	9.220	***	par_12
size3	<—	size	1.334	.146	9.133	***	par_13
size4	<—	size	1.497	.160	9.371	***	par_14
enviro3	<—	environment	1.000				
enviro2	<—	environment	1.022	.106	9.642	***	par_15
enviro1	<—	environment	1.110	.109	10.153	***	par_16
cost7	<—	coststability	1.000				
cost6	<—	coststability	.968	.081	11.948	***	par_17
cost2	<—	coststability	.998	.080	12.502	***	par_18
cost1	<—	coststability	1.047	.082	12.754	***	par_19
time1	<—	Timestability	1.000				
time3	<—	Timestability	1.082	.090	11.977	***	par_20
time4	<—	Timestability	1.105	.096	11.554	***	par_21
time5	<—	Timestability	.969	.087	11.104	***	par_22
productivity1	<—	proj_succ	1.000				
productivity3	<—	proj_succ	1.114	.094	11.912	***	par_23
productivity4	<—	proj_succ	.840	.080	10.493	***	par_24
riskaverse7	<—	Riskaverse	1.000				
riskaverse6	<—	Riskaverse	.974	.083	11.700	***	par_31
riskaverse5	<—	Riskaverse	1.030	.083	12.372	***	par_32
riskaverse2	<—	Riskaverse	.760	.069	11.064	***	par_33

In Table 13.26 all coefficients are significant, and the following table summarizes the main relationships in this part of the study:

TABLE 13.27
Summary of the Main Relationships in the Model of the Study

			Estimate	S.E.	C.R.	P	Label
Proj_complexity	<—	Riskaverse	.223	44.965	.005	.996	par_36
coststability	<—	Proj_complexity	−.704	141.916	−.005	.996	par_30
coststability	<—	Riskaverse	−.112	.064	−1.756	.079	par_34
Timestability	<—	coststability	.455	.070	6.536	***	par_35
Timestability	<—	Riskaverse	−.112	.053	−2.102	.036	par_37
proj_succ	<—	Timestability	.644	.079	8.119	***	par_25

In Table 13.27, the model indicates that there is significant positive impact of risk-averse knowledge creation on the level of project complexity. The more is the risk-averse knowledge creation, the higher is the level of project complexity. An increase in the level of risk-averse knowledge creation by 1% increases the level of project complexity by 22.3%. The interpretation of this relationship is due to the increasing of precautions and safety components in the system of the project in a way that may increase its system complexity. On the other hand, taking less safety precautions may simplify the project and make it less complex. The more is the complexity of the project, means the less is the cost stability of the project. That is when we have more system components and lots of variety and interdependency in the project. The cost tends to exceed the planned costs, and nonconformance appears. If the project manager wants to control the project costs and keep it conforming to the planned costs, he should keep an eye on the level of project complexity and should maintain the simplicity of the project. The more risk-averse knowledge creation is found in the project, the less is the cost stability of the project. That is, the increasing of the calculations and precautions against risks may increase the stability and conformance of the cost to the plan significantly, but at the same time weakly not strongly. The manager of the project in this case would prefer losing some cost conformance for the sake of increasing the creation of risk-averse knowledge because he believes that it is important to create this type of knowledge to keep the project successful. The model tells us that every increase by 1% in the level of project complexity means a decrease by 70.4% in the level of cost stability and that every increase by 1% in the level of risk-averse knowledge creation decreases both the time stability and the cost stability equally by 11.2%. Taking more time and cost by the project manager to create the needed risk-averse knowledge is slightly but significantly increasing the nonconformance in the project cost and time stabilities. A 1% increase in the cost stability of the project causes an increase in the time stability by 45.5%, and this is relatively moderate to strong impact of cost stability on time stability. Time stability is strongly and significantly impacting the project success. A 1% increase in the time stability means 64.4% increase in the level of project success.

Model Fit Summary

TABLE 13.28
CMIN

Model	NPAR	CMIN	DF	P	CMIN/DF
Default model	112	2179.414	553	.000	3.941
Saturated model	665	.000	0		
Independence model	70	9529.205	595	.000	16.015

In Table 13.28 the CMIN statistics add to the model fit.

TABLE 13.29
Baseline Comparisons

Model	NFI Delta1	RFI rho1	IFI Delta2	TLI rho2	CFI
Default model	.771	.754	.819	.804	.818
Saturated model	1.000		1.000		1.000
Independence model	.000	.000	.000	.000	.000

In Table 13.29 the baseline comparisons add to the model fit.

TABLE 13.30
Parsimony-Adjusted Measures

Model	PRATIO	PNFI	PCFI
Default model	.929	.717	.760
Saturated model	.000	.000	.000
Independence model	1.000	.000	.000

In Table 13.30 the parsimony-adjusted measures add to the model fit.

TABLE 13.31
NCP

Model	NCP	LO 90	HI 90
Default model	1626.414	1486.624	1773.720
Saturated model	.000	.000	.000
Independence model	8934.205	8621.342	9253.485

In Table 13.31 the NCP statistics of the SEM test add to the model fit

TABLE 13.32
FMIN

Model	FMIN	F0	LO 90	HI 90
Default model	5.435	4.056	3.707	4.423
Saturated model	.000	.000	.000	.000
Independence model	23.764	22.280	21.500	23.076

In Table 13.32 the FMIN statistics add to the model fit in SEM results.

TABLE 13.33
RMSEA

Model	RMSEA	LO 90	HI 90	PCLOSE
Default model	.086	.082	.089	.000
Independence model	.194	.190	.197	.000

In Table 13.33 the RMSEA statistics of the SEM analysis results add to the model fit.

TABLE 13.34
AIC

Model	AIC	BCC	BIC	CAIC
Default model	2403.414	2425.508		
Saturated model	1330.000	1461.178		
Independence model	9669.205	9683.013		

In Table 13.34 the AIC statistics of the SEM analysis add to the model fit.

TABLE 13.35
ECVI

Model	ECVI	LO 90	HI 90	MECVI
Default model	5.994	5.645	6.361	6.049
Saturated model	3.317	3.317	3.317	3.644
Independence model	24.113	23.333	24.909	24.147

In Table 13.35 the ECVI statistics of the SEM model add to the model fit.

TABLE 13.36
HOELTER

Model	HOELTER .05	HOELTER .01
Default model	113	117
Independence model	28	29

Minimization:	.025
Miscellaneous:	2.011
Bootstrap:	.000
Total:	2.036

In Table 13.36 the bootstrapping results as well as the HOELTER results of the SEM testing add to the model fit.

13.4.7 STRUCTURAL EQUATION MODELING FOR EXPERIENTIAL KNOWLEDGE CREATION STYLE OF PROJECT MANAGEMENT

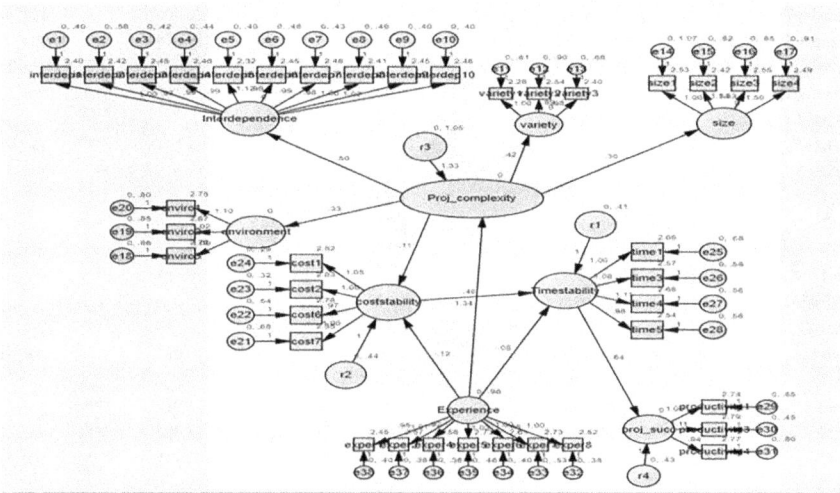

FIGURE 13.6 SEM model of the experiential knowledge creation style on the relationship between project complexity and project success.

In Figure 13.6 the minimum was achieved, Chi-square = 2358.779, Degrees of freedom = 657, Probability level = .000.

TABLE 13.37
Regression Weights

			Estimate	S.E.	C.R.	P	Label
Proj_complexity	<—	r3	1.333	134.444	.010	.992	par_39
Proj_complexity	<—	Experience	1.340	172.356	.008	.994	par_40
Coststability	<—	Proj_complexity	−.109	14.046	−.008	.994	par_30
Coststability	<—	Experience	−.119	.057	−2.107	.035	par_38
Timestability	<—	coststability	.456	.071	6.442	***	par_37
Timestability	<—	Experience	−.083	.043	−1.922	.055	par_41
proj_succ	<—	Timestability	.643	.079	8.099	***	par_25
Variety	<—	Proj_complexity	.415	53.427	.008	.994	par_26
Size	<—	Proj_complexity	.304	39.046	.008	.994	par_27
Interdependence	<—	Proj_complexity	.502	64.541	.008	.994	par_28
environment	<—	Proj_complexity	.330	42.476	.008	.994	par_29
interdep1	<—	Interdependence	1.000				
interdep2	<—	Interdependence	.968	.052	18.492	***	par_1
interdep3	<—	Interdependence	.991	.048	20.625	***	par_2

(Continued)

TABLE 13.37 (*Continued*)
Regression Weights

			Estimate	S.E.	C.R.	P	Label
interdep4	<—	Interdependence	.986	.049	20.168	***	par_3
interdep5	<—	Interdependence	1.123	.051	22.016	***	par_4
interdep6	<—	Interdependence	.980	.050	19.610	***	par_5
interdep7	<—	Interdependence	.994	.049	20.400	***	par_6
interdep8	<—	Interdependence	.983	.049	19.874	***	par_7
interdep9	<—	Interdependence	1.063	.049	21.502	***	par_8
interdep10	<—	Interdependence	1.023	.048	21.113	***	par_9
variety1	<—	variety	1.000				
variety2	<—	variety	.891	.075	11.811	***	par_10
variety3	<—	variety	.982	.072	13.618	***	par_11
size1	<—	size	1.000				
size2	<—	size	1.340	.145	9.230	***	par_12
size3	<—	size	1.329	.146	9.131	***	par_13
size4	<—	size	1.498	.160	9.389	***	par_14
enviro3	<—	environment	1.000				
enviro2	<—	environment	1.018	.106	9.589	***	par_15
enviro1	<—	environment	1.105	.110	10.087	***	par_16
cost7	<—	coststability	1.000				
cost6	<—	coststability	.972	.082	11.917	***	par_17
cost2	<—	coststability	1.004	.080	12.482	***	par_18
cost1	<—	coststability	1.051	.083	12.741	***	par_19
time1	<—	Timestability	1.000				
time3	<—	Timestability	1.076	.090	11.920	***	par_20
time4	<—	Timestability	1.110	.096	11.557	***	par_21
time5	<—	Timestability	.976	.088	11.132	***	par_22
productivity1	<—	proj_succ	1.000				
productivity3	<—	proj_succ	1.113	.093	11.915	***	par_23
productivity4	<—	proj_succ	.841	.080	10.496	***	par_24
exper8	<—	Experience	1.000				
exper7	<—	Experience	.853	.048	17.878	***	par_31
exper6	<—	Experience	1.003	.047	21.383	***	par_32
exper5	<—	Experience	1.031	.049	20.946	***	par_33
exper4	<—	Experience	.922	.044	20.968	***	par_34
exper2	<—	Experience	1.013	.047	21.482	***	par_35
exper1	<—	Experience	.946	.046	20.601	***	par_36

In Table 13.37 all coefficients are significant, and the following table summarizes the main relationships in this part of the study:

TABLE 13.38
Summary of the Significant Relationships in the Model of the Study

Proj_complexity	<—	**Experience**	1.340	172.356	.008	.994	**par_40**
Coststability	<—	Proj_complexity	−.109	14.046	−.008	.994	par_30
Coststability	<—	Experience	−.119	.057	−2.107	.035	par_38
Timestability	<—	coststability	.456	.071	6.442	***	par_37
Timestability	<—	Experience	−.083	.043	−1.922	.055	par_41
proj_succ	<—	Timestability	.643	.079	8.099	***	par_25

In Table 13.38, for the experiential knowledge creation model, there is a significant impact of the experiential knowledge creation on the level of cost stability, and this is logic because depending on the history of success for the project management saves costs of try and error and makes it easier for the project manager to repeat the same previously tried solutions in similar situations. The increase in the experiential knowledge creation by 1% decreases the cost stability by 11.9% at the level of confidence of 95% and the Alpha = 0.05. There is relatively weak relationship and impact of the creation of experiential knowledge on the stability of costs. More experiential knowledge may increase the nonconformance in the cost because of the need to copy and paste similar solutions again, and this is adding to the cost of implementation. The increase in the cost stability increases the time stability, and hence time stability will increase the project success, and this is a sequential positive relationship of impacts in which the time stability fully mediates the impact of the cost stability on the project success.

Model Fit Summary

TABLE 13.39

CMIN

Model	NPAR	CMIN	DF	P	CMIN/DF
Default model	122	2358.779	657	.000	3.590
Saturated model	779	.000	0		
Independence model	76	11489.456	703	.000	16.343

In Table 13.39 the CMIN statistics add to the model fit of predicting the variance in experiential rigidity knowledge creation impact.

TABLE 13.40

Baseline Comparisons

Model	NFI Delta1	RFI rho1	IFI Delta2	TLI rho2	CFI
Default model	.795	.780	.843	.831	.842
Saturated model	1.000		1.000		1.000
Independence model	.000	.000	.000	.000	.000

In Table 13.40 the baseline comparisons add to the model fit of the SEM testing results

TABLE 13.41

Parsimony-Adjusted Measures

Model	PRATIO	PNFI	PCFI
Default model	.935	.743	.787
Saturated model	.000	.000	.000
Independence model	1.000	.000	.000

In Table 13.41 the parsimony-adjusted measures of the SEM model add to the model fit.

TABLE 13.42
NCP

Model	NCP	LO 90	HI 90
Default model	1701.779	1557.265	1853.819
Saturated model	.000	.000	.000
Independence model	10786.456	10442.487	11136.839

In Table 13.42 the NCP results add to the model fit.

TABLE 13.43
FMIN

Model	FMIN	F0	LO 90	HI 90
Default model	5.882	4.244	3.883	4.623
Saturated model	.000	.000	.000	.000
Independence model	28.652	26.899	26.041	27.773

In Table 13.43 the FMIN statistics add to the model fit of testing the impact of experiential rigid knowledge creation style on project complexity.

TABLE 13.44
RMSEA

Model	RMSEA	LO 90	HI 90	PCLOSE
Default model	.080	.077	.084	.000
Independence model	.196	.192	.199	.000

In Table 13.44 the RMSEA statistics of the SEM test results add to the model fit.

TABLE 13.45
AIC

Model	AIC	BCC	BIC	CAIC
Default model	2602.779	2629.067		
Saturated model	1558.000	1725.851		
Independence model	11641.456	11657.832		

In Table 13.45 the AIC statistics show the model fit and that it can be used for prediction significantly.

TABLE 13.46

ECVI

Model	ECVI	LO 90	HI 90	MECVI
Default model	6.491	6.130	6.870	6.556
Saturated model	3.885	3.885	3.885	4.304
Independence model	29.031	28.173	29.905	29.072

In Table 13.46 the ECVI statistics add to the model fit of the SEM results.

TABLE 13.47

HOELTER

Model	HOELTER .05	HOELTER .01
Default model	123	127
Independence model	27	28

Minimization:	.156
Miscellaneous:	4.070
Bootstrap:	.000
Total:	4.226

In Table 13.47 the bootstrapping results as well as the HOELTER results add to the model fit of using the experiential knowledge creation in predicting the global level of project complexity.

13.5 CONCLUSION, RECOMMENDATIONS, AND SUGGESTIONS FOR FURTHER RESEARCH

The research answers the question of how to predict the level of project success at different levels of complexity and using different styles of knowledge creation by project managers. The study creates a validation methodology of the variables measurements and then tests the significance of three models. In each one of the three models there is one knowledge creation style of project management tested. The three styles of knowledge creation are found to have significant impacts on the cost stability of the project. The cost stability in the three models is positively impacting the time stability in the project, and the time stability of the project is positively and strongly impacting the level of project success in the three models. Before using the structural equation modelling, a multiple regression model illustrated that there is a positive significant and direct impact of the knowledge creation styles on the level of project complexity and the SEM models uncover the significance of indirect impacts as well. Number of full and partial mediation impacts was discovered in the models including the mediation of the cost stability, time stability, project complexity, innovative knowledge creation style, and risk-averse knowledge creation style. These mediation impacts can be used by project managers to understand, predict, and control the level of project success especially for the delayed projects and keep them successful despite of the delay and both time and cost nonconformities. The project manager can use the impact of the innovative knowledge creation to control the deviations in the cost and time. The level of complexity is also another significant factor in controlling the cost and the time of the project. The cost and time stability are the main reflecting constructs upon which the success of the project depends. They are mediators in the tested models and are directly affected by other variables. Studying the time and cost stabilities without studying the interrelationships between time, cost, and other variables like the complexity and the knowledge creation styles keeps the understanding of their impacts on the project success vague and not fully understood. Managers of projects will not only use one style of knowledge creation all the time. They may change the style they want to use according to the need of using it. In the research of Abousamra (Abousamra, 2017), it was found that the project manager may start by using the risk-averse style of knowledge creation and then follow it with the innovative one and not the opposite. In the current research both styles are found to have a significant impact on increasing the level of project complexity directly. At the same time they are not directly impacting the project success. The impact of using the knowledge creation style does not appear directly on the level of project success. They indirectly impact the level of project complexity or the time or cost stability of the project and then the time stability is the end impactor in the chain of impacts that has the direct impact on the project success. The mirror of the project success is the conformance between the planned time and the actual one; however having delays in the project performance does not mean that it is a failure. The delay in the project implementation is interpreted in this research by introducing more precautions or new innovative knowledge or even reusing successfully experience previous solutions and all these actions are related to the knowledge creation styles and are impacting the level of project complexity. An increase in the project complexity is reflected on consuming more time or more cost or even more of both of them and hence causing the delay in the implementation time of the project. The project manager will need to understand this sequence of significant impacts to predict and know how to take the right decision that can bring back the time to the conformance and by doing this increase the level of the project success. The results of the current research are contributing to the management of the delayed projects or how to manage the low levels of project success or even the project failures using the time and cost conformance control and by controlling the level of project complexity and using the right knowledge creation style for each project. The dealing with the KPIs and the keeping of the conformance as an isolated decision is no more enough to achieve the success and increase it for the project after considering the results of the current research. The conclusions of this research are generalized to the population of the study and all projects and project managers included in the population of the small- to medium-sized projects are possible to use the results of this study. This adds to the value of the current study. The results of this research are limited to the three knowledge creation styles and the used determinants of the level of complexity that are the size, variety, interdependency, and environmental complexity. There is possibility to have different results when using different measures of project complexity, and this is suggested to be included in the future research. It is also suggested to try to collect the same data from different contexts and fields of research and to apply the same research model on different industrial sectors to test if it can work again successfully to predict the level of the project success or not. Another suggestion for further research is to consider the use of a team of project managers to manage one project. In the current research the focus is on the projects that are managed by one manager only and not a team of managers. The use of teams is increasingly important, and that is why it is recommended to apply this study on the different combinations of project managers' knowledge creation styles in managing the project. For example, having innovative with risk-averse is better or having the risk-averse with the experiential is better. Also, the sequence of roles in managing the project needs further research. Like is it better to start with the risk-averse knowledge creation style and then follow it by the innovative one or vice versa. The impacts of the social, economic, financial, technical, and political factors are excluded from the scope of the current study and are recommended to be included in future research as they may have an impact on the results of the study but are avoided in the current research for the purpose of simplification.

REFERENCES

Abousamra, R. (2017). Qualitative Analysis of the Innovative Knowledge Creation Style of Project Managers and its Relationship with Performance Stability in IT Projects. *International Journal of Information Technology and Language Studies, 1.*

Abousamra, R. (2022, May). Quantitative Classification of Cognitive Behaviors for Industrial Projects' Managers in the MENA Region. In IEEE (Ed.), *2022 8th International Conference on Information Technology Trends (ITT)* (pp. 130–133). IEEE.

Asiedu, E. M. (2022). A comprehensive Assessment of Time Overruns in Ghanaian Telecom Cell Site Construction. *Cogent Engineering*, 9(1), 2109250.

Dao, B. K. (2022). Developing a Logistic Regression Model to Measure Project Complexity. *Architectural Engineering and Design Management*, 18(3), 226–240.

Dartey-Baah, S. K. (2022). *The Relationship between Project Complexity and Project Success and the Moderating Effect of Project Leadership and Roles in the Construction Industry of an Emerging Economy.* (Doctoral dissertation, Stellenbosch: Stellenbosch University).

Engström, A. J. (2022). Knowledge Creation in Projects: An Interactive Research Approach for Deeper Business Insight. *International Journal of Managing Projects in Business.* (ahead-of-print).

Hou, T. L. (2022). Linking Knowledge Search to Knowledge Creation: The Intermediate Role of Knowledge Complexity. *Management Decision.* (ahead-of-print).

Lalic, D. C. (2022). How Project Management Approach Impact Project Success? From Traditional to Agile. *International Journal of Managing Projects in Business*, 15(3), 494–521.

Nisula, A. M. (2022). Organizing for Knowledge Creation in a Strategic Interorganizational Innovation Project. *International Journal of Project Management*, 40(4), 398–410.

Samra, R. A. (2016, July). Exploring Chaotic Performance in Projects and its Relationship with Knowledge Creation Process. In *Proceedings of the The 11th International Knowledge Management in Organizations Conference on The Changing Face of Knowledge Management Impacting Society* (pp. 1–8). Association for Computing Machinery.

Taherdoost, H. A. (2022). Exploratory Factor Analysis; Concepts and Theory. *Advances in Applied and Pure Mathematics*, 27, 375–382.

Zaman, U. F.-P. (2022). A Stitch in Time Saves Nine: Nexus between Critical Delay Factors, Leadership Self-Efficacy, and Transnational Mega Construction Project Success. *Sustainability*, 14(4), 2091.

14 Deep Learning Method for Anomaly Detection in Cyber Physical Systems (CPS) and Social Networks

Reeja S R, Yaseen Ishfaq,
and Tamanna Jena Singhdeo

14.1 INTRODUCTION

The growth rate of CPS has increased due to the growth of Internet and technology. The application area of CPS is merged with applications of IOT. The waste areas are agriculture, manufacturing, medical, engineering, and technology, which expands the growth rate of CPS market. To avoid human error in technology, we use CPS in business. The main glitch in CPS market growth is attack in security. There are DL-based solutions for detecting intrusions that can be used to achieve a CPS with high availability. When compared to conventional intrusion detection techniques, DL-based intrusion detection solutions have a lower percentage of alerts, are capable of learning and adapting to changing attack patterns, and can discover new or unknown threats. Deep neural networks are conscience in unsupervised learning and can retain dynamic detection in response to countermeasures.

Deep neural networks use a limited number of datasets that are labelled with attack and normal states in semisupervised learning to dynamically learn from and react to situations in both familiar and unfamiliar settings with a low rate of erroneous successes. Trouble anticipation is the technique of anticipating an error in a component, which can be utilized to prevent system failure and, as a result, prevent unintended disaster and significant loss. Monitoring and evaluation is closely related to trouble anticipation, with the latter's results potentially serving as input data. Process optimization is related to cyber systems, in which each component's operation is broken down into a number of tasks and steps. To suit the real-time requirements of CPS, an artificial controller can be created for efficiently controlled job allocation and computation. A thorough analysis of numerous intelligence scheduling algorithms created utilizing ML/DL algorithms is given in this direction. To determine whether the appropriate QoS is being provided, QoS analysis entails accessing QoS-related data from multiple CPS subsystems and analyzing it.

DOI: 10.1201/9781003343332-14

Resource distribution and QoS analysis are closely related. Additionally, dynamic resource allocation takes into account different communication and computing needs for optimal CPS functioning. The widely used round-robin, decreased waiting time, and other scheduling algorithms are equivalent to the DL method, which also outperforms them in many situations.

As a positive shuttered network, CPSs incorporate the evolution of optimum phase instability and vibrant stages and fundamentally take such new forms into account while appealing for potential privacy protecting frameworks. For the purpose of studying the defense against assaults directed against CPSs, many monitoring mechanisms were presented. The evolution of confidentiality protection schemes and vulnerability scanning in CPSs made extensive use of data analysis, advanced analytics, and analytical measures. Therefore, gathering and analyzing the pertinent data is the main challenge in building an abnormality detection approach. Due to complexity in nature, the detector required more analysis and cognitive content. A new parametric method should be used to detect the anomaly in CPS.

Deep learning method is an efficient method to detect the anomaly in CPSs. To find the errors in various CPS, the survey shows various CNN architecture. Quality control is intimately associated to hazard forecasting, with the latter's results potentially serving as raw data. Strategy that makes it related to information systems, in which each component's operation is broken down into a variety of procedures and activities. Methodology can be used for unique systems integration, according to ML. In contrast, an automated dispatcher could be created to handle activity allocation and execution successfully in order to adhere to CPS's actual requirements. A thorough analysis of numerous cognitive optimization techniques created utilizing ML/DL procedures is given in this regard. To determine whether the appropriate quality is being provided, quality assessment entails accessing Quality-received information from multiple CPS modules and evaluating it.

Anomaly detection (AD) method is used to detect unwanted delusions. To identify regular levels of CPS, we can use regulation, circumstance assessment method, and mathematical models [1]. Due to the expansion of diverse applications in world norms and essential foundations for the development of the region, multiple scientists, legislative policymakers, professional workers, and practitioners are currently working toward CPSs. Additionally, since defense systems like military ships, remotely operated objects, and surveillance drones are crucial to CPSs, military officials in different nations depend greatly on the normal development of CPSs. Although investigation on protective measures like security features, identity verification, password management, security measures, and threat prevention for the safeguard of CPSs has advanced, there are still considerable hurdles of confidentiality because it is difficult to keep track of the computer system and natural assets. But the user should know, were they to use such models.

Cognitive content is not required for the model. It is not easy to detect the core parameter of CPS. Network security can be detected with the help of the intruder [2]. Perpetual behavior of CPS can be detected with the help of edibleness [3]. To scheme the CPS system the author used to describe the meaning of code by execution [4]. Several applications of artificial intelligence methods, such as algorithms with intelligence for sensors and controls, etc., have been extensively applied in the effective and responsive management plan of CPS system to be able to satisfy the require-ments of the present development surrounding. However, there are still a number of difficulties with time-series anomaly identification in CPSS. First, abnormalities are relatively uncommon in genuine CPSSs circumstances. Due to the vast volume of streaming data, it is extremely difficult and stagnant to label data, making it prac-tically unattainable for us to obtain information with a favorable and unfavorable classification proportion for feature learning work.

Another significant issue in the procedure of CPSS anomaly identification is the fact that the response variable data in CPS system gathered through tracking is gen-erally more complicated, including non-Gaussian disturbance or complicated data pattern. This form of data is typically more difficult to model than simple smoothed response variable and therefore attracts minimal study attention. Modelling complex response variable and researching the associated anomaly detection techniques are necessary to be able to enhance the entire plan of managing CPS system. Additionally, monitoring device failure, network transmission failures, etc. make the loss of mon-itoring data unavoidable. This is a significant issue that affects current CPSSs, and the influence that missing data has on data collection and model training represents a significant research challenge [5]. In particular, the prognostication channel method typically anticipates the value of the response variable for the following response as well as work with the forecast noise to find threads, however it should not be used with unreliable response variable. Dynamic infrastructure techniques, however, fre-quently employ reconstruction error to spot anomalies following training of neural networks to understand the allocation of typical response time.

14.2 LITERATURE SURVEY

As technology and science evolved in recent years, researchers began to use more than just data mining techniques to detect aberrant data traffic in commercial CPS. These methods have produced effective attack recognition rate in corporate CPS. Threats can be divided into assault and defect. The applications of anomaly detections are specified in [6]. In this paper the author narrates various deep learning method to detect anomaly in CPS systems. RNN, CNN, and other modified models are used. The change between the test and actual data is mentioned as anomaly gain. Anomaly gain is calculated with respect to the fault prediction, fault modification, and labels. This gain will give optimized outcome. The various techniques used in papers [7–10] are narrated in Table 14.1. Here we list out deep learning and CPS only.

Figure 14.1 shows the taxonomy diagram with various anomalies, methods to find the programming method and estimation methods. Various anomalies are classified into two, as discussed in the previous paragraph. Finding methods are classified into three. Programming can be done with respect to simulation, real data, and test data. We evaluate the assault and fault using situational analysis and extrapolation. A study of telemetry data is used to identify unknown in CPS activity. This could be utilized for technology malfunctioning, surveillance vulnerability scanning, etc. Cyber means determining different informational challenges which might lead the design meth-odology into dysfunction and then compromise actual device's intended intent. This paper presents a variety of ML strategies for identifying and defining vulnerabilities

TABLE 14.1

Various Techniques, Applications, CPS Used or Not and Gain Is Narrated

Related work	Techniques	CPS	Gain
[7]	Code scanning	Partial	App security and privacy
[8]	based on a model dendrogram	Yes	Cyberrisk evaluation
[11]	Physical mode	Yes	Anomaly detection (AD)
[9]	DL	Yes	AD
[1]	strategies should be adopted and interference identification	Yes	AD
[12]	Knowledge and behavior fault recognition approach	Yes	AD
[10]	Detecting fault, ML	Yes	Cyber security

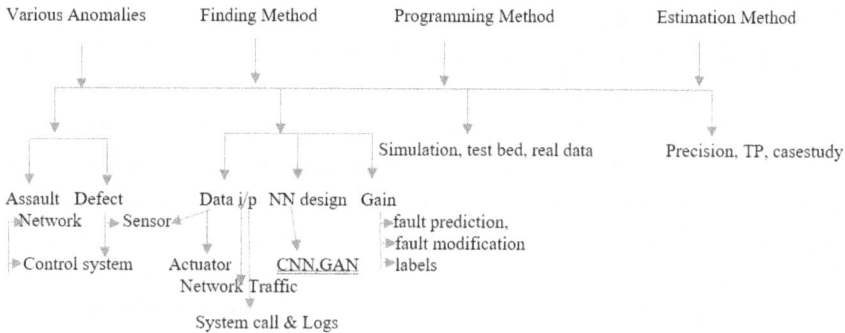

FIGURE 14.1 Taxonomy diagram.

that have been published in the latest studies. Trouble forecast is the technique of anticipating an error in a device, which can be utilized to prevent computer glitch and, as a result, prevent unintended emergency situations and significant reduction.

The main threats of data are private and financial-related data. The CPS data will be noticed as well as such type of data sent to data store and manages the assaultant entry. With the help of network layer, the sensor or actuator report are sent to the data stores by using control system commands. Through this, zero and level one assault can be managed. Well-established protection system lack will cause the assault in CPS. Some devices, the assaultant will place the malware in a control system which may cause damage to that device [2]. Failure in control signal also will cause the assault. During that time, the assaultant can enter into the system and get control over the devices. Unidentified, assault may occur in CPS. Examples are spacecraft [9] and PLC machines [10].

Finding methods are classified into three. They are data i/p, NN design, and gain. Data input is depended up on the type of anomaly. Mainly they are categorized into four. These types of findings use semi and unsupervised learning techniques. In a

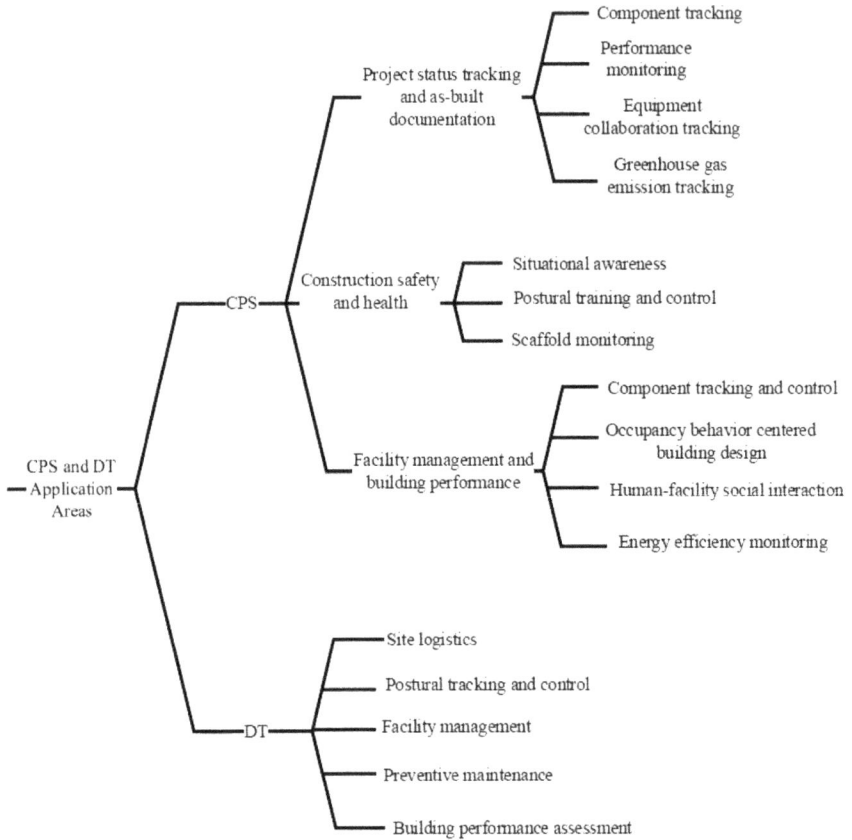

FIGURE 14.2   Areas of applications.

fault prediction, the anomaly data is taken from value and is predicted by the sensor. In fault modification, the anomaly data will be modified with the help of the threshold. We will be able to label the data with help of prediction mechanism. The areas of applications are mentioned in Figure 14.2.

The programming and estimation methods will give efficient anomaly detection method. During programming real and test bed data are collected and applied to CPS systems. Before applying to CPS, the real data will be tested with the help of the simulator. In health science and automated cars or in driverless cars, finding anomalies is a very difficult task. We can neither rely on simulated data because it may be harmful for the human nor the real data because cost is more. In such cases, we are identifying the anomaly manually. This can be done through the normal and real data [3] or follow some pattern to get the data [13] or test multiple varieties of data in the simulator [4]. Different types of faults can be evaluated with the help of the matrix fault finding operation [14]. The various anomaly detection survey is conveyed in this paper [15]. In industry-oriented applications, our method detects both assault [6] and defect [7]. Due to overloading and complexity of a machine, occurring of assault and defect will be more in industry machines. Such systems anomaly occurs through sensor and actuators. Change in anomaly can be detected with test bed [8].

Deep learning method will be a solution to find the anomalies. Through deep learning technique, we can improve performance, remove noise, modify feature extraction method and threshold. Various DLAD methods, techniques, various types of assault and disadvantages are narrated in Figure 14.3. The major disadvantages of various methods are noise, learning parameters, and false positive prediction. Our method improved all the previous disadvantages. Some show the smart grid application, others show the ITs DLAD application, and the remaining shows the Aircraft DLAD applications.

IDS has been proposed by authors in [16]. Utilizing statistical and pattern information extraction. For specific network aspects like the traffic layer, trans traffic layer, and packet layer, authors created a set of rules. Additionally, past system behavior is used in statistical analysis that is done to derive long-term data. These measurements were used to process the finding of abnormalities and patterns that don't match the recoded data. Authors in [17] used linear parameter varying (LPV) CPS to assess fake data injection attack situations. Based on the LPV's dynamical system, a framework model is developed. This can be used to forecast the anticipated measurement of future sensors. Anomalies are identified based on a particular threshold.

A novel network security strategy based on similarity learning was proposed by [18]. The suggested model enhanced the effectiveness of vulnerability scanning by

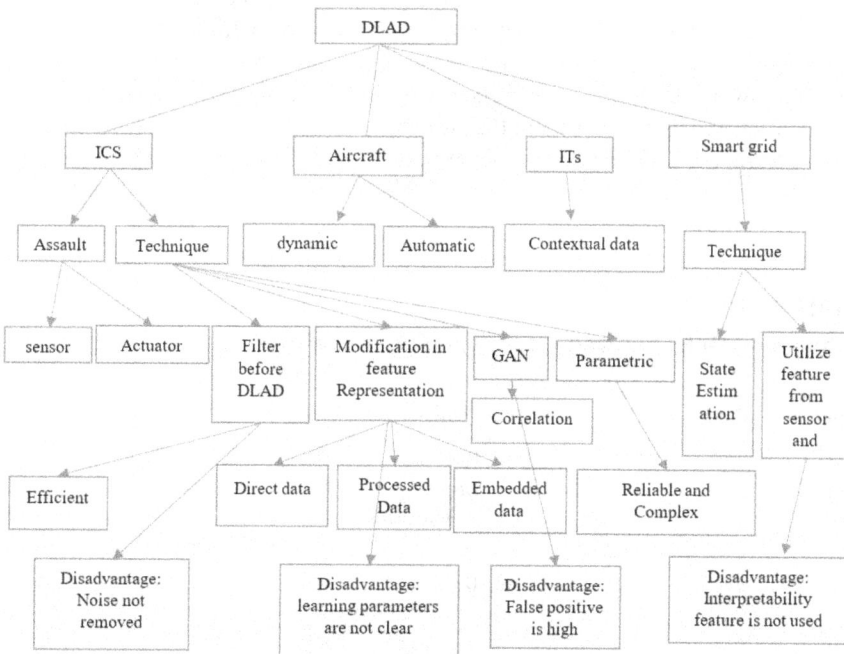

FIGURE 14.3 Various DLAD method, technique, assault, and disadvantages.

combining generative model with a trigonal deep net. During the training phase, both non-dependent generative models were retrained using both regular and irregular network. To generate characteristics descriptors for the congestion, the trigonal deep net was then trained. This depiction of features had the advantage of forcing the incoming congestion that will very certainly be retrieved in accordance with its reconstruction of the same category and far from its autoencoder-based compilation of the various categories. During the prediction stage, the autoencoder restored the latent space, and congestion in the validation set is classified into a group corresponding to the flow that was the most accurate rebuilding of the arrangement during the prediction stage after the autoencoder restored the latent space. The test data was assigned to the group associated with the stream that was the closest rebuilding of the fresh stream. By utilizing the embeddings acquired during the training phase, the forecast stage was successful in identifying fraudulent Internet usage with high predictive quality.

An overview of associated key techniques for spyware classification and identification in the IoT communication environment for healthcare CPS can be found in Table 14.2. The study of fuzzy-based edge computing spyware recognition and classification is presented by [19]. Even though the study is displayed using numerous extensive information, the scenario is based on machine code background subtraction. For the detection of Android and IoT malware, the CNN-based technique is suggested [20]. These characteristics are susceptible to obscuration. Extraordinary gradient bolstering was used to enhance the machine code feature with the goal of identifying later explored IoT ransomware [5].

The study's performance [21] for malware classification was lower than expected, and it demonstrates that longer byte sequences, more than 1,024 are needed to get better results. CNN uses CFG and printable string data from ELF files as features for IoT malware detection. The significance of IoT detecting attacks across platforms was also covered in the report. For the purpose of detecting IoT malware utilizing the CFG feature, the SVM-based technique is suggested [5]. Malware, which leverages a constrained virtualization environment for its operation, can go around the system.

TABLE 14.2
Summary—Detection and Classification of Malware

References	Platform	Data type	Length of byte	Objectives	Approach	Dataset size	Classes	Accuracy
[21]	IOT	CFG	–	Detection	CNN	11200	2	98.7
[20]	IOT	CFG	–	Detection, classification	CNN	6000	2, 3	99.7
[19]	IOT	opcode	–	Detection	Fuzzy	1207	2	99.8
[22]	IOT	Byte	128–1024	Detection, classification	SVM	4169	8	99.9
[5]	IOT	CFG	–	Detection	SVM	5476	2	99.1
[23]	IOT	opcode	–	Detection	XGBoost	1207	2	94.5

Support vector machine (SVM)-based IoT malware detection and CPU architecture categorization take into account the 1024-byte entry point sequences from a variety of CPU architectures in ELF files [22].

In conclusion, as time and technology developed, both machine learning techniques and deep learning approaches produced major advancements in the field of intrusion detection.

14.3 PHASES OF DLAD IN CPS

Choosing Dataset: Dataset is taken from iTrust lab water treatment and electric dataset [8]. Figure 14.4 shows the lab equipment. The data mentioned about the quantity of power measurement. Water level of home will be detected using sensors. The simulation of such system is shown in Figure 14.5. Due to the sensor activity, the assaultant can assault the entire system. They can add chemical to the system and impurify the water. The assaultant can assault the network through PLC control. The dataset consists of eleven days' data. Under these data four days data consists of assault that happened in sensors, actuator, as well as in network traffic. The assault happened every ten minutes of the interval.

During preprocessing, we considered sensor and actuator values. To normalize the features, we applied scaling and averaging to all the features.

Data Acquisition: Real data are collected with the assault and without the assault. During data acquisition, we used high performance computing to increase the speed and to improve the performance, which improved the packet processing during data acquisition. In order to identify and detect unusual activity in CPS, various techniques are used due to the complication of a huge CPS and the enormous number of respondents gathered. We gather information from the higher layer, intermediate layer, and lower layer for this purpose. We gather protection information from the

FIGURE 14.4 Lab equipment.

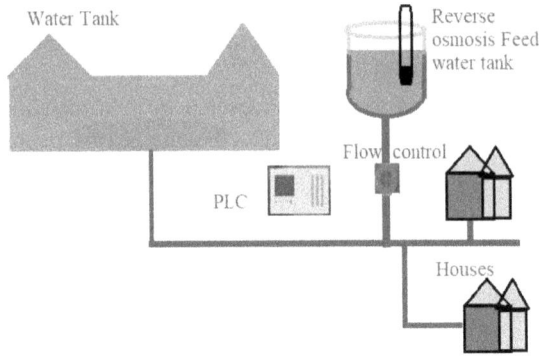

FIGURE 14.5   Simulation of the water distribution system.

sensor placed in the real world at the higher layers to acquire situational and psychological features. Data on datagrams and potentially dangerous network environments are gathered from the intermediate nodes. This data contains packet headers, SNR, broadcast network details, packet delivery frequency, and the average round-trip time from hardware to the software. Data about the computer system responsible for harming hardware and software systems, such as computation and storage, document storage privileges, and operation processing, are gathered at the lower layer. It is required to process well and integrate this knowledge before building the suggested framework when CPS include varieties of individual parameters such as actual as well as the Internet, that means datalogger characteristic, information transmission attribute, and smart phone function. While the environment of facts acquired over the website known as a cyber world. The workspace is the gathering of stored information from numerous sources.

Data Preprocessing: Before constructing the suggested skeleton, this data must first be preprocessed and integrated since CPS system have a variety of unique aspects from both terrestrial and virtual, comprising inertial sensors characteristic, data transmission attribute, and smartphone app function. While the workspace is the accumulation of local information from different resources, virtual world is the realm of information collected from the resource through the online. The gaussian mixed technique (GMT) is a probability-based method for representing information from probability parameters. Systems for generating different variables from terrestrial, virtual, and connection environment, giving them multivariate distinctive traits. The GMT is made up of a number of crucial components that combine data from features that were obtained from real, webbing, and online app systems as well as the real, transport, and smartphone app components; all details are regarded as being linear.

Learning and Training: We prepared the data for testing and training. The data is formulated into 70 percentage training data and the 30 percentage of testing data. To process the training occurrences, we used SVM method. For the training we have used CUDA cores-640, Base clock 1020MHz, Boost Clock 1085 MHz, and GTX 750 Ti GPU card.

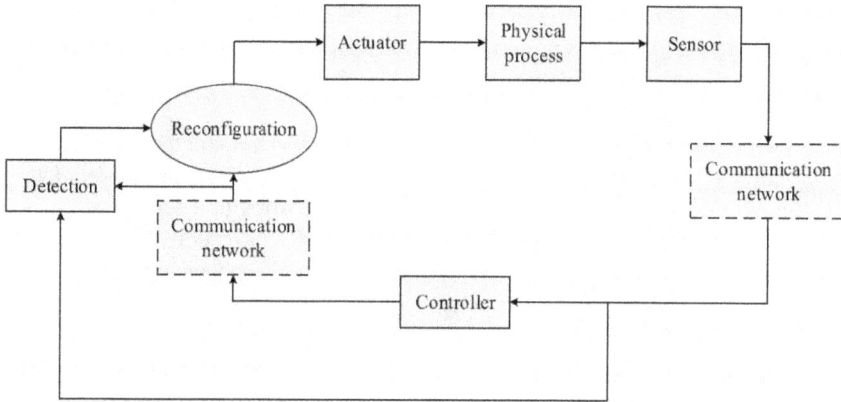

FIGURE 14.6 Structural platform for reliable management.

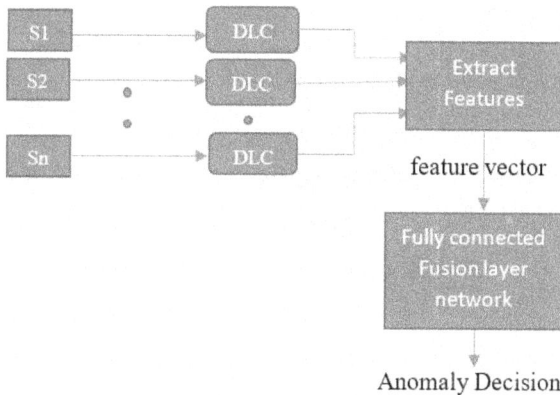

FIGURE 14.7 Process flow diagram.

Detection: Figure 14.6 shows the reliable detection management system, sensors, and actuator data are continuous variable. All the sensors make its own data $(S_1, S_2 \ldots S_n)$. These sensor data are trained by using deep learning classifiers (DLC). During the next steps, we extract required features from each sensor or actuator $(F_{e1}, F_{e2} \ldots F_{en})$. Figure 14.7 shows the process carried out. These extracted features are fed in to one feature Vector (F_v) with uniformity $(0,1)$. During the next step, we connect these vectors to fusion layer to produce the outcome. The outcome will give the anomaly detection score.

A confederal deep learning framework is the process model under examination, and it primarily consists of three types of firms: a confidence agency, a cloud provider, and K manufacturing units.

- Confidence agency: The confidence agency is tasked with setting up a confidential information flow between the data center and each manufacturing unit as well as rebooting the entire structure, creating public keys and private keys for the signature scheme to decipher the secret writing based on a highly secured mechanism.

- Cloud provider: The cloud provider is in charge of integrating the hyperparameters that were regionally acquired through each manufacturing unit's individual premises to create a complete intrusion detection model. The cloud provider must interact with each manufacturing unit several times in order to create an anomaly detection system that is ultimately "correct."
- Manufacturing units: Every manufacturing unit is responsible for developing a regional authentication scheme predicated on its own gathered commercial CPS information and assisting in refining the specifications of the anti-malware engine by periodically interacting with the cloud provider on behalf of the economic CPS proprietor.

Throughout this research, we evaluate the effectiveness of our fusion layer proposal against a number of cutting-edge investigations. We initially conducted tests to assess how well our suggested method performs in comparison to the aforementioned monitoring. Watched and mastered the ecological structure based as the dynamic source building automation at the beginning of this research. Additionally, according to the interaction activity consequence or results, researchers had identified and defined issues like equipment damage and vulnerabilities, which are amended and reviewed. Often demonstrate how irregularity references relate to ML frameworks for challenge delineation. The technology was conducted in a live showcase and technology demonstrator scenario using virtual, and the classifier was roughly 85% accurate. As a result, the recognition rate improves. This was supported by an empirical discovery, a comprehensive explanation of the types and characteristics used in this proposed strategy offered.

14.4 FUSION LAYER IMPLEMENTATION

In fusion layer, we intercorrelate the feature vector. Denoted the same by $(I_c(F_e))$ and can be calculated from eqn. (1).

$$I_c\left(F_e\right) = P\left(F_v \le F_{e1,} F_{e2,} \dots \dots, F_{en} \le f_{en}\right) \tag{14.1}$$

For any random sensor or actuator input F(S) is calculated from eqn. (2).

$$F\left(S\right) = I_c\left(F_{e1}\left(S_1\right), \dots F_{en}\left(S_n\right)\right) \tag{14.2}$$

Therefore

$$I_c\left(F_e\right) = P\left(S_1 \le F^{-1}{}_{S1}\left(S_1\right), \dots \dots F^{-1}{}_{Sn}\left(S_n\right)\right)$$

$$= F_s\left(F^{-1}{}_{Sn}\left(S_1\right), \dots \dots, F^{-1}{}_{Sn}\left(S_n\right)\right) \tag{14.3}$$

Eqn. (3) gives the closed form for intercorrelation feature vector.
In general $F_s(S_i)$ can be written as in eqn. (4) and (5).

$$F_s(S_i) = P(S \le S_i) = L_t \qquad (14.4)$$

Therefore, according to probability theory

$$1 - F_s(S_i) = P(S \ge S_i) = R_t \qquad (14.5)$$

Eqn. (4) and (5) will give a left (L_t) and right (R_t) tail probability value.
The next step is finding the skewness (S_k) to these tail values. This will evaluate, using eqn. (6).

$$S_k = \frac{\frac{1}{n}\sum_{i=1}^{n}\left(S_i - \bar{S}_i\right)^3}{\sqrt[2]{\frac{1}{n-1}\sum_{i=1}^{n}(S_i - \bar{S}_i)^2}} \qquad (14.6)$$

The entire inter-correlation value for non-parametric fitting of sensor or the actuator can be calculated by eqn. (7):

$$\Gamma_c(F_e) = \frac{1}{n}\sum_{i=1}^{n} I_c(S^{\wedge}_{i1}) \qquad (14.7)$$

To get a low probability of anomaly $A(S_i)$, we take $(-\log)$ probability of L_t, R_t, and S_k. While taking $(-\log)$ probability, we will get a high score. That is shown in eqn. (8).

$$A(S_i) = \text{Max}\{-log(L_t), -log(R_t), -log(S_k)\} \qquad (14.8)$$

From test cases, we fix the threshold value as $-\log(0.01)$. If the value exceeds, then we fix the anomaly.

The monitoring and evaluation of the prototype implementation is done by using both freely accessible benchmark information, the power system, and commercial dataset. These datasets are chosen for use in the scientific investigation because they have various kinds of feature character traits, such as perpetual or unambiguous. This dataset has ten separate types; single type one class describes typical events, and the other nine different classes specify system vulnerabilities. It moves between multiple network hops at a speed of approximately five to ten MBPS.

The algorithm 1 mentioned next is talked about, the workout stage.

Algorithm 1 Workout Stage
Input: Number of sensor or actuator inputs are given as $S_1, S_2 \ldots S_n$
Output: Workout Stage

1. For any random sensor or actuator input, calculate Feature Vector
   $F_{el}(S_1) \ldots F_{en}(S_n)$
2. Intercorrelate, the Feature Vector, $I_c(F_e)$
3. Calculate the probability of inverse Feature Vector to get the closed form
   $(F^{-1}_{Sn}(S_1) \ldots F^{-1}_{Sn}(S_n)$
4. Evaluate the left, right tail probability values, that is, L_t and right R_t
5. Find the skewness to those tail values S_k
6. The next step is for non-parametric fitting of sensor or the actuator, $I^\wedge_c(F_e)$
7. Then enumerate $(-\log)$ probability of L_t, R_t, and S_k
8. Quantify $\text{Max}\{-\log(L_t), -\log(R_t), -\log(S_k)\}$
9. Fix the anomaly if it exceeds, $-\log(0.01)$

The algorithm 2 mentioned next is talked about, the examine stage.

Algorithm 2 Examine Stage
Input: Get samples of data with abnormalities
Output: Eliminate deviation from the data samples

1. For all data samples do
2.   Compute fusion layer
3.   Compute PDF for fusion layer
4. Use the workout stage data to separate the examine stage dataset's routine from exceptional values
5. Enumerate maximum probability
6. For all examine sample do
7.   If Max(P(tail))> -log(0.01)
8.   Get Anomaly
9.   Else
10.   No Anomaly

The suggested work is developed in the R programming language and runs on a Windows 10 computer with an Intel Seven processor and roughly 16 GB of memory capacity. About 400,000 records, including both normal as well as exceptional information, are arbitrarily picked from every set for the research's workout and examine phase. The performance of the suggested and other approaches is attained by aggregating the five-fold out of sample validation results to adequately evaluate overall quality without making distinctions between benign or malicious classifications.

14.5   RESULTS AND ANALYSIS

With our proposed work, we evaluated the performance, which is shown with respect to probability identification and proportion and various quality parameter proportions (Q = 20%, 40%, 60%, 80% and 100%) narrated in Figure 14.8. The suggested work shows the probability identification is gradually increasing for the quality

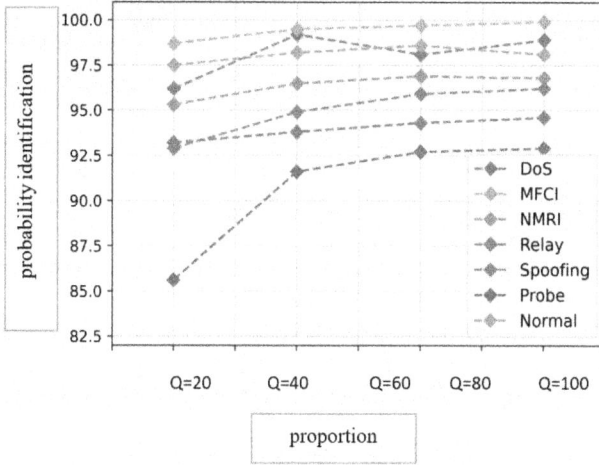

FIGURE 14.8 Various quality parameter proportions.

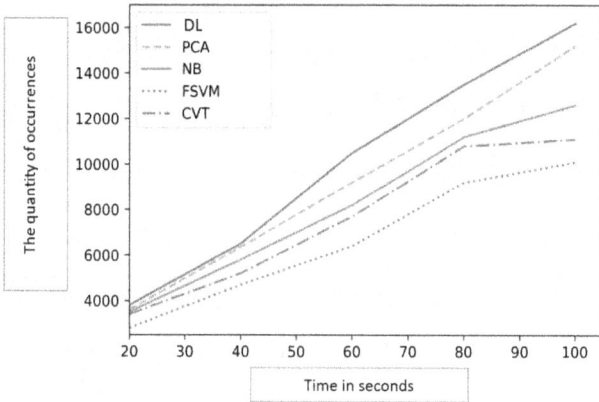

FIGURE 14.9 Calculation times for several methods currently in use are compared.

parameter proportion Q = 20 and Q = 40, whereas chart displaying is nearly identical to the quality parameter proportion Q = 60, Q = 80, and Q = 100.

The proposed framework outperforms, as shown by the conclusion presented in Figure 14.9, since that incorporates the capability of various approaches. The mix of different methods effectively restricts the typical zone, that can aid in effectively assessing key for computer security. Because we initially error check evidences, which can precisely analyze the PDF for every input vectors and accommodate the highest and lowest boundaries of sample points with probabilities, the research framework shows better results with criterion proportion configurations of Q = 80% and Q = 100%. Aggregately, harmful content can be distinguished from questionable content using the standard characteristic bounds in Figure 14.9. Evaluations of the proposed DLAD method narrates the probability identification for assessing the key assault. Figure 14.9 shows a comparison of calculation times for various methods that are currently in use, which indicated as the max and min extremes. Our selection of 80% of all components significantly improves the efficiency of the suggested framework. Our proposed method outperformed other squint methods, such as PCA, CVT, NB, RF, and F-SVM, in terms of identifying numerous different forms of computer attack in the shortest amount of calculation time. According to Figure 14.9, our suggested approach requires 65 seconds to work out 11,000 samples, whereas other existing approaches require, upon aggregate, 70 to 80 seconds.

The actual as well as differential data transmission is shown in Figure 14.10. This is done with respect to component count and lag. The red dot shows the actual, and blue dot shows a differential data transmission. When the component count increases, the lag will increase.

By using various model, we calculate the accuracy, precision, recall, and score. Table 14.3 displays the grade of Sensor-DLC, actuator, fully connected, fusion layer.

Our method compared with existing technique like KNN, CNN, and SVM. The comparison study is on Table 14.4. The relative study is done with respect to accuracy (A_s), precision (P_s), recall (R_s), and score (S_s).

The pictorial diagram Figure 14.11 narrates the comparison study. The graph shows the clear view of the performance of our approach. Additionally, it can be shown by contrasting the detection physical phenomena in Figure 14.11 that the viewing outcomes were mostly in line with those of an outlier detection activity.

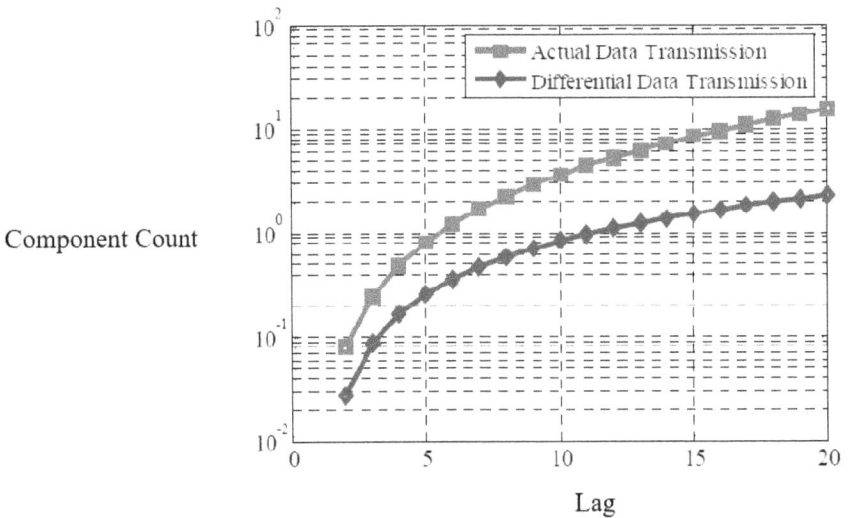

FIGURE 14.10   Component count vs. lag.

TABLE 14.3
Score for Each Step

Model	Score
Sensor- DLC	87.9%
Actuator- DLC	87.8%
Fully Connected	92%
Fusion Layer	88%
Anomaly Detected	99.91%

TABLE 14.4
The Comparison Study with Our Method

Methods	Accuracy (A_s)	Precision (P_s)	Recall (R_s)	Score (S_s)
KNN	68%	68%	67%	66%
CNN	65%	72%	64%	63%
SVM	75%	78%	75%	74%
Our Method (SWaT Data)	99%	99.9%	99.8%	99.91%

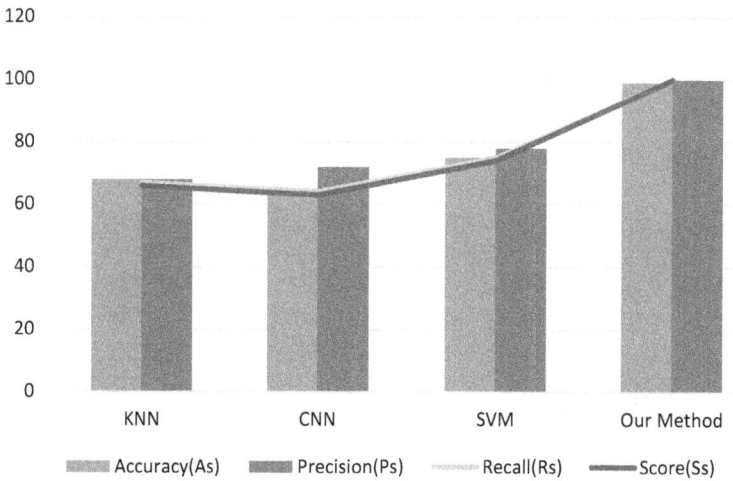

FIGURE 14.11   Graphical representation of performance evaluation.

This indicates that DL considers connection factors and more accurately distinguishes between healthy and pathological latent presentations, allowing for better categorization of pathological data as a necessary precondition for identification. The identification delay, or the separation between the starting point successfully identified in the abnormality region as well as the initial point in the abnormality region, is another significant issue that is brought up by the detection concept.

Therefore, to show that the proposed technique can conduct quick and efficient discovery, the tracking shutter lag is also assessed and compared during the trials. All in all, supervised modelling techniques include industrial and classical techniques, whereas unsupervised modelling techniques include prediction models and renovation models. The prediction model is more effective than conventional techniques since it concentrates on the temporal characteristics of the response variable. However, they don't perform as well on other types of data with weak or irrelevant cycles for the same reason when applied to seasonal response variable.

The study results are displayed and shows that the impacts of DL steadily diminished with an expansion on noise in every proposed methodology, but still the scope of impairment seemed constrained. The impacts were mostly rather minimal even when the noise intensity rose to 5%. This demonstrates that DL may continue to use correlated properties to reduce the effects of noisy outliers built on the work and likelihood estimation, maintaining performance even after the infusion of noise. Additionally, it can be observed by contrasting the research observations of respective recommendations that all these comparison frameworks had limited transferability to other implementations and were exclusively useful for the usage scenarios in the recommendations. Additionally, this shows that DL can demonstrate superior adaptability and scalability in a variety of real-world CPS situations.

14.6 CONCLUSION

This research confirms whether current trends and uncertainties could lead to various potential futures for CPS science research. Even if there are many alternative futures, the situations presented in this report are useful for academic, to absorbency, depict, and appreciate our tomorrow. We used a distinct strategy to find the anomaly present in CPS. The extracted features with the help of DLC and given to fully connected and fusion layer. With the help of three steps, we detected anomaly. The suggested model, which employs a DL, yields superior results when the asset proportions are set to $Q = 80\%$ and $Q = 100\%$. Our suggested approach successfully and rapidly distinguishes between various kinds of assaults. To increase the results and effectiveness of the system, extensive assessments in diverse circumstances will be carried out in subsequent study. Experimental results show, our method produce good accuracy, very good score, considered interpretability, and fastest method. The skewness gives the best performance.

REFERENCES

1. Jun Inoue, Yoriyuki Yamagata, Yuqi Chen, Christopher M. Poskitt, and Jun Sun, Anomaly detection for a water treatment system using unsupervised machine learning, International Conference on Data Mining Workshops, IEEE, 2017, pp. 1058–1065.

2. Li Yi, Xenofon Koutsoukos, and Yevgeniy Vorobeychik, Adversarial gaussian process regression in sensor networks, in *Game Theory and Machine Learning for Cyber Security*, 2021, pp. 149–159. IEEE.
3. Krishna H. Hingrajiya, and Ravi K. Sheth, Robust forgery detection method for copy-move and splicing forgeries in images using convolutional neural network, Proceedings IC-Recent Trends in Engineering Technology and Management, 6 and 7 May 2022.
4. N.C. Preethika, Swetha Sridevi, and T. Nachimuthu Subiksha, Novel cyber attack detection using hybrid deep learning model, Proceedings IC-Recent Trends in Engineering Technology and Management, 7 May 2022.
5. V. Ravi, T. D. Pham and M. Alazab, Attention-based multidimensional deep learning approach for cross-architecture IoMT malware detection and classification in healthcare cyber-physical systems, in *IEEE Transactions on Computational Social Systems*, 2022, doi: 10.1109/TCSS.2022.3198123.
6. S.K. Mazumder, A review of current research trends in power electronic innovations, in Cyber–Physical Systems, *IEEE J. Emerg. Sel. Top. Power Electron.*, Sp. Issue on Sustainable Energy through Power Electronic Innovations in Physical and Cyber Systems, 2020, *9*(5), 5146–5163.
7. Abiola A. Akanmu, and Chimay J. Anumba, Towards next generation cyber physical systems and digital twins for construction, *ITcon*, Vol. 26, 2021, pg. 505.
8. Shan Zhang, Baocheng Geng, Pramod K. Varshney, Muralidhar Rangaswamy, Fusion of deep neural networks for activity recognition: A regular vine copula based approach, EESS.SP, 21 November 2019.
9. P. Subbarayudu, B. Surya mohan, D.S. John Deva Prasanna, and G. Pavan Kumar, Detection of anomalous behaviour of a student in examination hall using deep learning techniques, Proceedings IC-Recent Trends in Engineering Technology and Management, 6 and 7 May 2022.
10. Montdher Alabadi, and Yuksel Celik, Anomaly detection for cyber security based on convolution neural network: A survey, *IEEE-Explor*, 26 January 2021.
11. N. Nandhini, J.P. John Bibith, S. Jaisuriya, and K. Fareeth Inzamam ul hak, Cyber security intrusion detection, Proceedings IC-Recent Trends in Engineering Technology and Management, 6 and 7 May 2022.
12. Giorgio Audrito, Roberto Casadei, Ferruccio Damiani, Volker Stolzc, and Mirko Viroli, Adaptive distributed monitors of spatial properties for cyber physical systems, *J. Syst. Softw.*, vol. 175, 2021, p. 110908.
13. J.M. Jose, and S.R. Reeja, Anomaly detection on system generated logs—a survey study, Lecture Notes on Data Engineering and Communications Technologies, vol. 68. Springer, 2022.
14. Mr. G. Madan, Miss. G. Madhu mitha, and Miss. B. Gowshika, Challenges and solution in smart grid cybersecurity, Proceedings IC-Recent Trends in Engineering Technology and Management, 6 May 2022.
15. Dr A. Pushpalatha, M. Kausik, R.R. Logeshwaran, and S. Mohamed Kamarudeen, Privacy assured clinical data findings in blockchain, Proceedings IC-Recent Trends in Engineering Technology and Management, 6 and 7 May 2022.
16. X. Tang, J. Wang, Y. Zhu, R. Doss, and X. Han, Systematic evaluation of abnormal detection methods on gas well sensor data, 2021 IEEE Symposium on Computers and Communications (ISCC). IEEE, September 2021, pp. 1–6.
17. A. Golabi, A. Erradi, A. Tantawy, and K. Shaban, Detecting false data injection attacks in linear parameter varying cyber-physical systems, 2019 International Conference on Cyber Security for Emerging Technologies (CSET), 2019, pp. 1–8.
18. Giuseppina Andresini, Annalisa Appice, and Donato Malerba, Autoencoder-based deep metric learning for network intrusion detection, *Inf. Sci.*, Vol. 569, 2021, Pages 706–727.

19. E. M. Dovom, A. Azmoodeh, A. Dehghantanha, D. E. Newton, R. M. Parizi, and H. Karimipour, Fuzzy pattern tree for edge malware detection and categorization in IoT, *J. Syst. Archit.*, vol. 97, August 2019, pp. 1–7.

20. H. Alasmary *et al.*, Analyzing and detecting emerging Internet of Things malware: A graph-based approach, *IEEE Internet Things J.*, vol. 6, no. 5, October 2019, pp. 8977–8988.

21. H.-T. Nguyen, Q.-D. Ngo, and V.-H. Le, A novel graph-based approach for IoT botnet detection, *Int. J. Inf. Secur.*, vol. 19, no. 5, October 2020, pp. 567–577.

22. T.-L. Wan et al., Efficient detection and classification of Internet-of-Things malware based on byte sequences from executable files, *IEEE Open J. Comput. Soc.*, vol. 1, 2020, pp. 262–275.

23. Weina Niu, Xiaosong Zhang, Xiaojiang Du, Teng Hu, Xin Xie, and Nadra Guizani. Detecting malware on X86-based IoT devices in autonomous driving, *IEEE Wirel. Commun.*, vol. 26, no. 4, 2019, pp. 80–87.

15 Industry 4.0
Intelligent Routing and Scheduling Algorithm for EV Charging Based on EV Battery Management System

Ashwin Kumar V

15.1 INTRODUCTION

The rapid growth of the smart cities and electric vehicle (EV) industry has led to an increase in demand for maintaining these new technological innovations. One such instance is to sustain the growth of EVs in the grid and to make efficient use of the same, and a significant issue is the lack of EV charging infrastructure. Although this can be overcome by building more charging infrastructure, it is impossible to do so overnight, and the cost involved is also great. One solution is to make efficient use of existing infrastructure through resource management principles until there is enough infrastructure to support the demand. The other major issue in the sustainability of the EV industry is the efficiency and life of the battery packs used for EVs. Multiple analytical researches [1] have shown that the Li-ion battery has high power density, high energy density, low maintenance requirement, low self-discharge, and no memory effect. But battery capacity degradation is still an issue. By combining routing algorithms with resource management to maintain consistency in battery charging and discharging, the battery's state of health (SoH) can be preserved for longer.

Resource management and scheduling algorithms are commonly used in the domain of computer science, particularly at the system architecture level for process scheduling and also in various other fields such as fleet management (logistics and transportation), human resource management, natural resource management, inventory resource management, and more. Applying these principles for efficient and improved routing and scheduling to help a user to make a trip from source to destination using an EV while maintaining the health and performance of the EV battery is the primary aim of this chapter.

The proposed solution involves a routing and scheduling algorithm for assisting users to charge their EVs based on their EV battery management system (BMS)

DOI: 10.1201/9781003343332-15

where EV battery parameters such as state of charge (SoC) and state of health (SoH), user's geolocation, and other parameters are used to calculate a cost to reach every EV charging station in a given network of charging infrastructures and the optimum charging station is recommended for the same.

From the user's perspective, this helps optimize the EV battery life and gives maximum range and less charging time. From the smart grid perspective, this helps optimize the grid power distribution and usage and efficiently handles the EV charging traffic.

15.2 LITERATURE REVIEW

For EV charging, it is important to keep in mind the charging cost per unit, which differs from user to user based on driving behavior, EVs full capacity and EVs SoC at the time of charging. [2] has suggested an approach for forecasting the best time of the day to charge EVs in a social community in a coordinated manner such that all users incur least charging cost. Comparing both coordinated and uncoordinated charging data (predicted using Latin Hypercube Sampling and Birnbaum Saunders distribution) the coordinated strategy reduces charging cost.

Another significant problem is the range anxiety for users traveling long distances using EVs, which [1] has simplified by an approach that minimizes the total travel time for each EV based on the A* algorithm with constraint verification and a peer-to-peer scheduling system.

To tackle the previous problem, [3] suggests another perspective which is to strategically place the charging stations at right locations to ensure charging availability at all points.

[4] proposes SoH estimation of lithium-ion batteries using a linear regression model between SoH and health parameter and also talks about why Li-ion batteries are used for EVs, what is a BMS and the three types of SoH estimation methods (direct method/Coulomb Counting, model based, data-driven based). The health parameter is extracted by considering a small range of voltage during battery charging and through simulation, data-driven approach using linear regression, low SSE, low RMSE, high R^2, and reducing estimation time helps in achieving online real-time SoH estimation. It is seen that the OCV curve shifts towards left as SoH decreases which shows that there is a co-relation between OCV and SoH.

Li-ion battery life cycle prediction is very important in terms of calculating the battery SoH. [5] proposes a method using a hybrid machine learning model that combines a shallow learning model, a relevance vector machine, and a convolutional neural network to predict the cycle life of a Li-ion battery cell.

A major requirement for EV charging services is maintaining fairness for service recipients while supplying chargeable electric power to multiple EVs. [6] discusses a priority-based method to control the required charge amount per parking time for EVs with relatively high urgency.

With the growth of large-scale EVs on road and their involvement in intercity transportation systems and power transmission systems, there is a growth in both traffic and power networks around the cities. [7] proposes a traffic-power integrated network model to convert a EV routing and charging problem into a shortest path problem.

In [8], the computation and monitoring of three key indices, namely, state of charge (SOC), state of health (SOH), and state of function (SOF) for EV using BMS (battery management system) is proposed.

Among many shortest-path algorithms, Dijkstra's is prominent. [9] compares both serial and parallel execution of this algorithm and its performances were measured on four different configurations, based on dual-core and i5 processors. The experimental results prove that the parallel execution of the algorithm has good performances in terms of speed-up ratio when compared to its serial execution.

Vehicular routing is always an optimization problem in any domain. The best solutions are inspired by nature. Hence [10] proposes a routing strategy based on an ant colony algorithm and then uses insert heuristic based on the time window classification.

There is a problem with how vehicles are used to transport goods to and from customers. [11] proposes an algorithm that would improve the way this is done by minimizing the number of vehicles needed and the travel time by using a Kohonen map algorithm, which is a type of unsupervised competitive neural network. This means that the network would be more likely to find different solutions to a problem, not just the one that it is biased towards.

[12] discusses a system in which electric vehicles are coordinated in order to provide frequency regulation service and that grid stability issue can be alleviated through the proposed model.

[13] discusses a new method for using electric buses to help restore power to the grid after a hurricane. The method includes figuring out the best way to route the buses and schedule their use so that they can help power the grid while still meeting the demand for public transportation.

[14] is about a model for electric vehicles that account for real-life demand behavior. The model minimizes travel time and energy costs for the electric vehicles. The electric vehicles are represented by a battery energy consumption model and the customer pick-ups and drop-offs were not certain, so they were modeled as stochastic parameters. The model is validated by computational results.

[15] looks at how electric vehicles can be used for ride-sharing services and specifically how to route the EVs to minimize customer waiting time and total distance traveled.

Load balancing requests to provide equal services to all users is very important.

In [16], a new load balancing algorithm called RTLB is proposed. The idea behind RTLB is to keep the load balancing of node so as to achieve the largest throughput of the system. RTLB uses the service load and node load as the decision-making measures and forwards requests to suitable nodes.

[17] is about how electric vehicles can be charged at work to help ease the burden on the central power system. The results of a study are presented, showing that there is a significant difference in EV charging capacity between peak winter and peak summer. Recommendations are made on how best to equip future distribution systems for increased EV charging.

In [18], a detailed analysis is done to understand the relationship between distance and charging power in a distribution network to better address the fairness in the proposed additive increase-multiplicative decrease EV charging algorithm.

[19] is about a study that was done using simulations to see what benefits there would be from using electric vehicles that can charge quickly, in combination with renewable energy sources like solar or wind power. The study found that there are definite benefits to using this type of system, and that it can improve the way power systems with renewable energy work.

In order to study how EV charging stations affect the load on the power grid, in [20] authors looked at how people use EVs by looking at data from the national household trip survey (NHTS). They classified different types of EVs based on where people were going and then looked at the probability of different variables (like how long people charge their EV) in a trip. After that, they used a method called Monte Carlo simulation (MCS) to establish a load model for different types of EVs. The results showed that with more EVs, the maximum load on the power grid increases. In [21], the authors created a model for how electric vehicles can be charged using the transportation network and the smart grid. They then created an algorithm to recommend when and where to charge the vehicles. The goal is to have the charging of the electric vehicles help balance the load on the power grid, instead of being a burden.

[22] discusses different ways to charge electric vehicles, and then presents a new way to charge them. It uses coils that are energized depending on the position of the receiver coil mounted on the vehicle. A reduced scale demonstrator will be presented. Numerical 3D simulations are used to calculate the parameters of the inductive charging system and their variation with the EV position as well as the energy transfer efficiency. Some of the numerical results are experimentally validated.

15.3 PROPOSED SOLUTION

This chapter proposes a solution to improve EV battery performance and health by calculating a route schedule between the source and destination of a user's trip based on various parameters. The solution algorithm, as illustrated in Figure 15.1, includes three sub-algorithms, namely, the cost function algorithm to find the SoC cost from the current location to any location, the shortest path algorithm to find the shortest path between any two points, and the load balancing algorithm to reroute or distribute charging request, and these sub-algorithms are interdependent of each other. Each sub-algorithm's output decides whether the route schedule needs to be updated to give the optimal route through iterations. Finally, when all sub-algorithms find their optimal solutions, the final route is generated.

The input decision parameters are a combination of route parameters that are independent of the EV and user parameters that are dependent on the EV. The final route may or may not include one or more charging allocations for the user to stop by at EV charging stations based on the SoC required for the trip and the EV's SoC at the start of the trip. The proposed solution also ensures that the user's EV has double the SoC required for the trip to travel to and fro from the destination. The charging allocated for the EVs is preemptive scheduling and also depends on the EV's SoC and need not charge up to 100% SoC all the time.

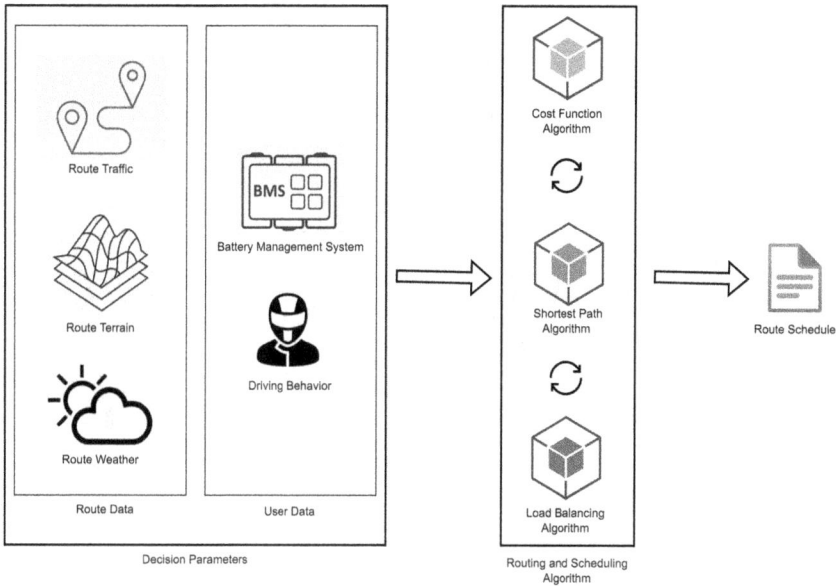

FIGURE 15.1 Architecture diagram.

15.3.1 Cost Function Algorithm

Generally, the battery's discharge rate along with the time to reach between two points is measured to calculate the SoC cost to travel between the respective two points, but the other BMS and route parameters are neglected which results in an error rate in SoC cost estimation that affects the battery performance slowly over a long time. The EV's BMS as the main control component of the EV's battery pack monitors and provides more vital battery state information such as state of charge (SoC), state of health (SoH), safe operating area by monitoring temperature, battery faults, and alerts through sensors, etc. These parameters in combination with route parameters such as route weather, route terrain, route traffic, and driver behavior from past trip data can be measured and fed to an advanced cost function to calculate a more accurate SoC cost to reach any location from the current location geographically.

Consider a set of coordinates of five locations, namely, A, B, C, D, and E, and four charging stations, namely, CS1, CS2, CS3, and CS4 represented in a weighted undirected graph data structure to evaluate the working of the algorithm, as illustrated in Figure 15.2. Each vertex represents either a location or charging station and each edge represents the path that connects two locations, and its weight represents the time cost to travel that respective path. Every vertex in the graph need not be connected to every other vertex.

The cost function algorithm in addition to the base time cost adds the decision parameters cost calculated for each edge of the graph based on the cost values described in Table 15.1. For example, if the base time cost between A to B is 5 and

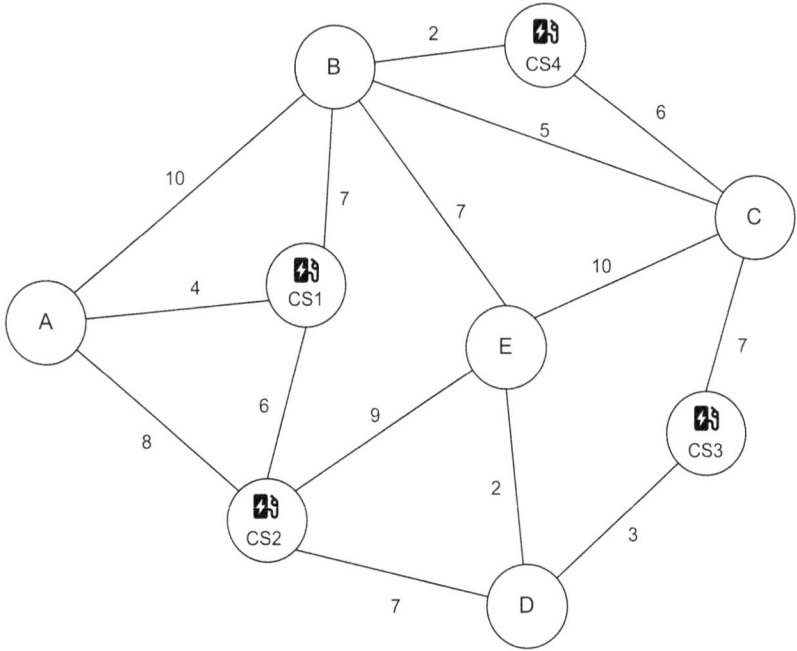

FIGURE 15.2   Coordinates graph visualization.

TABLE 15.1
Decision Parameter Cost Table

Parameter						
Route Traffic	Low	0	Medium	1	High	2
Route Terrain	Smooth	0	Rough	1	Very Rough	2
Route Weather	Cold	1	Normal	0	Sunny	1
Driver Behavior	Slow	2	Average	1	Fast	0

route traffic is medium, route terrain is rough and route weather is sunny, and the driver usually drives slowly between those two points based on previous trips, then the final time cost to travel between A to B will be $5 + 1 + 1 + 1 + 2 = 10$. The SoC cost to travel from A to B will be the product of final time cost and the EV battery discharge rate.

The cost function updates the edge cost dynamically every time it executes and every trip might have unique costs for every edge based on real-time route status. But the permutations of edge costs is finite, meaning there is always maximum cost and minimum cost between two points on the graph and all the other possible edge costs lie between these two extremes. The algorithm always uses the combination of edges with optimal cost to reach the destination. The cost function is executed every time the driver reaches a location or vertex on the graph and the route parameters are recalculated at every execution. This makes sure that the algorithm gives the optimal path for every instance of the trip and not just at the start of the trip since parameters like route weather and route traffic can change frequently and even roadblocks and other unpredictable obstacles cannot be determined at an early stage. The cost function algorithm is also executed before each execution of the shortest path algorithm and load balancing algorithm to make sure even those two algorithms give optimal output at any particular instance.

The cost function algorithm is also responsible to calculate the heuristic cost for each vertex in the graph given one source and one destination vertex. Heuristic cost is necessary as the shortest path algorithm uses A* search, and here, the heuristic value is measured by the Euclidean distance between every vertex and the destination vertex. This heuristic cost is only calculated once between the source and destination and is not recalculated until a new destination has been updated. Here both the time and heuristic cost are measured on a scale of 1 to 10 to standardize the computational values. After each execution, the final calculated cost for every edge (time cost and heuristic cost) in the entire graph is passed on to the next shortest path algorithm.

15.3.2 Shortest Path Algorithm

The main aim of any routing and scheduling algorithm is to find the optimal path between every source, destination, and stops on the trip. Similarly here, the shortest path algorithm is responsible for finding the shortest path between a given source and destination at every iteration of execution. Although Dijkstra's algorithm and breadth first search are good algorithms to find the shortest path, they are more time-consuming when it comes to an infinitely large graph of locations, and repeated iterations make it heavily compute intensive. Also these algorithms are better suited for a graph whose weights don't change often, which is contrary in this scenario.

Hence A* search algorithm is used here as it uses a heuristic approach to find the shortest path faster than Dijkstra's and breadth first search. The heuristic value for every vertex is calculated using the Euclidean distance measure between every vertex and the given destination. When the output from the cost function algorithm is fed into the shortest path algorithm, it first figures out the shortest path between the source and the destination. Then from the edges of this shortest path, it aggregates the required SoC for this path. Then it decides if the EV needs a charging allocation or not based on the EV's current SoC and the required SoC for the trip. To ensure sufficient SoC is available for the driver, the required SoC for the trip is always double the required SoC between source to destination.

$$SoC_{trip} = 2 \times SoC_{path}$$

Eq 15.1 SoC required for the trip

If the EV has a SoC greater than or equal to the SoC for the trip, then there is no charging allocation, and the already calculated shortest path is produced as the final optimal path. Else if the EV's current SoC is less than SoC required for the trip then a charging allocation is allotted based on certain criteria and load balancing protocols.

To allocate an EV charging station stop on the trip, the shortest path algorithm first finds three nearby charging stations from the current location using breadth first search. Here we use breadth first search rather than A* search because it searches for charging stations in all directions simultaneously. Once three nearby charging stations are found, the path cost to reach the destination from source through each of these stations separately is calculated, and the three stations are ordered from least path cost to highest path cost in order. The priority is initially given to the route with least path cost and charging allocation and this route is produced as the output of the shortest path algorithm and passed on to the load balancing algorithm which further decides if there is an unavailability of charging ports at the allocated charging station, then the other two nearby stations are allotted based on priority.

From the previous figures we can notice that the heuristic value for each vertex is represented right next to the node inside the rectangle and that the shortest path on the first iteration uses the heuristic A* search to not just traverse the nodes based on least path cost but also on how close the node is to the destination node based on the heuristic value.

$$f(x) = g(x) + h(x)$$

Eq 15.2 A* search function

Here f(x) is the final path cost, g(x) is the time cost, and h(x) is the heuristic cost. From Figure 15.3 in an EV trip from source A to destination C requires no charging allocation since it has sufficient SoC (assuming to be greater than or equal to 30) to reach the destination so it goes from A to B and then to C directly with least path cost of $((10 + 5) + 0 = 15)$ where $(10 + 5)$ is the time cost and 0 is the heuristic cost at destination C. This implies a trip cost of 30.

From Figure 15.4 considering that the EV has insufficient current SoC for the trip (assuming to be less than 30), the shortest path algorithm first finds the three nearby charging stations, namely, CS1, CS2, and CS4 using breadth first search. In order to find the overall shortest path from source to destination with charging allocation, the algorithm breaks the path into two shortest paths considering the charging station as the destination in the first half and then as the source in the second half and the combination of both halves aggregates as the total trip. So first it finds the shortest path from source to the charging station and then the shortest path from the charging station to the destination. In case more than one charging station is to be allocated in one single trip, the same method is implemented to break the trip into individual parts for each allocation stop and find the shortest path for each part and aggregate them in the end. So for each of the three charging stations, first it calculates the shortest path from A to station CS1 with a path cost of $(4 + 8 = 12)$ and the shortest path from CS1 to C, which is through B with a path cost of $((7 + 5) + 0 = 12)$, so final path cost from A to C through CS1 will be $(4 + 7 + 5 = 16)$. Then, it calculates the shortest path from A to station CS2 with a path cost of is $(8 + 9 = 17)$ and the shortest path from CS2 to C which is through D and CS3 with a path cost of $((7 + 3 + 7) + 0 = 17)$,

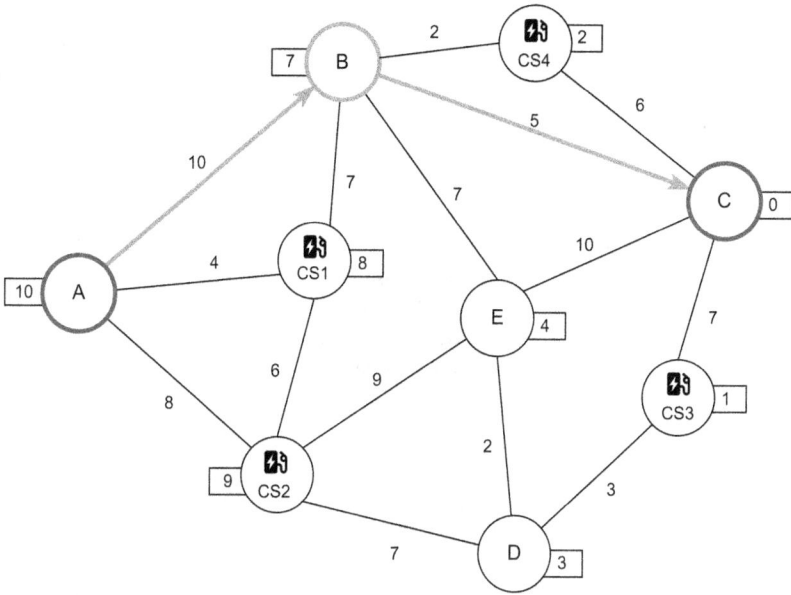

FIGURE 15.3   EV trip from A to C (no allocation).

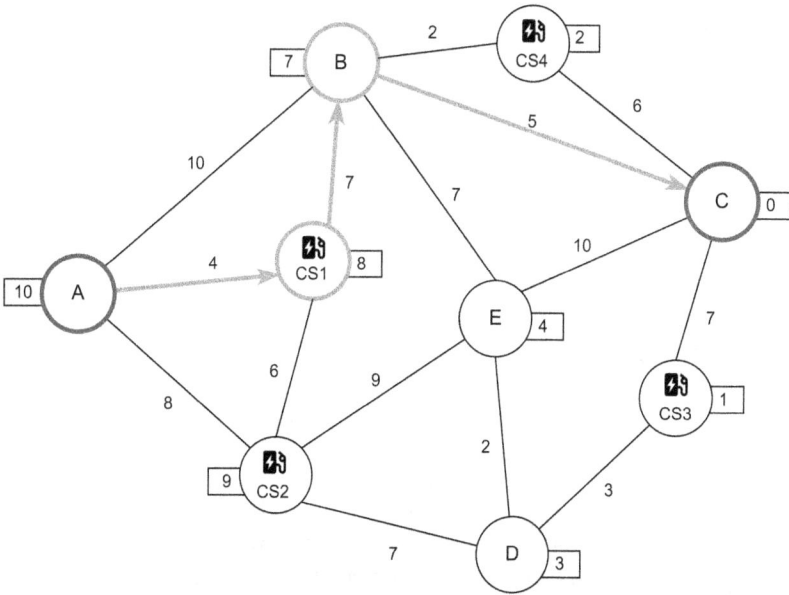

FIGURE 15.4   EV trip from A to C (with allocation).

so final path cost from A to C through CS2 will be (8 + 7 + 3 + 7 = 25). Similarly, it calculates the shortest path from A to station CS4 with a path cost of ((10 +2) + 2 = 14) and the shortest path from CS4 to C with a path cost of (6 + 0 = 0), so the final path cost from A to C through CS4 will be (10 + 2 + 6 = 18). From this, the algorithm ranks the charging stations as CS1, CS4, and CS2 in order of least path cost from source to destination through the respective charging station and allocates CS1 as the EV charging station stop for the trip routing from A to CS1 to B to C.

The other two charging stations and their paths are not neglected and are used in further iterations of the shortest path algorithm based on the availability of the charging ports at CS1, which is handled by the load balancing algorithm. The same three stations need not be used throughout the trip and can be dynamically updated based on the location of the driver and change in destination.

15.3.3 LOAD BALANCING ALGORITHM

In any commercial business, treating all customers equally is very important for customer retention. This includes providing equal services to all users at all times. When in terms of petrol pumps or gas stations, there is a limited number of pumps for users to fuel up their vehicles and when there is an increase in demand for fuel, there is a huge crowd at gas stations and a lot of chaos and imbalance for providing fuel services to all users. This chaos is resolved by means of pre-booking slots and maintaining time periods for each user. But when all slots are already filled and yet there is a demand for services, then comes the importance of load balancing algorithms. These algorithms basically balance incoming load across various terminals. Similarly, here in terms of EV charging, when there is an increase in demand for charging allocations at a particular charging station, then the load balancing algorithm will either decide to redirect the user to a different charging station or distribute the charging slot equally to both parties.

The output from the shortest path algorithm only allocates the user with the charging station with the least path cost. But it does not resolve to see if there are any available charging ports or not at that allocated charging station. One charging station can have one or more charging ports based on the infrastructure of the station, and also not every charging port are of the same type. So availability of charging port also includes compatibility of the port type, meaning there can be charging ports available but not of the compatible type for the EV. Also every charging station based on the number of ports and its power rating should not be allowed to consume more power from the main power grid at point of time, which might result in a rise in electricity bill of the station owner or service provider. The imbalance in power consumption from the main power grid by charging stations is also a major issue in terms of managing EV charging infrastructures. All these issues are addressed by the load balancing algorithm whose main responsibility is to allocate charging ports to user equally with no bias in turn ensuring that there is no over usage of power consumption at any charging station at any point of time. The algorithm does this by means of two actions (redirect or distribute) based on the SoC of both the user who is already allocated a charging port and the user who is about to be allocated the same charging port and also based on the trip cost of both users. A priority-based preemptive scheduling method is implemented to calculate the trip cost of both users for both actions, and the action which results in least trip cost for both the users is chosen. This ensures that both users reach their destinations with least total path cost.

Consider two EVs, starting from different sources and traveling to different destinations are allocated the same charging station. By means of first come first serve, the EV which was first allocated the charging station will continue its trip as usual unless the cost to distribute the charging station between both the EVs is lesser than the cost to redirect the second EV to its second nearest charging station using the shortest path algorithm. After redirecting, if the second charging station is also clashing with another third EV, then the same distribute or redirect calculation is done for the second and third EV. If redirected again to the third nearest charging station and there is again another clash with a fourth EV for a charging port, then the first charging station is forcefully distributed between the first and the second EV. These iterations of calculations involve both the cost function algorithm and shortest path algorithm also to be executed again and again until both the EVs have been allocated charging stations equally.

From Figure 15.5 we notice EV1 travelling from source C to destination B through charging station CS4 with a path cost of (6 + 2 = 8) and from Figure 15.6 we notice EV2 travelling from source B to destination C also through charging station CS4 with a path cost of (2 + 6 = 8). Since both the EVs are clashing for charging ports at CS4, we calculate the trip costs for both EVs for both actions. To redirect EV2, the path cost of EV1 remains the same (6 + 2 = 8) and the path cost of EV2 through B to CS1 to B to C is (7 + 7 + 5 = 19). And to distribute CS4 between EV1 and EV2, the path cost for EV1 through C to CS4 to C to CS3 to C to B is (6 + 6 + 7 + 7 + 5 = 31) and the path cost for EV2 through B to CS4 to B to CS1 to B to C is (2+ 2 + 7 + 7 + 5 = 23). So the cost to redirect EV2 is lesser than to distribute CS4 between EV1 and EV2. So by means of first come first serve EV1 will continue to charge at CS4, while

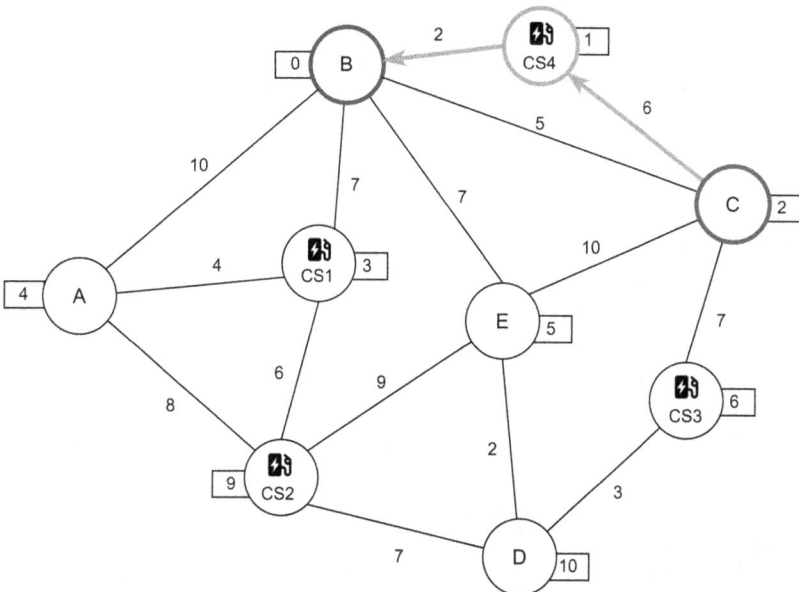

FIGURE 15.5   EV1 trip from C to B (CS4 allocated).

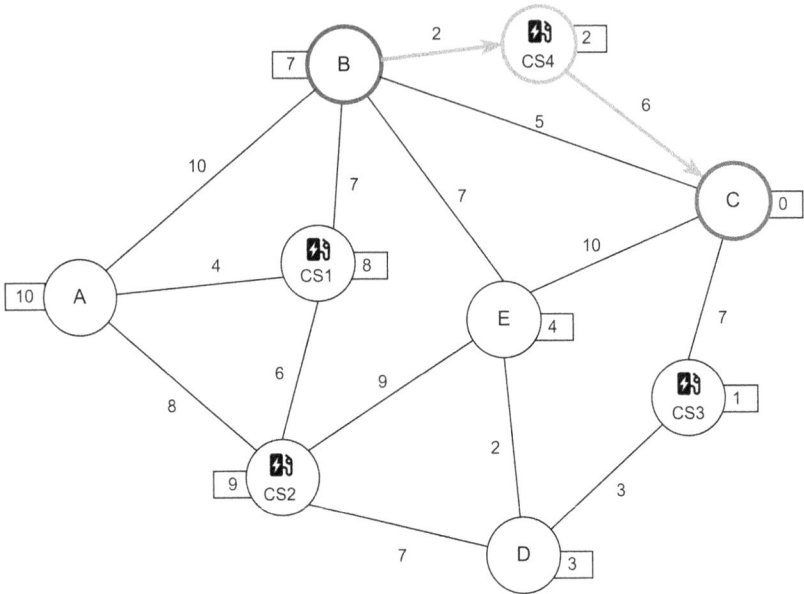

FIGURE 15.6   EV2 trip from B to C (CS4 allocated).

EV2 will be redirected to its second nearest charging station CS1 from its list from the shortest path algorithm. We also notice that for distributing CS4, both the EVs are allotted another charging station to fuel up. This is because by distributing, one charging station is not sufficient to charge up the EV, so the second nearest charging station is allotted for both EVs respectively.

From Figure 15.7 we can notice EV2 being redirected through CS1 and from Figure 15.8 we can notice another EV3 travelling from source A to destination B through charging station CS1 with path cost of (4 + 7 = 11). Since both EVs are again clashing, we recalculate the trip costs for both EVs for both actions. To redirect EV2, the path cost of EV3 remains the same (4 + 7 = 11) and the path cost of EV2 through B to E to D to CS3 to C is (7 + 2 + 3 + 7 = 19). And to distribute CS1 between EV2 and EV3, the path cost for EV3 through A to CS1 to CS2 to CS1 to B is (4 + 6 + 6 + 7 = 23), and the path cost for EV2 through B to CS1 to B to E to D to CS3 to C is (7 + 7 + 7 + 2 + 3 + 7 = 33). So the cost to redirect EV2 is lesser than to distribute CS1 between EV2 and EV3. So by means of first come first serve EV3 will continue to charge at CS1 while EV2 will be redirected to its third nearest charging station CS3 from its list from the shortest path algorithm.

From Figure 15.9 we can notice EV2 being redirected through CS3 and from Figure 15.10 we can notice another EV4 travelling from source C to destination D through charging station CS3 with path cost of (7 + 3 = 10). Since both EVs are again clashing, we cannot redirect EV2 anymore. So we forcefully distribute CS4 between EV1 and EV2 since that has the least path cost that affects all EV1, EV2, EV3, and EV4. This is the final resort after iterating through all three nearby charging stations of EV2, and after three clashes, we preemptively distribute the first nearest charging station.

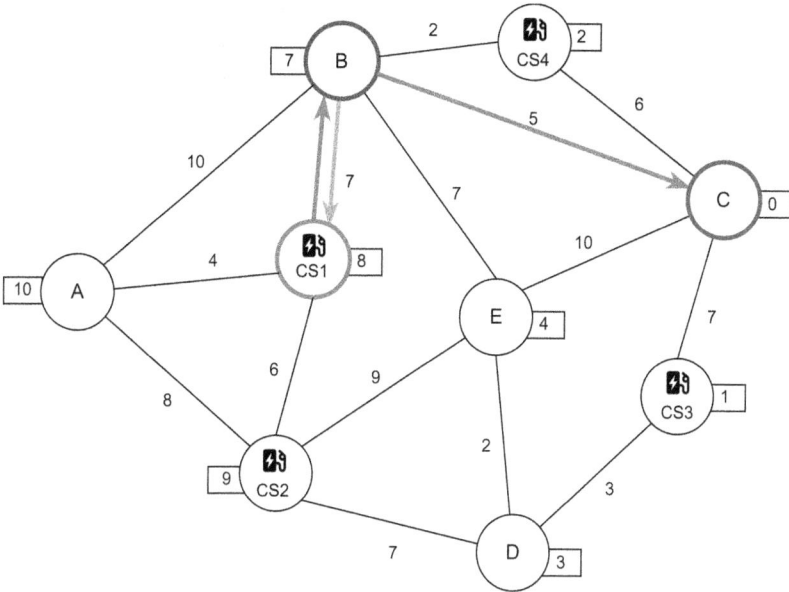

FIGURE 15.7 EV2 trip from B to C (redirected to CS1),

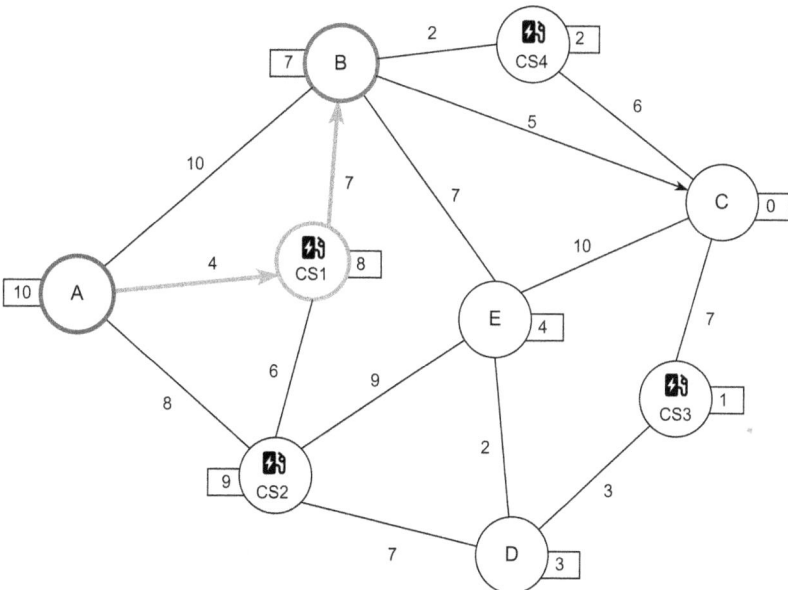

FIGURE 15.8 EV3 trip from A to B (CS1 allocated).

As mentioned earlier, we can notice in Figure 15.11 and Figure 15.12 that both EVs are allocated two charging stations to fuel up throughout the trip since the initial clashing charging station has been preemptively distributed between both the EVs. This almost doubles the path cost for both the EVs, but it is the optimal method

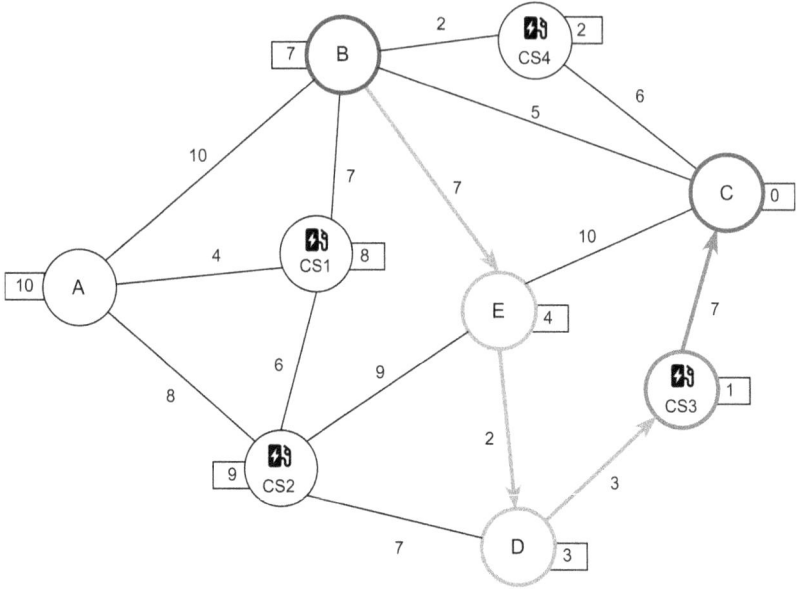

FIGURE 15.9   EV2 trip from B to C (redirected to CS3).

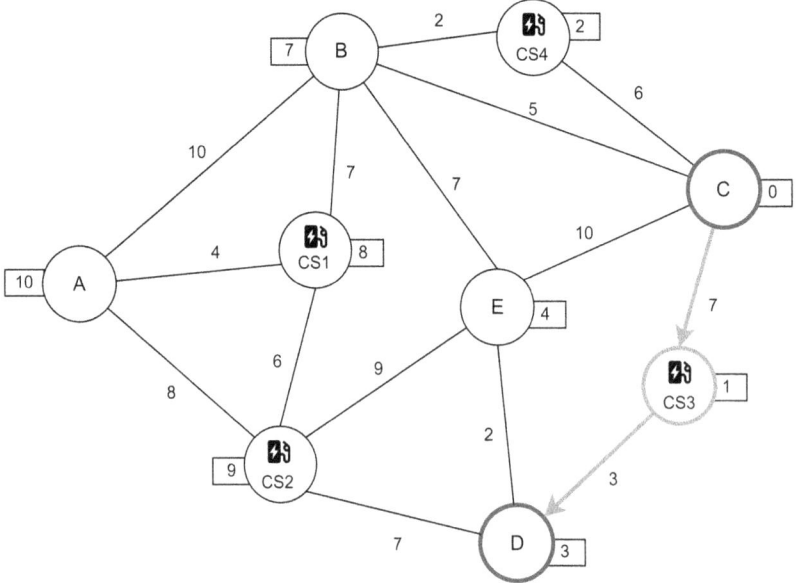

FIGURE 15.10   EV4 trip from C to D (CS3 allocated).

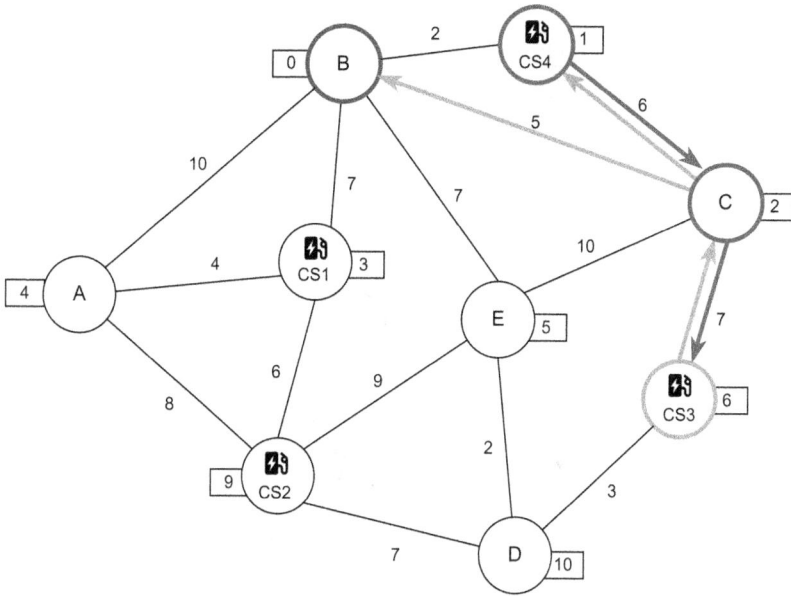

FIGURE 15.11   EV1 trip from C to B (CS4 distributed).

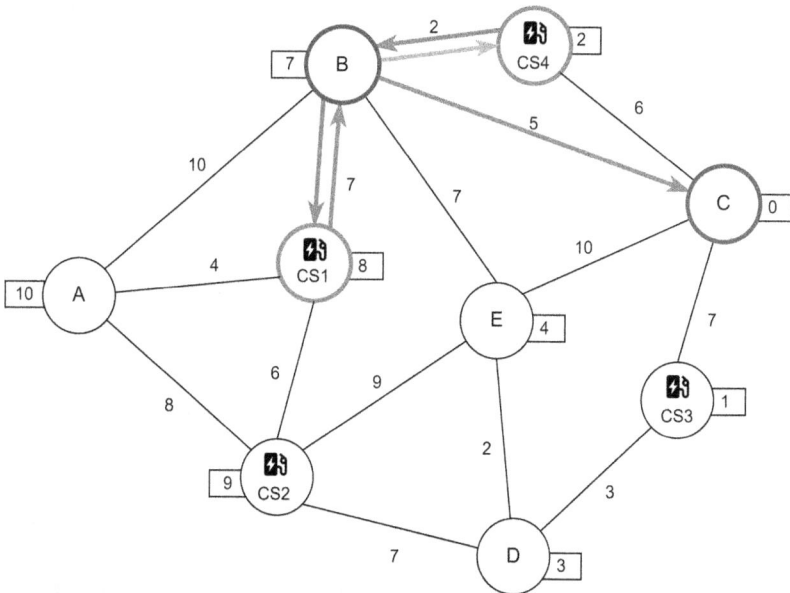

FIGURE 15.12   EV2 trip from B to C (CS4 distributed).

to equally balance services between all EV charging station allocation requests. In areas of high demand for charging or high traffic, the end result will mostly be distribution over redirection, but with shorter time slots for charging, clashes itself can be avoided.

15.4 FLOWCHART

The entire routing and scheduling algorithm as discussed until now executes in iterations of the sub-algorithms based on various changing parameters. Figure 15.13 illustrates in great detail of the overall flow of the execution of the algorithm.

The process begins when the input decision parameters are fed into the algorithm, then the cost function algorithm is executed first to calculate the time cost of every edge in the graph and the heuristic cost of every vertex given a source and destination for the trip. This is then fed into the shortest path algorithm which first finds the shortest path between the source and the destination and then its output is a feedback into the cost function to compare the current SoC of the EV and the required SoC of the shortest path and based on the need of charging allocation either the final output is produced, or the shortest path algorithm is run again with a charging allocation stop in between the trip. Once allocated the output is then fed into the load balancing algorithm which balances the charging requests and ensures all users are equally allocated charging resources. The output from this is sent back to shortest path algorithm to find new shortest paths for every new charging station that is allocated as part of redirect or distribute from the load balancing algorithm. This cycle of allocate and reroute continues until the best path is figured out for all EVs involved and the final optimal path is produced.

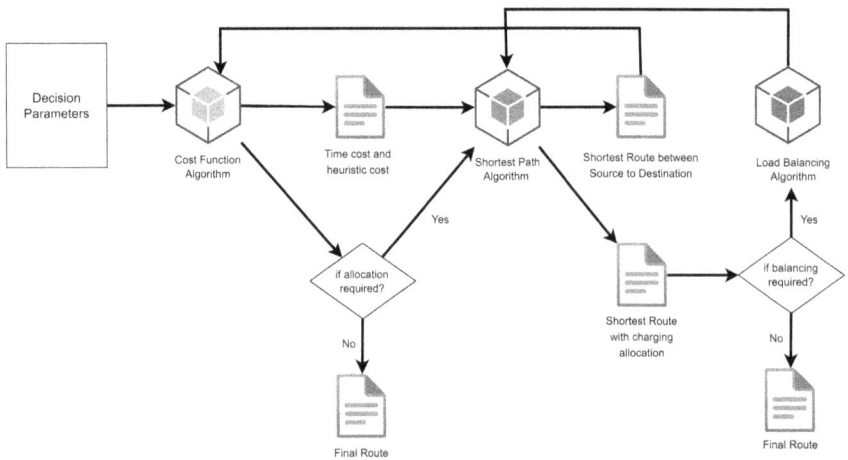

FIGURE 15.13   Flowchart.

15.5 IMPLEMENTATION

The proposed solution was partially implemented through a web application prototype to accommodate EV users to register and manage their respective EVs. It includes a map view where the algorithms once integrated with the web application could render its output to help registered users to navigate through the optimized trip route. It also allows the various power supply service providers in the grid to register and manage their EV power stations and balance their energy reserves in the grid through various business strategies.

The dashboard from Fig 15.14 included the overview details of the EV User's EVs including battery SoC, estimated distance that can be covered before running out of charge and other details. The map view from Fig 15.15 was only partially implemented

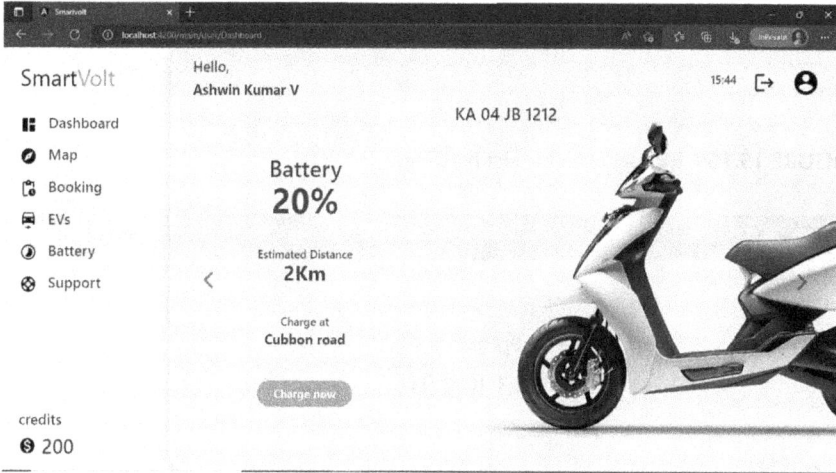

FIGURE 15.14 EV user dashboard.

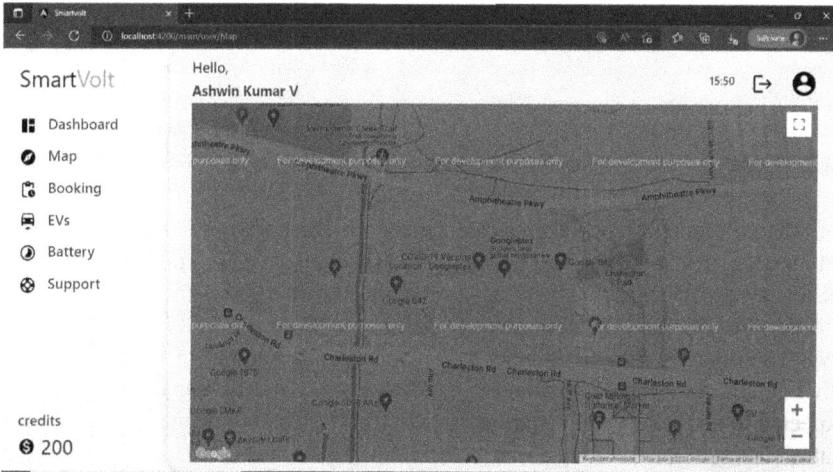

FIGURE 15.15 EV user map view.

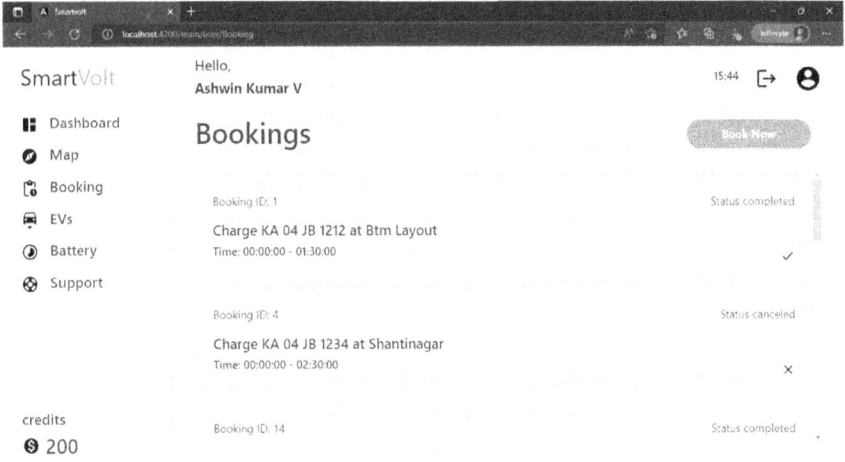

FIGURE 15.16   EV user charging slot bookings.

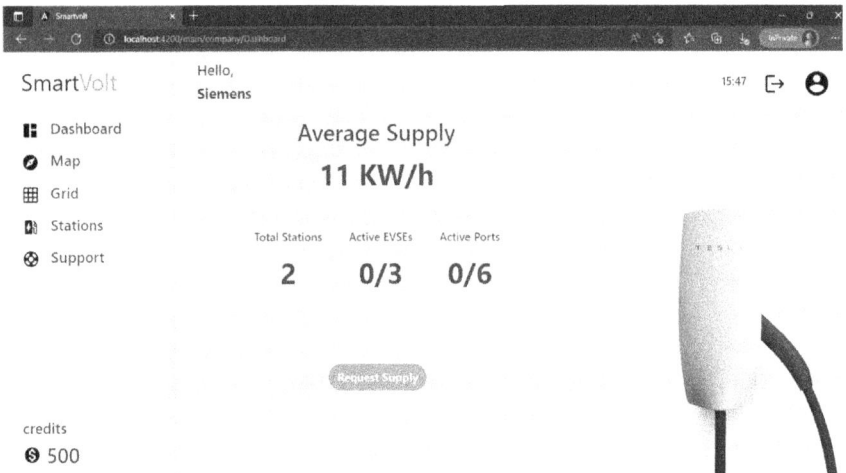

FIGURE 15.17   Power supply service provider dashboard.

and was expected to show the final calculated trip route once fully implemented along with terrain, traffic details, and EV charging schedules. The bookings from Fig 15.16 include all the EV charging bookings ever made and allotted for the user.

The power supply service provider dashboard from Fig 15.17 includes all the EV station details such as number of active stations, number of active ports, and other details. The map view from Fig 15.18 was only partially implemented and was expected to show all the EV stations owned in the grid and its energy reserves and status. The stations from Fig 15.19 include all the stations owned, its EVSEs, and its ports and other relative details.

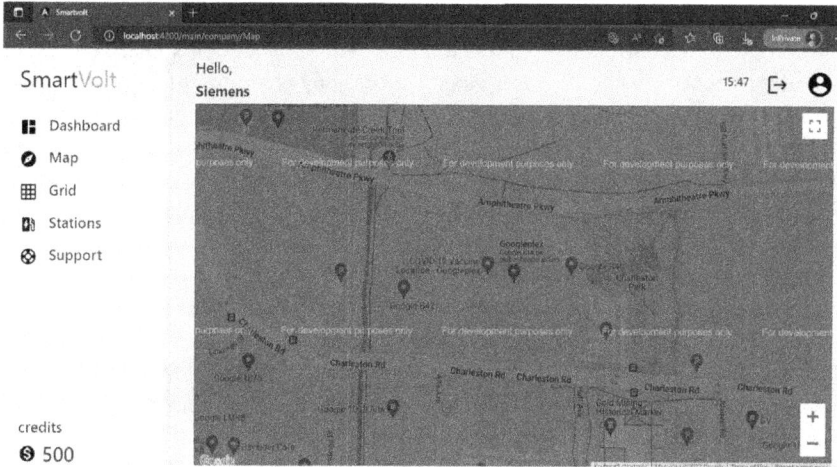

FIGURE 15.18   Power supply service provider map view.

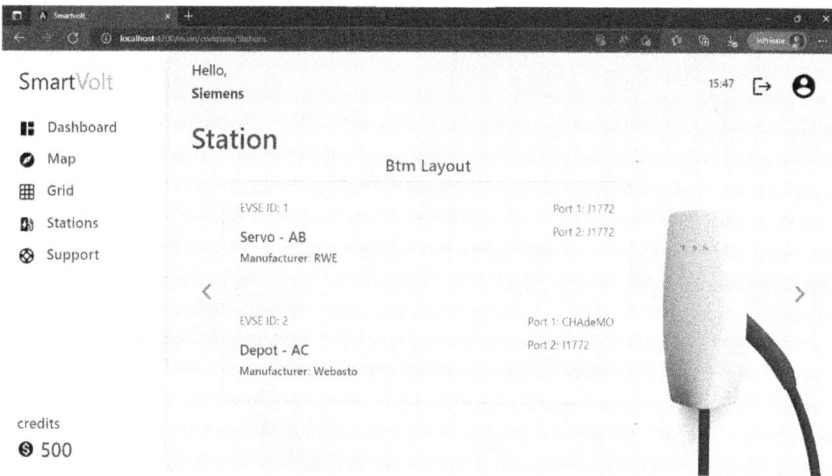

FIGURE 15.19   Power supply service provider EV stations.

15.6   RESULTS AND DISCUSSION

The proposed solution discusses an algorithmic design for dynamic routing and scheduling for EV users to travel from source to destination and also make stops at EV charging stations. Considering a hypothetical implementation of the proposed system, one should be able to effectively travel between any two locations with sufficient EV charging infrastructure to accommodate an abundant audience of EV users.

The worst case can be a user whose source and destination are at either corner of the city and the user's current SoC is less than 5% with no nearby EV charging station, meaning a scenario where the user cannot access any charging infrastructure at all. This is not considered part of this algorithm's use cases, and it is the responsibility of the user to charge or request charging allocation as early as possible by monitoring the EV's SoC. There are other exceptional cases where failure in charging infrastructure and infinite distribution clashes in load balancing algorithm might occur which needs to be handled in future enhancements.

15.7 CONCLUSION

The routing and scheduling algorithm helps users effectively travel from source to destination in the shortest and most optimum path with or without one or more charging allocation based on the user EV current SoC and the required SoC for the trip. The algorithm makes sure once charging is allocated there is surplus SoC to travel to and fro from the destination. Also the algorithm makes sure all charging requests are handled efficiently with no bias keeping a balance in the microgrid power exhaustion and the load at every charging station. The algorithm has three sub-algorithms, namely, cost function algorithm, shortest path algorithm, and load balancing algorithm. The cost function algorithm is responsible for calculating the edge time cost and heuristic cost for all edges of the graph dynamically for every iteration. The shortest path algorithm is responsible for finding the shortest path between the source and the destination with or without charging allocation. And the load balancing algorithm is responsible to ensure equal balance of charging requests by redirecting requests or distributing charging resources equally to both parties. All these three sub-algorithms are executed in iterations based on the input decision parameters until the optimum path is produced.

REFERENCES

1. V. del Razo and H. -A. Jacobsen, "Smart Charging Schedules for Highway Travel With Electric Vehicles," *IEEE Transactions on Transportation Electrification*, vol. 2, no. 2, June 2016, pp. 160–173, doi: 10.1109/TTE.2016.2560524.
2. S. Ayyadi, M. Maaroufi and S. M. Arif, "EVs Charging and Discharging Model Consisted of EV Users Behaviour," 2020 5th International Conference on Renewable Energies for Developing Countries (REDEC), 2020, pp. 1–4, doi: 10.1109/REDEC49234.2020.9163594.
3. A. Pal, A. Bhattacharya and A. K. Chakraborty, "Allocation of EV Fast Charging Station with V2G Facility in Distribution Network," 2019 8th International Conference on Power Systems (ICPS), 2019, pp. 1–6, doi: 10.1109/ICPS48983.2019.9067574.
4. P. Jain, S. Saha and V. Sankaranarayanan, "Novel Method to Estimate SoH of Lithium-Ion Batteries," *2021 Innovations in Energy Management and Renewable Resources*, vol. 52042, 2021, pp. 1–5, doi: 10.1109/IEMRE52042.2021.9386881.
5. S. Shen, V. Nemani, J. Liu, C. Hu and Z. Wang, "A Hybrid Machine Learning Model for Battery Cycle Life Prediction with Early Cycle Data," 2020 IEEE Transportation Electrification Conference & Expo (ITEC), 2020, pp. 181–184, doi: 10.1109/ITEC48692.2020.9161647.

6. S. Miyahara et al., "Charging Prioritization of Electric Vehicles under Peak Demand in Commercial Facility: Destination Charging as a Service," 2020 International Conference on Smart Grids and Energy Systems (SGES), 2020, pp. 226–231, doi: 10.1109/SGES51519.2020.00047.

7. Y. Shi, T. Sun and D. Feng, "The Economic Impact of Electric Vehicle Routing and Charging Strategy on Traffic-Power Integrated Networks," IECON 2017–43rd Annual Conference of the IEEE Industrial Electronics Society, 2017, pp. 453–458, doi: 10.1109/IECON.2017.8216080.

8. Zong-You Hou, Pang-Yen Lou and C. -C. Wang, "State of Charge, State of Health, and State of Function Monitoring for EV BMS," 2017 IEEE International Conference on Consumer Electronics (ICCE), 2017, pp. 310–311, doi: 10.1109/ICCE.2017.7889332.

9. N. Jasika, N. Alispahic, A. Elma, K. Ilvana, L. Elma and N. Nosovic, "Dijkstra's Shortest Path Algorithm Serial and Parallel Execution Performance Analysis," 2012 Proceedings of the 35th International Convention MIPRO, 2012, pp. 1811–1815.

10. L. Meng, C. Lin, H. Huang and X. Cai, "Vehicle Routing Plan Based on Ant Colony and Insert Heuristic Algorithm," 2016 35th Chinese Control Conference (CCC), 2016, pp. 2658–2662, doi: 10.1109/ChiCC.2016.7553766.

11. M. Moumou, R. Allaoui and K. Rhofir, "Kohonen Map Approach for Vehicle Routing Problem with Pick-Up and Delivering," 2017 International Colloquium on Logistics and Supply Chain Management (LOGISTIQUA), 2017, pp. 183–187, doi: 10.1109/LOGISTIQUA.2017.7962895.

12. S. Zhang and K. -C. Leung, "Joint Optimal Power Flow Routing and Vehicle-to-Grid Scheduling: Theory and Algorithms," *IEEE Transactions on Intelligent Transportation Systems*, vol. 23, no. 1, Jan. 2022, pp. 499–512, doi: 10.1109/TITS.2020.3012489.

13. B. Li et al., "Routing and Scheduling of Electric Buses for Resilient Restoration of Distribution System," *IEEE Transactions on Transportation Electrification*, vol. 7, no. 4, Dec. 2021, pp. 2414–2428, doi: 10.1109/TTE.2021.3061079.

14. O. Oladimeji, A. Gonzalez-Castellanos, D. Pozo, Y. Dvorkin and S. Acharya, "Impact of Electric Vehicle Routing With Stochastic Demand on Grid Operation," *2021 IEEE Madrid PowerTech*, 2021, pp. 1–6, doi: 10.1109/PowerTech46648.2021.9495092.

15. J. Shi, Y. Gao and N. Yu, "Routing Electric Vehicle Fleet for Ride-Sharing," 2018 2nd IEEE Conference on Energy Internet and Energy System Integration (EI2), 2018, pp. 1–6, doi: 10.1109/EI2.2018.8582518.

16. C. Yan, M. Zhu and Y. Shi, "A Response Time based Load Balancing Algorithm for Service Composition," 2008 3rd International Conference on Pervasive Computing and Applications, 2008, pp. 13–16, doi: 10.1109/ICPCA.2008.4783561.

17. O. Alatise, A. Karlsson, A. Deb, R. Wu and J. Ortiz-Gonzalez, "Expanding EV Charging Capacity in Distribution Networks: A Case Study for Charging EVs at Work," CIRED 2021—The 26th International Conference and Exhibition on Electricity Distribution, 2021, pp. 2945–2949, doi: 10.1049/icp.2021.1919.

18. E. Ucer, M. Kisacikoglu and M. Yuksel, "Analysis of an Internet-Inspired EV Charging Network in a Distribution Grid," 2018 IEEE/PES Transmission and Distribution Conference and Exposition (T&D), 2018, pp. 1–5, doi: 10.1109/TDC.2018.8440252.

19. Y. Chen, H. Z. De La Parra and F. Hess, "Dynamic Simulation of EV Fast Charging with Integration of Renewables," 2012 IEEE International Electric Vehicle Conference, 2012, pp. 1–5, doi: 10.1109/IEVC.2012.6183187.

20. L. Gong, W. Cao and J. Zhao, "Load Modeling Method for EV Charging Stations Based on Trip Chain," 2017 IEEE Conference on Energy Internet and Energy System Integration (EI2), 2017, pp. 1–5, doi: 10.1109/EI2.2017.8245572.

21. X. Chen, H. Wang, F. Wu, Y. Wu, M. C. González and J. Zhang, "Multimicrogrid Load Balancing Through EV Charging Networks," *IEEE Internet of Things Journal*, vol. 9, no. 7, 1 Apr. 2022, pp. 5019–5026, doi: 10.1109/JIOT.2021.3108698.
22. T. Tudorache, A. Marinescu and I. Dumbrava, "On-road Charging System Demonstrator for EVs," 2019 Electric Vehicles International Conference (EV), 2019, pp. 1–4, doi: 10.1109/EV.2019.8892887.

16 Design of Eco-Friendly Batteries Using PCM

Lina Momani, Chanyalew Belachew,
Ahmed Abdulla Qasem, Hamad Hareb,
Jasim Yousif, Ali Ahmed, and Abdullah Ali

16.1 INTRODUCTION

As the world's population grows, the global energy demand grows as well. Recent projections predict that primary energy consumption will rise by 48% in 2040 (Doman, 2016). However, the fear of fossil fuel depletion has urged several organizations and entities around the world to explore new methods of obtaining energy that has less impact on the environment and is sustainable as well. According to the international renewable energy agency (IRENA), the total renewable energy generation capacity was 2,351 GW at the end of 2018 (IRENA, 2020; Petrova, n.d.), which is a third of the total installed capacity. Solar power generation has a share of nearly 24% of the world's total installed capacity (C2ES.org, n.d.).

To meet the world's energy needs sustainably, energy storage is becoming an increasingly important piece of the puzzle. While energy production technologies like solar and wind have come a long way in recent years, being able to store that energy for use when it's needed most can make all the difference in creating a reliable and resilient energy system. With advances in energy storage technologies, there is a potential to transform the way energy is generated and used, making it more efficient, cost-effective, and environmentally friendly.

Energy storage devices are systems that store different types of energy. The most common method of energy storage in solar power generation is battery systems. Energy storage devices are systems that store energy in various forms such as electrochemical, kinetic, pressure, potential, electromagnetic, chemical, and thermal.

The International Energy Agency stated that lithium-ion batteries are the most utilized technology globally (Sani et al., 2020). However, batteries have a limited life span, and they can negatively impact the surrounding environment if not disposed of properly. Such depletable batteries contradict the purpose of using renewable energy, which strives to provide energy without harming the environment. Current systems lack high efficiency and have lower capacities. The cost of these systems is high. Therefore, there is a need to develop efficient thermal energy storage systems. Such systems can be more efficient and environmentally friendly energy use. Meanwhile keeping the costs low, small form factor, and portable.

DOI: 10.1201/9781003343332-16

16.1.1 PROBLEM STATEMENT

Solar energy is adequate and has the highest potential for a sustainable future, and many companies are exploring and developing new technologies to obtain and use this energy. However, the use of batteries is contradictory to the principle of sustainability. Thermal energy storage is therefore a crucial bottleneck solution toward the widespread and integrated use of renewable energy systems. The challenge is to find an appropriate solution that can fix the issue of storage with a proper design and low cost.

To solve the issue of depletable batteries that can cause harm to the environment, whether from manufacturing or disposal, a thermal energy storage system is proposed that can store energy by using the physical properties of a phase-changing material. The thermal storage system is composed of relatively cheap materials compared to widely used alternatives such as lithium-ion batteries. Also, the system's components use materials that are widely available in the market in terms of the metals used, storage material, and insulation, and those materials have little to no harm to the environment, which makes them easier to maintain and replace if needed. Therefore, such a battery indeed implies the concept of sustainability and completes the missing link in clean energy, which is clean and sustainable storage.

16.1.2 MOTIVATION

The global energy demand and consumption are tremendously increasing due to rapid growth in the world population and economies. According to the International Energy Agency, primary energy production and CO_2 emissions have increased by 49% and 43%. There is a growing necessity for energy storage due to the intermittent nature of renewable energy sources (Mohan, 2018). Such batteries will have a wide market, it would be of interest to energy storage industries, governments, self-employed individuals, or project owners and individuals.

There are some challenges associated with using PCMs in batteries, including finding the right combination of materials and designing an enclosure that can accommodate the phase change. However, with careful design and testing, a PCM battery could be a promising alternative to traditional batteries for certain applications.

Phase-changing materials (PCMs) are materials used to store and release thermal energy during the processes of melting and freezing. They release/absorb energy upon freezing/melting in the form of latent heat to/from the immediate environment. This enables thermal energy storage from one process or period and is used at a later time or transferred to a different location. Another application for PCMs, they are used to provide thermal barriers or insulation, for example, in temperature-controlled transport.

Unceasing global economic development leads to a continuous increase in energy demand. Considering the limited conventional resources of energy as well as the impact on the environment associated with its use, it is important to focus on the rational management of energy resources and on supporting the development of new technologies related to both conventional and renewable energy resources. The use of PCMs turns out to be a reasonable solution.

16.2 THERMAL ENERGY STORAGE (TES)

16.2.1 DEFINITION

Thermal energy storage (TES) is the temporary storage of high- or low-temperature energy for later use, bridging the gap between requirement and energy use. A storage cycle might be daily, weekly, or seasonal depending on the system design requirements, and whilst the output will always be thermal, the input may be thermal or electrical (PCM Limited, n.d.).

TES deals with storing energy by heating, cooling, melting, solidifying, or vaporizing material, and the thermal energy becomes available when the process is reversed. Energy demands vary on daily, weekly, and seasonal bases. These demands can be matched with the help of TES also. TES is a significant technology in systems involving renewable energies and other energy resources as it can make their operation more efficient, particularly by bridging the period between periods when energy is harvested and periods when it is needed (Abedin & Rosen, 2020; Dincer & Rosen, 2002; Enescu et al., 2020).

16.2.2 TYPES OF TES

There are several types of TES systems. Each type of TES system has its advantages and disadvantages, and the choice of system depends on the specific application and requirements. In the following section three types are described.

16.2.2.1 Sensible Heat Storage (SHS)

Responsive heat storage is a device where the energy is stored in any temperature of the storage material and operates based on the heat power and the related temperature difference of the respective storage material during the charging and discharge phases. The specific heat of the medium, the temperature variation, and the amount of storage content determine how much thermal energy can be retained. The medium of storage is available in two forms, solid and liquid. Water is a common commercial heat storage medium with a range of applications in the residential and industrial sectors. It is also one of the most widely used mediums since it is the cheapest. SHS has two major advantages in that it is relatively inexpensive and has low danger as fewer dangerous materials are used (Socaciu, 2012; Whiffen & Riffat, 2013).

16.2.2.2 Latent Heat Storage (LHS)

In the LHS system, the heat containment process takes place during the phase change that occurs at a relatively stable temperature and corresponds proportionally to the fusion heat of the product. The best choices for the LHS system are high thermal conductivity phase change materials (PCMs) and high fusion heat. It also presents itself as a material that exhibits high energy containment and maintains a uniform temperature during the process. The PCMs represent themselves as materials that display high energy content as well as maintain a uniform temperature during the process. It should also have a high density and relatively low volume variance as conversion between phases takes place to reduce containment volume, low cost, and high availability for large-scale applications (Abedin & Rosen, 2020; Socaciu, 2012).

TABLE 16.1

A Comparison between Different Types of TES (Socaciu, 2012)

Performance	Sensible TES	Latent TES	Thermochemical TES
Media materials	Water, rock, brick, soil, etc.	Paraffin, ice, salts, etc.	Porous materials, hydrated salts, composite sorbents, etc.
Storage density	Small (54 kWh/m³)	Moderate (84–140 kWh/m³)	Large (200–840 kWh/m³)
Lifetime	Long	Limited	Depends on the reactant degradation
Technology status	Available commercially	Available commercially for some temperatures and materials	Generally not available, but undergoing research and pilot project tests
Advantages	• Low-cost, reliable, simple application with available materials	• Medium storage density • Small volumes • Short-distance transport possibility	• High storage density • Low heat losses (storage at ambient temperature) • Long storage period • Long-distance transport possibility • Highly compact energy storage
Disadvantages	• Significant heat loss over time (depending on the level of insulation) • Large volume needed	• Low heat conductivity • Corrosivity of materials • Significant heat losses (depending on the level of insulation)	• High capital costs • Technically complex

16.2.2.3 Thermochemical Heat Storage (THS)

Thermochemical reactions typically produce more heat than the process of phase change and sorption. The reversible thermochemical reaction can be used for heat storage with high energy density. All possible reactions are the hydration reaction of salt hydrate, except for the coordination of ammonia suggested in sorption heat storage (Abedin & Rosen, 2020). A comparison between different types of TES is shown in Table 16.1.

16.3 PHASE CHANGE MATERIAL

16.3.1 Definition

Materials capable of absorbing or releasing large amounts of energy, at certain periods and under specific operating conditions are called phase change materials (PCMs). These materials are capable of storing 5–14 times more energy per unit of volume than materials that store energy via sensible heat, such as water, concrete, or rocks, and present specific phase-changing temperatures that tend to remain constant during the transformation of matter (Sharma & Chen, 2009).

16.3.2 TYPES OF PCM

There are many types of PCMs (phase change materials), which can be classified based on their chemical composition, melting/freezing point, and other properties. The following section elaborates on the most two common types of PCMs.

16.3.2.1 Organic PCM

Organic PCMs, which are typically made from materials like paraffin, fatty acids, and sugar alcohols, are commonly used in building applications. Paraffinic PCMs are particularly popular due to their ability to adjust their melting temperature and phase change enthalpy based on the length of their carbon chains. When the number of carbon atoms in the paraffin molecule falls between 13 and 28, the melting temperature typically falls within a range of −5 to 60°C, making them suitable for use in buildings in a wide range of climates around the world. They also possess other desirable qualities such as chemical inertness, non-toxicity, reliability, and biocompatibility, and have a minimal subcooling effect. Additionally, fatty acids like capric acid, lauric acid, and palmitic acid have similar storage densities to paraffin, and their melting temperatures increase with the length of the molecule. However, they may react with the environment due to their acidic nature (Ravikumar, 2005; Zhou et al., 2012).

16.3.2.2 Inorganic PCM

Inorganic PCMs offer a broad range of temperatures. They have similar latent heat per unit mass as organic PCMs but generally have a higher latent heat per unit volume because of their higher density. One example of inorganic PCMs is salt hydrates, which are composed of one or more water molecules combined with inorganic salts. They are known for their non-toxicity, non-flammability, low corrosiveness, and high thermal conductivity, which is higher than organic PCMs (Patil et al., 2020; Ravikumar, 2005; Zhou et al., 2012).

16.4 INCORPORATING PCM IN TES

Thermal energy storage (TES) is one of many technologies that are used to save energy and consume it later. TES is storing thermal energy by heating or cooling a storage medium and saving it for later use. TES system using phase change materials (PCMs) as storage medium offers advantages such as high TES capacity, small unit size, and isothermal behavior during charging and discharging when compared to the sensible TES. PCMs suck up energy during the heating process as phase change takes place and release energy to the environment in the phase change range during a reverse cooling process. In this article, we will demonstrate a model that we produced for experimentation purposes. The model is a composite PCM that combines paraffin wax and metal foam to enhance thermal conductivity with minimum cost and system complexity.

There are several methods for incorporating phase change materials (PCMs) in thermal energy storage (TES) systems, depending on the specific application and requirements. The choice of the PCM incorporation method depends on several factors, such as the operating temperature range, heat transfer rate, thermal stability, mechanical strength, and cost. Some common methods are illustrated in the following sections.

16.4.1 ENCAPSULATION

The encapsulation process is a crucial aspect of using PCMs. Different companies have developed various techniques for encapsulating PCMs. One example is EPS, which specializes in versatile macro-encapsulation solutions. The technology for encapsulating PCMs continues to evolve to improve performance. The PCM and container material mustn't chemically interact with each other to ensure long-term stability and prevent corrosive mixtures (Whiffen & Riffat, 2013),(Cabeza et al., 2011).

16.4.2 DIRECT APPLICATION

Early research on PCM focused mainly on direct integration and direct immersion methods. In the direct integration method, PCMs in either liquid or powder form is mixed into the bulk material during processing. This approach is attractive due to its simplicity, but it has limitations such as leakage issues and inconsistencies in the material. Due to these challenges, it is not widely used in practice (Zhou et al., 2012).

16.4.3 BULK STORAGE

Bulk storage is another method of incorporating PCM in TES, which involves storing the PCM in large tanks. This approach is cost-effective as it requires minimal material processing. However, one issue that has been encountered is poor thermal conductivity, which leads to fluctuating efficiency. To overcome this, methods such as agitation and increasing the surface area have been explored as ways to improve heat transfer and enhance the efficiency of bulk storage (Regin et al., 2008).

16.4.4 MACRO-ENCAPSULATION

PCM can be encapsulated in containers that range in size from 10 cm to a few meters in length (EPS Limited, n.d.). Standard units are incorporated into the technology to optimize heat transfer. These macro-encapsulation units are anti-corrosive and can be used to encapsulate inorganic PCMs. One issue that can arise when using these containers is poor heat transfer within the PCM, which can cause solidification on the edges and decrease efficiency (Zhou et al., 2012).

16.4.5 MICROENCAPSULATION

Microencapsulated PCM (MPCM) is a type of PCM that consists of small particles (core) of PCM, a few microns in diameter, encased in a sealed membrane (shell).

MPCM can be integrated into various applications through methods such as uniform sealing in a solid matrix or suspension in a heat transfer fluid (HTF). MPCM is a preferred method for modern PCM applications due to its high surface area-to-volume ratio, which minimizes the effects of poor thermal conductivity (Tyagi et al., 2011). The small scale of the capsules also means that the total thermal conductivity of the shell is greatly influenced by the thermal conductivity of the shell material. MPCM also allows for internal volume changes without affecting bulk volume changes. MPCMs can be produced using various physical and chemical methods such as pan coating, air suspension coating, centrifugal extrusion, vibration nozzle, spray drying, interfacial polymerization, in situ polymerization, and matrix polymerization. MPCMs are classified based on the core structure and shell deposition process, they can be mononuclear, polynuclear, or matrix encapsulation. In mononuclear, a single nucleus with a single shell around it; in polynuclear, several cores are surrounded by a single shell. In matrix encapsulation, the core material is homogeneously dispersed in the shell material. The production process of MPCMs usually follows four stages (Neto, 2020): emulsification, coacervation, cross-linking, and filtration.

16.4.6 SHAPE-STABILIZED PCM

PCM is dispersed homogeneously over a supporting material, such as high-density polyethylene (HDPE) to form a composite material that has several advantageous properties. These properties include high thermal conductivity, stability, and the ability to retain stable shapes during the phase-change process (Zhou et al., 2012).

16.4.7 SLURRY

There are two main methods for transporting phase change materials (PCMs) in thermally active fluids (slurries): microencapsulation (MPCM) and direct mixing of paraffin in water using immiscible fluids. MPCM involves encapsulating the PCM in a molecular vinyl coating, which allows for expansion without leakage into the slurry. In contrast, direct mixing uses sufficient surfactants to allow for the solubility of paraffin in water. Initial investigations suggest that MPCM slurries may be more cost-effective, faster to process, and more efficient in terms of heat transfer, but further research is needed to fully understand the process of melting and solidifying water for broader use of paraffin water slurry (Lu & Tassou, 2012).

16.4.8 HEAT TRANSFER

PCM technology is a method of controlling heat transfer by utilizing materials that change phase at specific temperatures. The performance criteria for PCM technology are determined by factors such as thermal conductivity, the exchange area between the PCM and the heat transfer medium, and the temperature difference between them. During operation, heat transfer occurs through convection, conduction, and radiation. In models of packed bed systems, the efficiency of the system is often expressed using the effective conductivity and the total coefficient of heat transfer (Cabeza et al., 2011).

16.4.9 ENHANCEMENT

PCMs in bulk systems can have poor heat transfer characteristics due to their low thermal conductivity (Zhou et al., 2012). This can make it difficult for heat to be transferred efficiently through a bulk system that uses PCM, resulting in slower heat transfer rates and lower overall performance. Additionally, the low thermal conductivity of PCMs can also lead to temperature gradients within the bulk system, which can cause uneven heating and cooling. Research has been conducted to improve the heat transfer capabilities of PCMs in bulk systems. One approach has been to improve the thermal conductivity of the PCM itself by incorporating materials such as metal foams, Al powders, carbon fiber, and extended graphite (EG) structures (Agyenim et al., 2010; Lu & Tassou, 2012). Another approach has been to improve the heat transfer through the container by increasing its emptiness and surface area to volume ratio (Regin et al., 2008). Additionally, the addition of metal fins and additives has been shown to improve conductivity, but the expense and incompatibility of such PCMs can be a barrier to advancement (Zhang et al., 2012). EG has been found to have great potential for the development of paraffin-based systems. It has high thermal conductivity, high stability, high compatibility, and lower density than metal promoters. Studies have shown that using EG paraffin with 92 weight percent PCM resulted in a minimal loss of latent heat while speeding up the conduction rate by more than 250 min (Whiffen & Riffat, 2013).

16.5 MODEL PCM FOR TES

Undergraduate students of the Capstone Design project have been working on this topic which resulted in a prototype that could be the future of PCM batteries. In the next section, an introduction to the design of those two prototypes is summarized. Both of the projects implemented the same structure with minor differences. Figure 16.1 illustrates the work breakdown structure for the proposed systems.

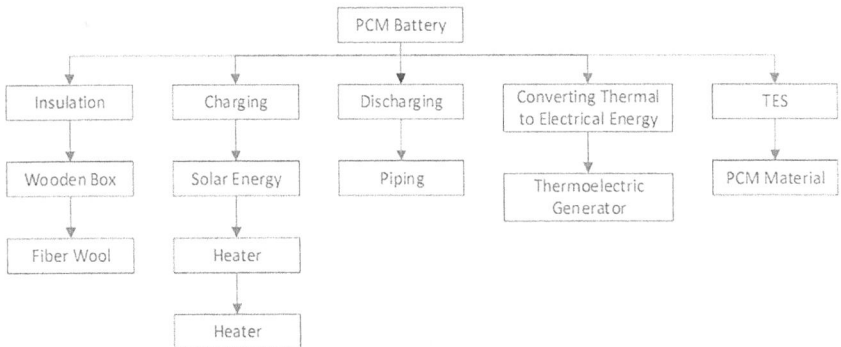

FIGURE 16.1 Work breakdown structure for the model of PCM thermal energy storage.

16.5.1 PROJECT DESCRIPTION

The purpose of the project (Qasem et al., 2021) is to design a model that is safe and eco-friendly for thermal energy storage by using PCM that allows charging and discharging when the phase changing takes place. An organic material was used that can be easily disposed of without harming the environment, with minimum cost and system complexity, and the battery is relatively light. Such batteries will be accessible to users and easy to dispose of or recycle with a small and convenient size to ensure wider use. Paraffin wax was used as PCM material due to its low melting point and other advantages. Various methods were applied to improve the conductivity and thermal storage capacity of paraffin.

16.5.2 MODEL DESIGN

Figure 16.2 represents the proposed system. The system consists of a wooden box to ensure low conductivity of heat and provide proper insulation for the battery. The box would be internally covered with three to four layers of fiber wool to insulate the heat of the internal system. Copper pipes will feed the water through the battery while being surrounded with paraffin wax. Initially, the paraffin is in a solid state. Solar energy is used to heat the paraffin (another alternative would be heating the paraffin with a concentrated solar heating system). In addition, a copper mesh was used to improve the thermal connectivity of paraffin and ensure the uniform distribution of the heat. During the charging stage, the temperature of the paraffin will increase and convert it from a solid state into liquid, and energy stored as latent heat in the paraffin. During the discharge stage which takes place at the night, water is passed through the copper pipes in the heated paraffin, which cools down the paraffin, and this is accompanied by a discharge of latent heat, thus increasing the temperature of the water. This heated water is discharged from the battery and harvested either as thermal energy or electrical energy through the thermoelectric generator to be used for different applications such as USP, LED, and lamps. At the same time, the hot water coming out of the battery can be used for bathing or heating a building. The system is run by a controller box that consists of a switch button and three lights; when the battery is fully charged, the green light will turn ON, when the battery is empty, the red light will turn ON, and if the battery is in process of charging, the yellow light will turn ON (Qasem et al., 2021).

16.5.3 COMPONENTS

Figure 16.3 illustrates the main components of the proposed system. The list of components is detailed here:

16.5.3.1 Phase Change Material

Three alternatives for PCM were investigated: paraffin wax, honey wax, and coconut oil. The properties of these PCM materials are shown in Table 16.2. The paraffin wax was chosen for our model because of its features such as low cost, recyclability, ability to melt congruently, and chemical stability. It is classified as organic material.

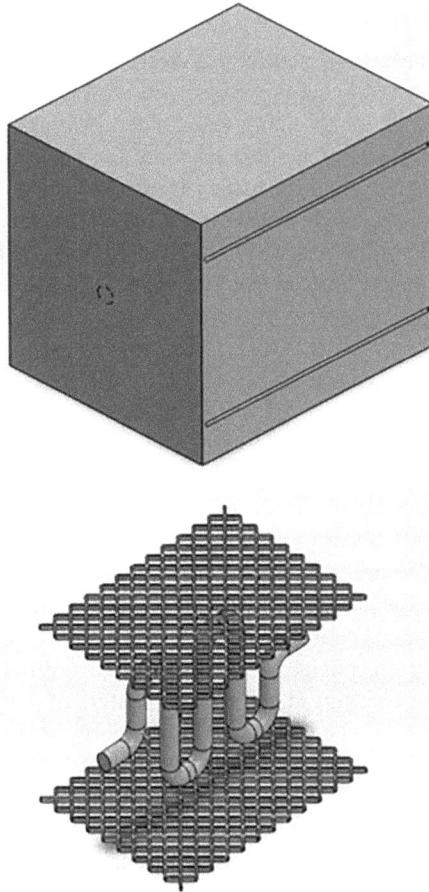

FIGURE 16.2 Schematic drawing of the PCM battery: external view and internal view (Qasem et al., 2021).

Paraffin wax possesses several properties that are suitable for making a battery, a far better alternative than lithium-ion, being cheaper and more energy dense. This material utilizes the temperature difference between day and night for storage and releases thermal energy for multi-energy storage applications such as cooling, heating, and producing power. Paraffin wax is an organic material that can be acquired easily from local stores, is relatively less expensive than other PCMs, and is harmless for the environment. In addition, it is chemically inert, non-toxic, reliable, biocompatible, and shows negligible subcooling effect.

16.5.3.2 Thermoelectric Generator (TEGs)

Thermoelectric generators represent a reliable, robust, and compact way of directly converting heat into electricity. A TEG is a solid-state device that converts

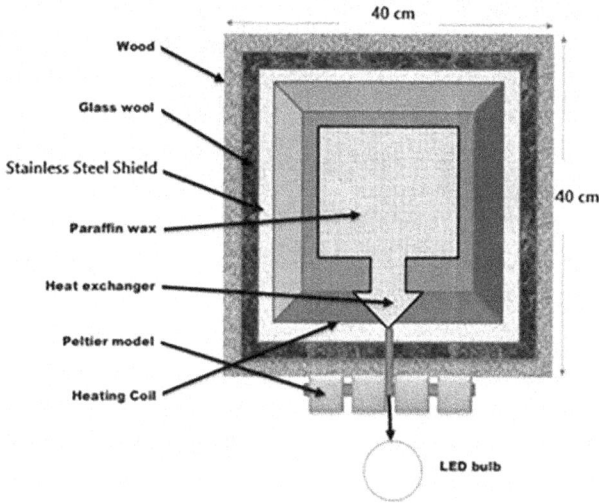

FIGURE 16.3   Main components of the PCM battery (Qasem et al., 2021).

TABLE 16.2
Properties of PCM Materials

PCM	Specific heat	Melting point (°C)	Evaporation point (°C)
Paraffin wax	2.14–2.9 kJ/kg.K	46–68°C	> 370°C
Honey wax	0.54–0.6 kJ/kg.K	40–50°C	
Coconut oil	114.6 kJ/kg.K	17.38°C	204°C

temperature differences into electrical energy or TEG Power, using a phenomenon called the "Seebeck effect." Their typical efficiencies are around 5%–8%, it is possible to generate 56 mV of electric voltage and 3 W/cm² of electrical power density under 37°C of body heat temperature and 15°C of environment ambient temperature. The thermoelectric (TE) effect is the direct conversion of a temperature difference between two dissimilar electrical conductors or semiconductors to an electrical voltage. When the sides of TE materials are exposed to different temperatures, then a voltage is created across the two sides of the material. Conversely, when a voltage is applied, a temperature difference can be created, known as the "Peltier effect" (Coherent, n.d.).

Thermoelectric power generation provides reliable, renewable energy for a variety of industries and applications. Benefits include ease of installation, safe storage (safer than most existing batteries), can operate in any geographic location, cost-effectiveness, and low maintenance (Enescu, 2021). Figure 16.4 shows different types of TE generators used in the project. As illustrated in Figure 16.4, two heat sinks and four thermoelectric generators were used to convert heat flux (temperature

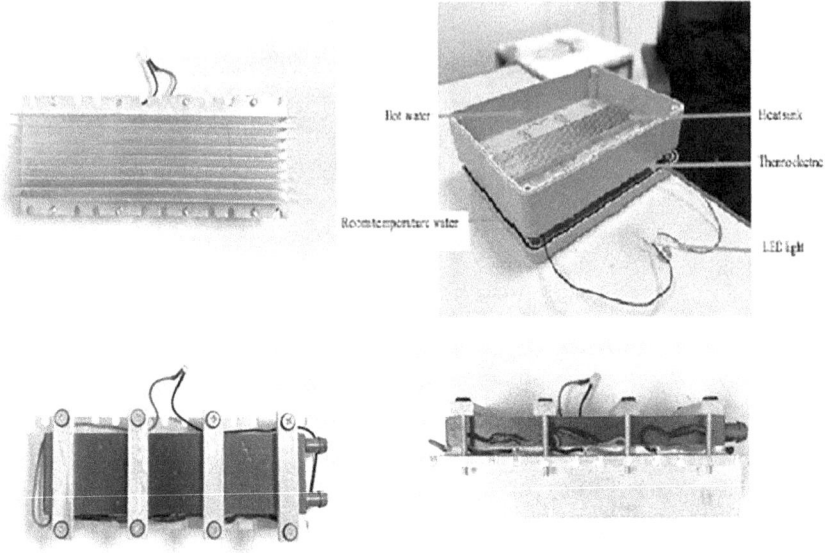

FIGURE 16.4   Four thermoelectric generators with heat sinks.

differences) directly into electrical energy by adding hot water from the tap at the top and room-temperature water at the bottom.

16.5.3.3   Insulation

16.5.3.3.1   Mineral Wool

Mineral wool is porous with an open fiber structure. Mineral wool traps air, making it highly efficient as an insulation material. The wool's porous nature also makes it a good noise absorber. It is also incombustible and does not fuel the spread of flame and even it can slow it down.

Mineral wool also has low thermal conductivity and resists the passage of sound. It is highly resistant to expansion and shrinkage, which results in joints that stay as close as possible. It is also relatively inexpensive. There is also zero water absorption, environmentally friendly product, and fire resistance—the material belongs to a class G1 (Širok et al., 2008). Mineral wool was used to thermally insulate the inner parts and wood from the outer box, thus reducing the thermal losses. At the same time, it protects any user from burns thus improving the safety aspects.

16.5.3.3.2   Glass Wool

Glass wool is an insulator that is globally used in thermal insulation, sound absorption, and noise reduction in construction, the chemical industry, electronics, and many other industries. Glass wool is used in several industries and applications due to non-combustibility, non-toxicity, corrosion-resistance, small bulk density, low thermal conductivity, strong chemical stability, low moisture absorption, and good water repellency. The thermal conductivity of glass wool is 0.04 W/m.K and its thermal diffusivity is 0.023×10^{-6} m^2/s.

16.5.3.3.3 Wood Box

Wood exhibits a low thermal conductivity (high heat-insulating capacity) compared with materials such as metals, marble, glass, and concrete. Thermal conductivity is highest in the axial direction and increases with density and moisture content; thus, light, dry woods are better insulators. That's why wood has been chosen to insulate the proposed system. The advantages of using wood are long-term sustainable construction, high strength concerning weight, shorter construction times, eco-friendly summer heat protection and winter cold protection, without harmful additives, sustainable and certified products, and ease of use (PUUINFO, n.d.).

16.5.3.4 Digital Temperature Controller (Thermostat)

A digital temperature controller is an essential instrument in the field of electronics, instrumentation, and control automation for measuring and controlling temperatures. It can be used in home and industrial applications. Different types of analog and digital temperature controllers are readily available in the market, but they are generally not only expensive, but their temperature range is also usually not very high (Rathore, 2022). The digital temperature controller (Thermostat) was used in the projects to set the required heating temperature and display the instantaneous temperature of the water and PCM material.

16.5.3.5 Controller Box

A control box provides the physical interface to allow an operator to control a piece of equipment and monitor its performance. Control boxes typically contain a variety of instruments such as switches, knobs, sliders, and buttons. These are connected to the equipment and are used to control it, allowing operators to start, stop, or adjust various functions. The controller box will control the proposed system. The power switch button (silver) is used to turn the ON/OFF of the system. In this model, the LED light will be used as an indicator of the charging status. The green color refers to a fully charged battery, the yellow color refers to the charging process is in process, and the red color means the battery is empty

16.5.3.6 Charging the PCM

In the actual prototype, solar panels will be used to charge (heat) the PCM by converting the solar energy into heat. However, in this demonstrative model, a heating coil shown in Figure 16.5 is used. The 2,000 watt coil ensures uniform and fast charging of the PCM for the planned experimental work. The heater will heat the PCM to reach the required temperature to change its phase to store the heat energy as latent heat.

16.5.3.7 USB Controller

USB ports shown in Figure 16.6 allows USB devices to be connected and transfer digital data over USB cables. In addition, they can supply electric power across the cable to devices that need it. In the proposed system, it will be used as an output port of the battery to charge any device such as a mobile or laptop. The USB controller is the main output port of the proposed system. The power which will be generated by the thermoelectric generators will be used to run out the USB controller. The USB controller consists of a display screen that will show the output current and voltage of the system.

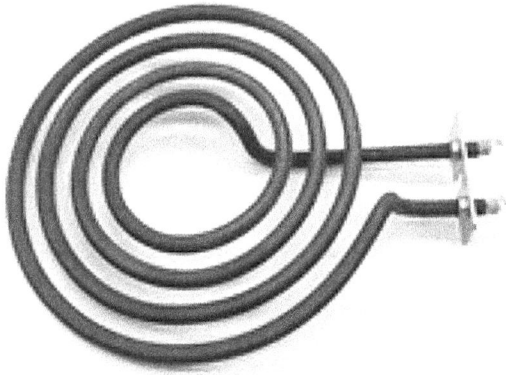

FIGURE 16.5   Heating coil.

FIGURE 16.6   USB controller/ports and display screen.

16.5.3.8   Copper Tubes

Copper is a material that is durable, ductile, and malleable. These properties make copper perfect for the formation of tubes. Copper has many features and specification that makes it a great material for tubes, some of the advantages that we are considering are excellent heat conductivity, corrosion resistance, good machinability, and nonmagnetic, which are very crucial when installing it near an electrical circuit because copper will not hinder it and the ability to retain mechanical properties when exposed to low temperatures. Copper tubing is most often used for heating systems and as a refrigerant line in HVAC systems. Copper offers a high level of corrosion resistance. Copper is a tough, ductile, and malleable material, with excellent heat conductivity, excellent electrical conductivity, good corrosion resistance, and good biofouling resistance. These properties make copper extremely suitable for the proposed system. The copper tubes in this model are used to carry the coolant liquid (water) within the system during discharging.

16.5.3.9   Heat Transfer Enhancing Fins

Most PCMs have very low thermal conductivities and require features to enhance heat transport to effectively utilize all the latent heat potential. This can be achieved

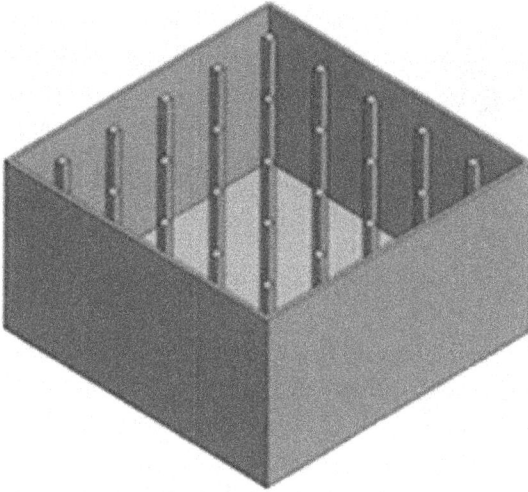

FIGURE 16.7   Addition of fins to battery design.

easily by incorporating meshes or fins or even heat pipes into the PCM to better distribute the heat and evenly melt or freeze the PCM. Metal mesh can be applied to PCM since metal is a good conductor of heat. Such an addition would improve the thermal conductivity of the PC material and increase the speed of heat transfer. The material of the mesh should have a high melting point, high electrical resistivity relative, corrosion and oxidation resistance, and non-magnetic and high thermal conductivity. Examples of such materials are copper and nichrome. Nichrome has better properties in this respect compared to copper. The appropriate number of meshes, their orientation, and the material of the meshes should be considered when designing such PCM batteries in addition to the type of PC material to store the highest amount of energy with the least size and cost.

Alternatively, mixing 10% of copper powder with paraffin can improve thermal conductivity. However, in this model fins are used instead of meshes and copper powder to ensure uniform distribution of energy within the PC material as shown in Figure 16.7.

16.5.3.10   Resistance Wires

The resistance wires will be used as connection wires to resist heat from the hot water and the hot coconut oil to avoid damaging of wires.

16.5.3.11   Heat Exchanger

Heat exchangers are devices designed to transfer heat between two or more fluids. Its design is an exercise in thermodynamics, which discusses heat energy flow and temperature. Heat exchangers are utilized in cooling or heating depending on the application. The heat exchanger design we selected is a hybrid flow heat exchanger that has a combination of cross and counter flow. By applying this flow, we can get

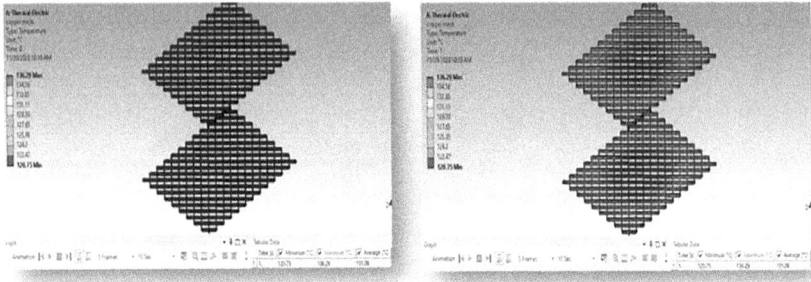

FIGURE 16.8 Thermal simulation of the meshes at the start and the end of the charging process.

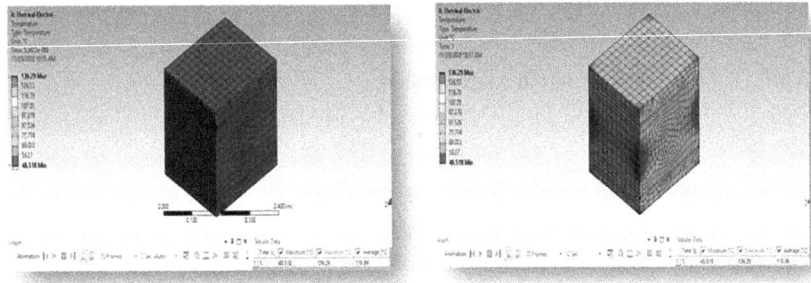

FIGURE 16.9 Thermal simulation of heating paraffin wax using the meshes before and after the heating process.

the desired output while overcoming several obstacles such as cost and available space. Also, this orientation allows the fluid to have a larger heat transfer area.

16.6 SIMULATION

Ansys software was used for simulations to observe and analyze the behavior of the system. The main components, that is, meshes, paraffin wax, and tubes were simulated in their proposed geometry, dimensions, and assigned material. The operating conditions were defined, that is, input electricity through meshes.

Figure 16.8 illustrates the thermal simulation of the mesh at the start and end of the PCM charging process (heating process) when electricity flowed through it. It is observed a rise in the internal temperature of the copper mesh with the highest temperature at the center of the mesh (red color), 136.29°C, and the minimum temperature at the edges of the mesh (green color), 120.75°C.

Figure 16.9 shows the thermal simulation of the paraffin wax. It can be observed that thermal changes happen when the meshes are heated. The temperature of the wax ranged between 107°C to 115°C (yellow color). The minimum temperature was at the top and bottom surfaces ranging from 87.5°C to 97.2°C. Nevertheless,

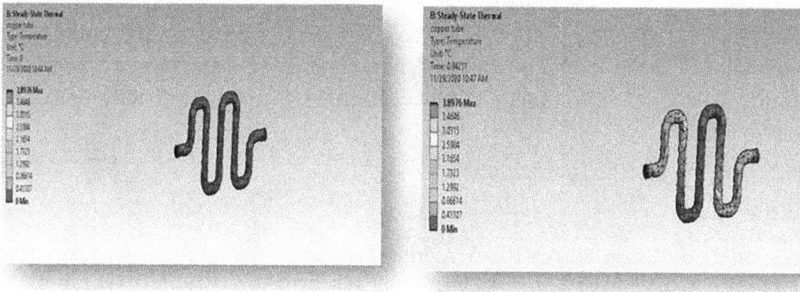

FIGURE 16.10 Thermal simulation of heating copper tube using the water; before cooling and after cooling.

the temperature range surpasses the melting point of paraffin wax, which is ranged between 46°C to 68°C. This indicates that the design succeeded in changing the phase of the material using heat that is generated from the electrical resistivity of the mesh metal.

Figure 16.10 illustrates the thermal simulation in the water tubes, where the cold water flows through them and the cooling process starts at the ends of the copper tubes. Then observed a gradual decrease in temperature is distributed throughout the rest of the tube. Such simulations can help in determining heat loss, the number of meshes included in the design, their orientation, and the material of the meshes. In addition, different PCM materials can be tested to select the best PCM that will ensure the highest stored energy possible with minimum size and cost.

16.7 APPLICATIONS

Currently, PCM batteries used in large-scale solar energy collectors and for heating buildings are salt. Such batteries can be used in different applications to replace non-degradable current batteries, especially when used with solar sources where the outcome can take two forms. The first form is hot fluid, for example, water or air, or oil that can be used for heating/cooling applications which, if used along the traditional heating systems, would reduce energy consumption. The second form is the conversion of thermal energy into electrical energy that accordingly can be used in different applications, for example, providing the required grid for domestic applications.

16.8 FUTURE SCOPE AND CONCLUSION

There are challenges and issues facing current batteries: no established system for recycling, less life cycle, and it's difficult to dispose of. Developing an eco-friendly battery (that stores clean renewable energy in the form of thermal energy by using PCM) is a future target that might come true soon. The biggest challenge is reducing the size of the batteries and utilizing the best option of phase change material that is available in the local market and would improve the efficiency of the battery storage. In addition, thermal energy storage can be combined into a smart integrated energy system that can improve the efficiency of the overall system and provide the required power for domestic applications (heating, lighting, hot water, electricity) in different output forms.

REFERENCES

Abedin, A. H., & Rosen, M. A. (2020). A critical review of thermochemical energy storage systems. *Open Renew*, *4*, 1. https://benthamopen.com/ABSTRACT/TOREJ-4-42

Agyenim, F., Hewitt, N., Eames, P., & Smyth, M. (2010). A review of materials, heat transfer and phase change problem formulation for latent heat thermal energy storage systems (LHTESS). *Renew. Sustain*, *14*(2), 615–628. https://doi.org/10.1016/j.rser.2009.10.015

C2ES.org. (n.d.). *Renewable energy. Solutions, center for climate and energy*. www.c2es.org/content/renewable-energy/

Cabeza, L. F., Castell, A., Barreneche, C., de Gracia, A., & Fernández, A. I. (2011). Materials used as PCM in thermal energy storage in buildings: A review. *Renew Sustain*, *15*(3), 1675–1695. https://doi.org/10.1016/j.rser.2010.11.018

Coherent. (n.d.). *How does a thermoelectric cooler module work?* Retrieved February 18, 2023, from https://thermalbook.wordpress.com/how-does-a-thermoelectric-cooler-tec-work/

Dincer, I., & Rosen, M. (2002). *Thermal energy storage: Systems and applications*. Wiley.

Doman, L. (2016). *EIA projects 48% increase in world energy consumption by 2040*. U.S. Energy Information Administration. www.eia.gov/todayinenergy/detail.php?id=26212

Enescu, D. (2021, January 21). *Thermoelectric energy harvesting: Basic principles and applications*, 2. www.intechopen.com/chapters/65239

Enescu, D., Chicco, G., Porumb, R., & Seritan, G. (2020). Thermal energy storage for grid applications: Current status and emerging trends. *Energies*, *13*, 2. https://doi.org/10.3390/en13020340

EPS Limited. (n.d.). *Thermal energy storage (PCM and ice-based applications)*. Environmental Process Systems Ltd. Retrieved February 18, 2023, from www.epsltd.co.uk/

IRENA, R. E. S. (2020). *International renewable energy agency*. Abu Dhabi.

Lu, W., & Tassou, S. A. (2012). Experimental study of the thermal characteristics of phase change slurries for active cooling. *Applied Energy*, *91*(1), 366–374. https://doi.org/10.1016/j.apenergy.2011.10.004

Mohan, M. (2018). Perovskite photovoltaics. *Perovskite Photovoltaics: Basic to Advanced Concepts and Implementation*, 447–480. https://doi.org/10.1016/B978-0-12-812915-9.00014-9

Neto, R. C. (2020). *Micronal ® PCM intelligent temperature management for buildings a broader base for your success*. Retrieved November 6, 2020, from www.academia.edu/34694086/Micronal_PCM_Intelligent_Temperature_Manageme

Patil, D., Tech, M., Raisoni, G. H., & Karale, S. (2020). Design and analysis of phase change material based thermal energy storage for active building cooling: A review. *Undefined*, *2012*.

PCM Limited. (n.d.). *Integrated thermal management applications*. PCM Products Ltd. Retrieved February 18, 2023, from www.pcmproducts.net/Thermal_Energy_Storage.htm

Petrova, V. (n.d.). *Global renewables capacity reaches 2.35 TW in 2018*. Renewables Now. Retrieved February 18, 2023, from https://renewablesnow.com/news/global-renewables-capacity-reaches-235-tw-in-2018–649118/

PUUINFO. (n.d.). *Thermal properties of wood.* https://puuinfo.fi/puutieto/wood-as-a-material/thermal-properties-of-wood/?lang=en#:~:text=The thermal conductivity of wood is perpendicular to the grain.

Qasem, A. A., Hareb, H., Yousif, J., Ahmed, A., & Ali, A. (2021). Thermal energy storage components: Design & simulation. *Capstone Design Project, 2021, 22.* https://thermalbook.wordpress.com/how-does-a-

Rathore, A. (2022). Digital temperature controller: Full circuit diagram with explanation. *Electronicsforu.Com, 13.* www.electronicsforu.com/electronics

Ravikumar, M. (2005). Phase change material as a thermal energy storage material for cooling of building. *Journal of Theoretical and Applied Information Technology, 10.*

Regin, A. F., Solanki, S. C., & Saini, J. S. (2008). Heat transfer characteristics of thermal energy storage system using PCM capsules: A review. *Renew. Sustain, 12*(9), 2438–2458. https://doi.org/10.1016/j.rser.2007.06.009

Sani, S. B., Celvakumaran, P., Ramachandaramurthy, V. K., Walker, S., Alrazi, B., Ying, Y. J., Dahlan, N. Y., & Rahman, M. H. A. (2020). Energy storage system policies: Way forward and opportunities for emerging economies. *Journal of Energy Storage, 32.* https://doi.org/10.1016/j.est.2020.101902

Sharma, A., & Chen, C. (2009). Solar water heating system with phase change materials. *International Reviews in Physical Chemistry, 1,* 297–307.

Širok, B., Blagojevic, B., & Bullen, P. (2008). *Mineral wool.* Woodhead Publishing. https://www-sciencedirect-com.ezproxy.hct.ac.ae/book/9781845694067/mineral-wool#book-info

Socaciu, L. (2012). Thermal energy storage with phase change material. *Leonardo Electron, 11*(20), 75–98.

Tyagi, V. V, Kaushik, S. C., Tyagi, S. K., & Akiyama, T. (2011). Development of phase change materials based microencapsulated technology for buildings: A review. *Renew. Sustain, 15*(2), 1373–1391. https://doi.org/10.1016/j.rser.2010.10.006

Whiffen, T. R., & Riffat, S. B. (2013). A review of PCM technology for thermal energy storage in the built environment: Part I. *International Journal of Low-Carbon Technologies, 8*(3), 147–158. https://doi.org/10.1093/ijlct/cts021

Zhang, Z., Zhang, N., Peng, J., Fang, X., Gao, X., & Fang, Y. (2012). Preparation and thermal energy storage properties of paraffin/expanded graphite composite phase change material. *Applied Energy, 91*(1), 426–431. https://doi.org/10.1016/j.apenergy.2011.10.014

Zhou, D., Zhao, C. Y., & Tian, Y. (2012). Review on thermal energy storage with phase change materials (PCMs) in building applications. *Applied Energy, 92,* 593–605. https://doi.org/10.1016/j.apenergy.2011.08.025

17 Synthesizing User Comments to Prospective Viral Interests on New Media Using Sentiment Analysis for Effective Data Visualization Premeditated for Journalism 4.0

Monikka Reshmi Sethurajan and Natarajan K

17.1 INTRODUCTION

Since its inception in 1997, social media has experienced a meteoric rise in popularity. Furthermore, social media's importance has consistently shown that it cannot be ignored. Since technology is always changing, social media is one of the public communication channels that is growing, developing, and having the most effect. Given the huge changes social media endured in the last ten years, predicting its future is rather challenging. However, it's reasonable to predict that social media will continue to exist for a very long time, perhaps even until the extinction of humanity.

In India, Facebook has a market share of 77.26% in the social networking space. India, as already mentioned in the article, has the highest Facebook users, with over 300 million members as of January 2020. YouTube has a market share of 9.03%, whereas Pinterest has a share of 7.15%. The data pertains to the months of December 2019 and December 2020.

Since the year 2020, rent life has taken on a radically different trajectory and, in many respects, differs significantly from years before. Countries being kept under lockdown for months has affected our needs, priorities, and manner of life. Before the year 2020, we had quite different ways of earning money, spending it, learning, and even interacting with others. We couldn't leave our homes for days or even months, so the Internet and social networking sites were our only options. We've only just realized the actual importance of social media and its impact on our lives.

 DOI: 10.1201/9781003343332-17

Statistical data may not be very fascinating to you if social media is not handled properly and in practice. If you wish to benefit from or establish relevance for your business with regard to all of the themes discussed before in the previous section, you must understand where your audience spends the bulk of their time.

By interacting with your audience on these popular websites, you can help your business get greater visibility while assuring the success of your campaign. A somewhat different approach would be required for other websites; for instance, LinkedIn would call for a more formal and professional demeanor, while Instagram would place more focus on aesthetics.

Over the last few years, data journalism has tended to focus more on stories based on structured data rather than unstructured data (e.g., text). Various studies conducted by the Data Journalism Awards from 2012 to 2016 indicated that geographic and financial information were the most common sources, in addition to sensor, self-reported data, socio-demographic data, and data from polls and other sources (Loosen et al., 2020); however, according to projections made by "Data Management solutions review" and "IDC," the Internet will be filled with 80% unstructured data for the next five years. Since media organizations are going to deal with a growing range of social media postings, conferences, mails, and lengthy government statistics, computer methods for evaluating and evaluating these resources have become more crucial. Sentiment analysis, which automatically classifies text by sentiment (positive, negative, neutral) and reads for the writer's viewpoint and emotion, is one of the most prominent field of application for unstructured data gathering, according to several studies engineers, and data scientists with skills in natural language processing used to be the only ones who could perform sentiment analysis. In recent years, however, the AI community has developed astounding tools to foster deep learning access. Some reporters would have seen the value in categorizing comments or papers as good, bad, or neutral (or such rating schemes) depending upon the speaker's opinion towards the subject at hand. Examples include evaluating the sentiment over a problem by analyzing a topic, a hashtag, or posts from a Twitter user, including doing analogous computations on news articles or live comments on a site. Discussing the importance and principles behind text data work, including such information extraction, appropriate theories and algorithms, and textual data visualization, can assist you in contributing to Journalism 4.0.

17.2   LITERATURE SURVEY

The AI field has made some efforts to read text employing Sentiment Analysis as a medium. In this mapping research, we highlight some of these efforts, solutions, and suggestions. The assumptions embedded proved to show that there are many constraints to sentiment analysis method, according to this article. Constraints that are frequently overlooked in public dialogue. [1] The method outlined in this paper combined the need for an unstructured MLA in which already dataset was not accessible at initially with a lexicon-based approach that trained and tested the algorithms using data directly from the Twitter API. To determine the sentiment of each tweet, a lexicon-based classifier employed a manually built vocabulary. [2] This study and application as a declaration for people curious regarding sentiment analysis and other forms of SM monitoring. In order to generate deep understandings of such modern occurrences, the essay recommends that a variety of views should be brought together. The perspectives of sentiment analysts, crucial observations deduced from Marxist-influenced structural systems, a moral economy point of view investigating morality and agency in sentiment analysis inference, as well as the outlooks of Internet users—because without events media platforms cognition just wouldn't occur—have been recommended. Putting these varied perspectives altogether enables us all to understand the activity of any and all popular social media sites (monitor) users, as well as the restricting structural contexts under which these behaviors occur. [3] A lot of work has been done in the area of opinion mining and sentiments of customer evaluations to mine opinions in the form of documents, sentences, and feature level sentiment analysis, which addresses a basic difficulty in sentiment analysis: sentiment polarity classification. [4] Social media has risen to prominence in sociopolitical discussions, influencing our thoughts and behaviors in a variety of ways. Along with extensive usage of SMP around the globe, as well as the ability to express oneself, expression that comes with it, various vices have grown in recent years, with racism being one of the most prominent. Racism and related stress appear to be flourishing in a new environment represented by SMP like Twitter. [5] Currently, 22% of individuals in the United States (US) use Twitter [6], with there being 1.3 B profiles and 336 M active users in the world, 90% of whom have a public profile, resulting in 500 M tweets every day. [7] Unless they are made private, tweets are publicly accessible until they are retweeted, tagged, liked, or replied to by Twitter users. [8] A sentimental analysis is the analysis of the opinions, feelings, and attitudes expressed on Twitter. [9] As SMP have grown in popularity, to widespread usage of them for a variety of old and new types of racism. [10] Racist statements and tweets have indeed been made on social networks linked to a variety of mental and physical illnesses, as well as poor health outcomes. [11–13] Within the wake of a global epidemic, the web's relevance was never bigger, and so as contemporary society moves to a much more digital world of living, socializing, teaching, and generating, the use of various SMP has exploded. This research is based on inciting violence, bigotry, and misogyny, and the researchers made a good algorithm to recognize racist and sexist tweets. Studying public information online data can lead to important innovations and morsels of information about nearly every item, organization, or individual within the field of public attitudes. Information extraction is the scientific method of presumptions, ideas, and subjective in data gathered from a database utilizing NLP and ML to discover a range of feelings. [14]

17.3   EXISTING METHODOLOGY

17.3.1   COMPARISON BETWEEN EXISTING METHODS

This research seeks to give readers a full and in-depth overview and analysis of how to extract and display feelings from a text. Current surveys generally rely on methodologies that remove social media from its original settings, resulting in misconceptions and misinterpretations that are harmful to both its use and our larger social knowledge of the medium. In his discussion of sentiment analysis, he implies that traits that were never part of its original design or use are now assigned to it, which may have implications for our understanding of social reality. [1] Our suggested methodology combined supervised and unsupervised modelling in a unique way. As a result, the forecast improved significantly compared to similar research that used label data. To identify the most suitable for our information, the model employed a number of techniques. [2] Sentiment analysts' concerns give insight into the field's empirical realities. Their considerations of analytical accuracy, data quantity, and data quality highlight the challenges of effectively assessing sentiments, as well as the restricted availability and dependability of sentiment in social media. As a result, unless we can properly deploy visualization approaches, the dissemination and influence of sentiment analysis may be rather restricted. [3–5] A sentiment polarity classification and POS procedure has been developed, as well as full explanations of each stage. Preprocessing, pre-filtering, biassing, data accuracy, and other qualities that need machine learning understanding are included in these processes. [6–10] As a result, within the realm of NLP, sentiment analysis is a very well discussed topic. We found that XGBoost with word2vec produced the best results, so we fine-tuned it even more. We also discovered which the F1 score = 0.690285. As a result, one might achieve a 69% accuracy in our classifier. [11–13] The aim of this review is to look into SMP in terms of learning more about racism, its causes, and the first steps toward combating racist ideology. Trends in cognitive, challenge, character traits, beliefs, and coping skills were discovered through a qualitative assessment that included content analysis of 600 American Facebook postings. [14–15] In the past, news reporters often attended meetings of local governments around the country to closely monitor city councils and other organizations on behalf of the general people. But during the previous 15 years, more than 2,000 publications have shut down. [16] Although text-mining techniques are widely used in academics and business, they are still fresh in journalism. [17] Following tokenization, bigram extraction, stop word removal, and part of speech filtering, only nouns, adjectives, verbs, and adverbs are kept in the text. Since we want our model to identify subjects across government agencies rather than just inside one agency, we also utilize entity extraction to remove the names of people who are mentioned in the documents. [18] In addition to learning word vectors, Moody's lda2vec model also learns Dirichlet-distributed latent mixes of topic vectors at the document level. It incorporates the Skipgram Negative-Sampling method created to train word embeddings while utilizing the interpretability of subjects produced by LDA. [19] Utilizing distributed word representations, Nikolenko suggests new coherence measurements, concluding that these metrics more closely resemble human judgement. [20] We may think of the issue of finding news items in our corpus as an event detection challenge by referring to the paper and finding meeting document snippets close to one another in vector space by using techniques like paragraph vector on our corpus is another potential strategy. [21]

17.3.2   LSTM And Topic Modelling

The topic model can be thought of as a methodology to present the enormous volume of data produced as a result of improvements in computer and web technology in low dimension and to present the concealed concepts, prominent features, or latent variables of data, depending on the application context, should be identified effectively. The first approach to dimension reduction was an algebraic one that divided the original matrix into a factor matrix. So in this survey, our topic modelling classification technique is probabilistic model, non-probabilistic topic model (algebraic model). [22] The idea of latent semantic analysis and non-negative matrix factorization was brought to non-probabilistic techniques, which are algebraic matrix factorization methodologies. Both LSA and NNMF employ the bag of words technique, in which the corpus is transformed into a term document matrix, and the order of the terms is completely disregarded—only the terms themselves matter when it comes to documents. To enhance the algebraic model, such as latent semantic analysis, probabilistic models were created using generative modelling techniques. [23] The latent semantic analysis, non-negative matrix factorization, probabilistic latent semantic analysis, and latent Dirichlet allocation are the most well-known topic modelling methods that have contributed to every area of text analysis in several disciplines. [24–25]

This news headlines dataset for sarcasm detection is compiled from two news websites in order to get over the restrictions associated with noise in Twitter datasets. TheOnion tries to produce satirical interpretations of current events, thus we gathered all the headlines from the satirical News in Brief and News in Photos categories. From HuffPost, we pulled actual (and unsarcastic) news headlines. [26] In comparison to the current Twitter datasets, this new dataset has the following advantages: (1) News headlines are written by professionals in a serious way, thus there are no spelling errors or casual language. This decreases sparsity and raises the likelihood of discovering pre-trained embeddings. (2) In addition, because TheOnion is solely dedicated to publishing satirical news, we obtain high-quality labels with far less noise than Twitter datasets. (3) In contrast to tweets that are answers to other tweets, the news headlines we collected stand alone. This would make it easier for us to identify the true sarcastic components. (4) The more subjects there are in a dataset for topic modelling, the more specialized those topics become. After preprocessing the data, we used LDA to generate three out of five out of ten themes. LDA for ten themes aids with our understanding of the subjects. Topic 8 for LDA10, for instance, is typically speaking about social media. Politicians, specifically Donald Trump, are the subject of Topic 9. Children, violence, and climate change are discussed in Topic 2. The ratio of subjects appears to be similar for the sarcastic and non-sarcastic datasets, and there is no discernible difference. Each topic's sizes are comparable. All of this data demonstrates that sarcastic headlines are common and may be seen in all news genres. [27] Text in a document is categorized to a certain topic using LDA. It develops a Dirichlet distribution-modeled topic per document and words per subject model. Every document is represented by a multinomial distribution of themes, and each subject by a multinomial distribution of words. LDA makes the assumption that each piece of text we give it will include words that are connected in some way. Choosing the appropriate corpus of data is essential as a result. It also presumes that papers are created using a variety of subjects. Based on the probability distribution of those themes, words are then generated. According to LDA, each headline (i.e., document) is made up of a variety of themes (multinomial distribution), and each topic is made up of a variety of words (multinomial distribution). [28–29]

17.3.3 NLP Based Data Visualization and Manipulation

In the age of big data, extracting information from text remains a challenging but vital problem. The sheer number of data to be analyzed might overwhelm information to be retrieved, whether it's from consumer feedback, social media posts, or the news. Modern natural language processing (NLP) techniques can help with this. They can analyze prevalent emotions regarding a topic or product (sentiment analysis), summarize/classify essential topics from texts (summarization/classification), and remarkably, even answer context-dependent questions (like Siri or Google Assistant). Individuals and businesses now have access to consistent, sophisticated, and scalable text analysis tools thanks to their development. Even yet, characteristics of languages that are unique to them might make it difficult to examine data for NLP or explain the results. Metrics that apply to the numerical realm, for example, may not be available for NLP. (For instance, what is the mean or standard deviation of a set of word tokens?) Even if they could be computed, getting the facts in front of an audience can be difficult.

NLP is a branch of AI that studies whether systems communicate using different languages, particularly how or when to construct systems which can analyze and interpret large quantities of natural language data. Textual data is routinely classified using natural language processing (NLP). Text analysis is the division of categorizing text material depending on its quality. A most important part of text classification is features design, which is the process of creating characteristics for an ML model from raw text input, as shown in Figure 17.1.

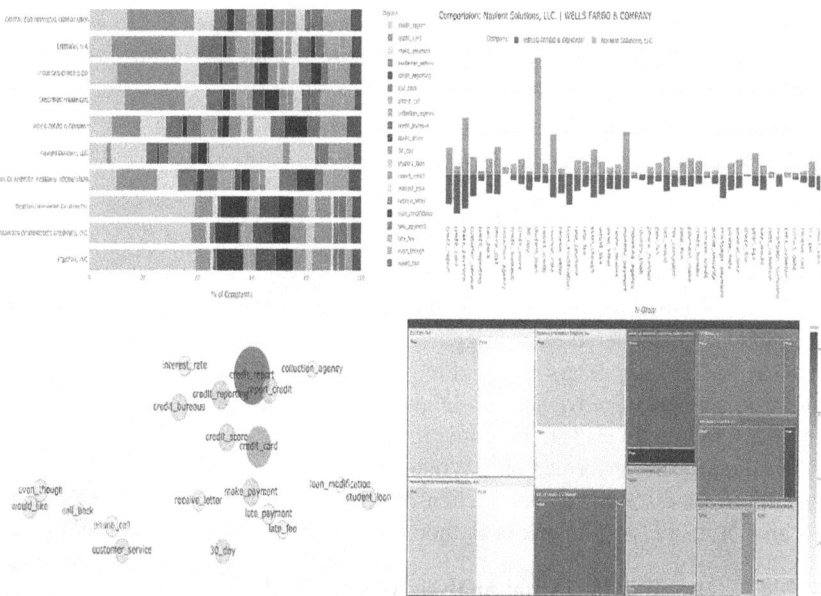

FIGURE 17.1   Sample of NLP visualization.

17.4 RESEARCH METHODOLOGY

The three-stage framework suggested in this study will be used to identify and categorize nasty remarks and associate the inciters' characteristics. The Tweepy API will be used in the initial stage to stream and gather Twitter comments in order to construct a dataset. The second level of preparation for the dataset including Twitter comments will ensure that it is ready for mining and feature extraction. The dataset including Twitter comments will be subjected to a number of data cleaning processes. As part of the cleaning process, formatting elements like punctuation, hyperlinks, and white spaces will be eliminated. Training features will be taken out, and the data will be converted into an array using TF-IDF vectorizer.

In the third stage, classifiers will be trained using a training subset of the preprocessed dataset, and they will then be applied to test data to determine whether Twitter comments are positive or negative. The data will be divided into training and testing sets using Python's Scikit-learn package and the train-test-split methodology. The training data will be used to invoke the algorithm and apply it. Predictions are generated using a range of measures, including as F1 score, precision, and recall on the testing set, after models have been trained using training data. At this stage, we utilize the pickle library to preserve the model so that it may be used again later. The testing information kept in a different file will be utilized to create the model.

The categorization stage's goal is to divide the comments into two categories: one comprises the remark or tweet's content, and the other characterizes the attitude as normal or hostile. The Scikit-learn is the chosen classification method because it offers strong support for multinomial Naive Bayes classifiers. It also provides a range of tools that may be used to enhance the model, including classification, clustering, regression, and visualization methods. The categorization results will be saved in a text file after completion. The percentage of good and negative comments in the file will be calculated, and the results will be shown, using a tool that has been built specifically for this model.

17.5 PROPOSED METHODOLOGY

17.5.1 ANALYSIS OF NEW DATA VISUALIZATION TOOLS

17.5.1.1 Experimental Setup

The data are evaluated using a secondary qualitative technique in the study. The information was gathered from Kaggle and assessed for its potential to spread hate ideas on the Internet. Users of social media have the ability to publicly voice their hate, gain support from friends and followers, and experience some kind of validation which indirectly leads to improper news spread.

The model was performed on a Windows 11 computer with an Intel Corei7 11th generation processor. Scikit, Kara's, Tensor-flow learn frameworks are used to create ML and DL models on Jupyter in Python.

17.5.1.2 Use of POS Textual Analysis and Word Cloud

When visualized, it is established that categorizing tweets would be an intriguing subject of research. Opinion mining, which is based on a collection of identified

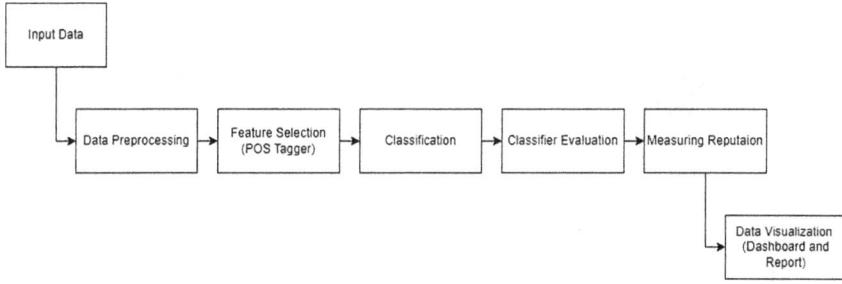

FIGURE 17.2   Sentiment analysis and data visualization.

feature expressions collected from reviews and user comments, has emerged as a fascinating study topic. We may apply this to the field of news categorization based on Internet comments. Caused by SM's prominence in the social economic arena, a variety of old and new forms of bigotry has emerged just on site. Racial prejudice has manifested itself along social networks in a number of forms, either disguised and open, concealed through the use of meme, and allowed access via racist utterances posted behind false names in order to spread hate, murder, and social upheaval. Racism has flourished in recent years based on race, ethnicity, languages, culture, and, most crucially, religion. Hatred is commonly associated with nationality, and now it is flourishing on the basis of race, ethnic background, languages, history, and very importantly, faith. Views and comments on social networks that incite racial divisions are seen as a serious threat to public, economic, stability, and the security for so many nations. As a result, social media, which is the most common medium for the transmission of racist ideas, should be closely watched, and racist statements should be recognized and blocked as soon as possible. The goal of this paper is to use sentiment analysis to identify Tweets that include racist language.

For some time now, data visualization has been an element of both print and television journalism. The interactive graphics used in today's digital journalism are different since they start with a database and continue through presentation. In contrast to infographics, which convey a (or more) particular message through a graphic representation, data visualization, as seen in Figure 17.2, enables the reader to study, explore, and discover information.

17.5.1.2.1 Word Cloud

This idea proposes to categorize hate speech data using text processing and machine learning methods. Using statistical metrics, the outcomes were assessed, and comparisons between the outcomes and the principal methods thought to be state-of-the-art were made. In light of this, a word cloud that provides a visual representation of text data is created. Word clouds are frequently used to display keyword information on websites or to represent free-form text. Tags are often single words, and the font size or color of each one indicates its significance. In contrast to hateful remarks on social media, a word cloud can be created that shows that the most often used terms were those for hate comments. The delicate line between freedom of expression and hate speech is a topic of intense debate at the moment. The first is necessary for a democracy to function; it stands for discourse that is intolerant and sympathetic. So it is important to comprehend what constitutes hate speech and how damaging it can be to a democratic society. Word cloud helps for such projection to happen much easier.

17.5.1.2.2 Tweet Corpus

The definitions of hate speech vary, but they are all comparable. The expression of hate speech refers to beliefs that encourage racial, social, or religious prejudice against particular groups, most frequently minorities. From this perspective, however, it only discusses issues of racial, social, or religious discrimination, ignoring factors like gender, sexual orientation, weight, a particular type of disability, and class, among others. In addition, expressions of prejudiced hatred, contempt, or intolerance against particular groups might be considered a component of hate speech. We may thus infer that hate speech is a series of behaviors with an intolerance-based content directed against groups, most often social minorities, based on these two definitions and the common sense that exists regarding the phrase. Digital journalism is hugely affected with this spread of hate.

It was discovered that the tweet corpus data specified the number of occurrences. Communication has altered as a result of the iIternet. Anyone may share their opinions and leave comments on the seemingly endless topics and materials by only holding a cell phone. But not everything is flawless. Easy was also used for bad. Hate speech and discrimination have become more prevalent, largely due to fake profiles. Based on theoretical ideas of violence and how hatred is positioned in the virtual world, the work aims to explore the link between hate speech and the development of the profile of haters and their statements on social networks. To put an end to this, they examined the speech in a blog post and comments made in a Twitter post in Figure 17.3.

17.5.1.2.3 Tweet Labels

The results demonstrated that the environment of social networks is a motivating factor and is capable of amplifying the aggression of hateful people through the spread of hatred and the advancement of ideology that make up the discourse. This keeps the

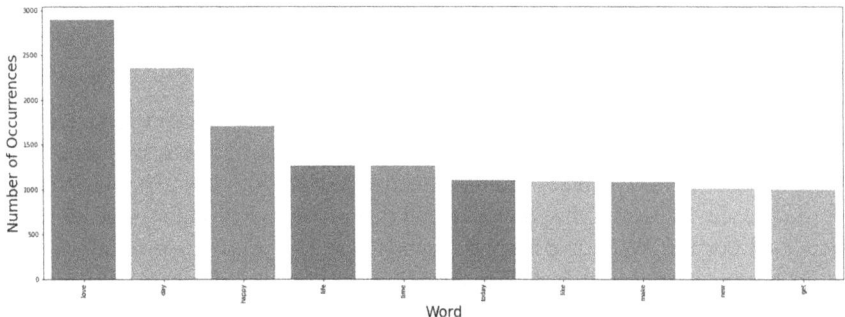

FIGURE 17.3   Occurrences of tweet corpus.

profile design alive and creates a following of biased and prejudiced people. Social media posts that duplicate messages of hatred, racism, intolerance, and violence are monitored by an online program. The tool, which will go live this month, will enable user identification and reporting. The researchers claim that hate speech has become more prevalent online and that human rights are seen negatively. It is essential to stop this process. The idea is that by making the data available, public policies may be made to assist and empower victims. The Internet will look for discussion starters that promote sexual assault against women, racism, and prejudice against black people, Native Americans, immigrants, homosexual people, lesbians, transgender people, and transsexuals. The labels for tweets are shown in the next figure. A score of 0 indicates non-hateful tweets, whereas a score of 1 indicates hateful tweets, in Figure 17.4 and Figure 17.5.

Recently, one has been distinct from the other. One example is the beating of a street seller who attempted to assist transvestites and came from a demographic that was particularly vulnerable to collective violence. In this case, the act truly qualifies as a hate crime motivated by hatred of difference. The notion that "I murder the homeless person or the transvestite that he might become and in such a way this will never become that same person" inspires virtually all of the crimes. There is terror and anxiety of being able to empathize with the victim. It serves as the primary justification for many racist attitudes and acts of ethnic cleansing. This is one form of violent mechanism; another, like in the case of the nightclub in Istanbul, is the urge to demolish the location where Westerners congregate for their infidel parties because "I don't want to be lured by it and kill my own temptation to go insane."

On the other side, a trans woman who lived in the interior and believed she was a monster, the only one of her types, and was doomed for a secret life learns unexpectedly that there are others like her across the world, as well as groups and people eager

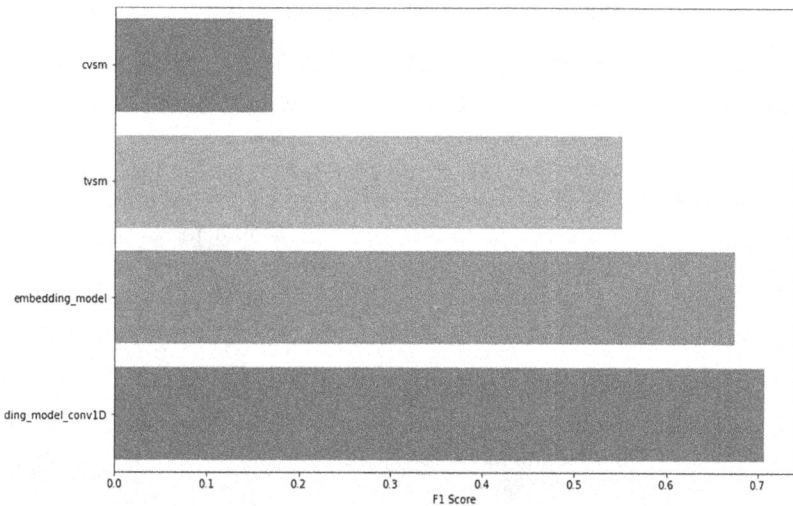

FIGURE 17.4 Corpus lab.

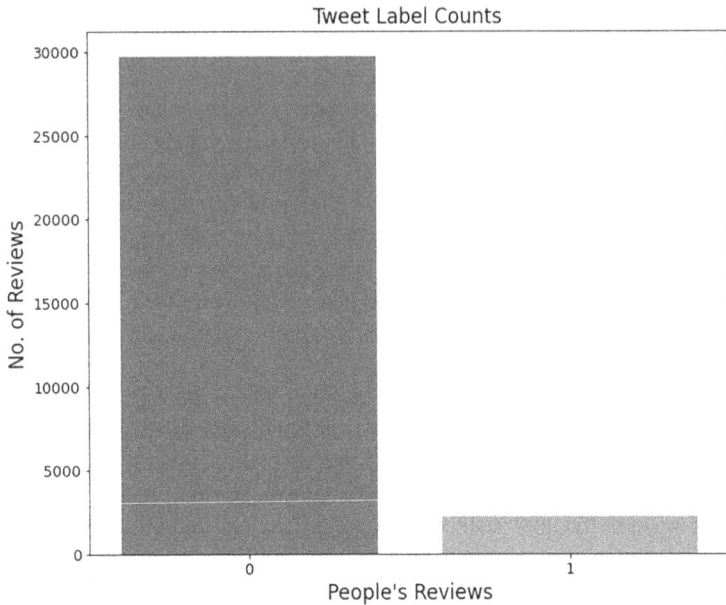

FIGURE 17.5   Tweet count.

to listen and offer assistance. This is another advantage of networks. It is true that social networks are basically constructed on the premise that people are deserving of the respect that they generate in modern society. Or in this instance, the potential amount of likes a post can acquire. Even without social networks, this would still occur. That is, a person's value in today's culture is defined by who and how many other people think the same way about them, not by their history or level of education. This is how modern society functions, whether or not we like it. The issue is that because people thrive when others appreciate them, it's quite simple to get caught up in very great group dynamics.

More intercultural engagement has been made possible by the Internet, particularly social networks. After all, we now have access to a variety of communication tools like Facebook, Instagram, Twitter, and instant messaging application like WhatsApp that let us make new acquaintances and even develop love relationships. Nonetheless, these platforms have consistently been used to spread violent and hateful discourse, which often happens in an anonymous manner. The Internet has brought about a change in mankind that, although on the whole favorable, has also given rise to fierce debates about bias and anti-discrimination ideologies.

17.5.1.3   Steps Involved in Using Text Sentiment Analysis

Step 1: **Data collection**

This research will look into racist trends based on Twitter tweets in particular as shown in Table 17.1. Sentiment analysis relies on this as a crucial factor. Data quality and annotation and labeling will ultimately determine how everything is handled from here on out. Twitter was used to collect the racist tweets dataset. As Twitter has been the most often utilized medium by

TABLE 17.1

Text Sample

User	Text
@_LeBale racism	@_LeBale racism is good
tonyhasanidea	@manoutdoors4 @AJ_Lady_Liberty @FBIWFO @TheJusticeDept ©FBI it is clear to hundreds of millions of people of all walks that this country has a severe problem with systemic racism, your denial is discussing.

a huge number of participants to share their sentiments, opinions, remarks, and thoughts, the majority of the empirical studies have selected this as their main tool for textual and sentiment analysis. This research will look into racist trends based on Twitter tweets in particular.

Step 2: **Data preprocessing**

The data type in issue will dictate how it should be processed: image, text, video, audio. The sentiment analysis procedures in Restate IQ also feature the ability to handle video content analysis with the same simplicity as text analytics. The sub-tasks are listed in the next section.

For ensuring whether any audio and video file (for example, a webcast) within the dataset is just not overlooked, the audio from the video sequences is translated utilizing voice to text software.

Caption overlay—If the movie contains any captions, Restate IQ extracts them and analyzes them for any people, traits, or issues you've highlighted as important. Likewise, the system employs OCR to identify and gather any images embedded in audio or textual information. Any logos that appear in the video backdrop are promptly recognized by Restate IQ's sophisticated data scanner. This includes films that appear on the presenter's clothing or on a desk object such as a pen or cup. It even recognizes logos on posters in the backdrop. When the platform does sentiment research of your brand, it does it in such a way that not even the tiniest element goes undetected.

Text extraction—Only those text is focused on extracting in the very same way during and after the sentiment analysis process. Emoticons and hashtags are included as they're a big part of SM sentiment analysis. Unlike other sentiment analysis solutions, Restate IQ never excludes emojis in data acquisition, as this could contribute to erroneous affirmations or exclusions.

Step 3: **Data analysis**

The following is a list of the several steps that are involved in sentiment analysis at this level.

- Preparing the dataset for training the model—The dataset for training the model must be preprocessed and manually labelled. In order to train the model, the correctly categorized data will be compared to the incorrectly classified data. This will allow a bespoke model to be developed for a specific brand.

TABLE 17.2

Text Before and After Preprocessing

Before Preprocessing	After Preprocessing
@_LeBale racism is good	racism good
@ manoutdoors4	clear hundr million people walk
@ AJ_Lady_Liberty @FBIWFO ©TheJusticeDept	country sever problem system racism denial
@ FBI it is clear to hundreds of millions of people of all walks that this country has a severe problem with systemic racism, your denial is discussing, the world is changing, get on board or gel left	

- Multilingual data—The data for every language has been individually tagged and tested in sentiment analysis processes that involve international processing of data as part of Restate IQ. This is because the platform does not rely on translations at all because enormous disparities in particular languages, such as Spanish and Korean, might cause information to be misinterpreted or nuances to be lost. As a result, Restate gets top accuracy rate as contrasted toward other systems.
- Custom tags—Throughout this phase of the design, customized tags for qualities and topics including such mentions, new product, etc. will be created. When you've already taught the algorithm, this should intelligently split statements based just on tags you've given.
- Topic classification—The topic classifier determines the topic of a document. This line, for instance, would be classified with subcategory "clothes" because it contains the word "Clothes." "The dresses were wonderful, and I discovered some pretty excellent scarves as well."

Step 4: **Data visualization**

After all phases of the sentiment analysis process have been completed, the insights are immediately translated into actionable reports as graphs and charts. After that, the data can be shared across groups. In addition to providing you with detailed and aspect-based reports, these visuals are extremely useful. Once you get an aggregate rating of the business, for instance, one could use sentiment analysis platform to sort through the results to see if aspects got high marks or what got low marks. This should assist you in determining that areas demand increasing emphasis.

As shown in Table 17.2, only those text is focused on extracting in the very same way during and after the sentiment analysis process.

17.5.1.4 Process of Sentiment Analysis and Data Visualization for Twitter API

The tool separates every feature or subject at around this level, which then is analyzed for emotions. The sentiment rating scale runs between −1 to +1. A neutral remark is denoted by the number zero. Whereas the tool assigns specific ratings to numerous aspects, such as efficiency, quickness, cleanliness, capabilities, drinks, atmosphere,

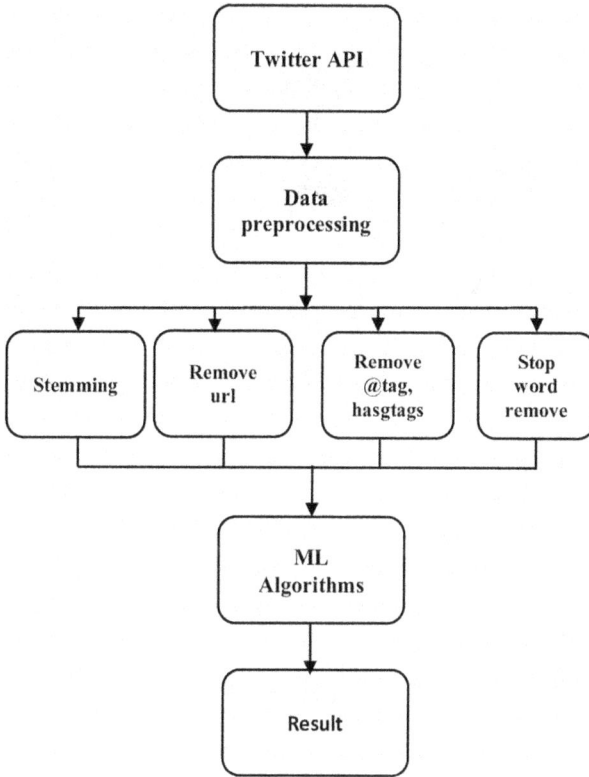

FIGURE 17.6 Flowchart of sentiment analysis.

etc., it's really the overall rating that has been used to assess the viewer's emotional response to a business. Thus, if three of the elements are scored negatively (−.65) and four were evaluated favorably (+.5), the sentiment analyzer would give a total brand sentiment an average rating as shown in Figure 17.6.

17.6 IMPLEMENTATION

17.6.1 Sentiment Analysis by Training Tweets

We first solve using dataset from Kaggle wherein we use to find the Sentiment analysis and later visualize the data to know the topic of interests of users. Firstly, we begin with importing several libraries that are necessary for text analysis, including regular expressions, the natural language toolkit (NLTK), and string. It then filters out warnings related to deprecated features. Finally, it plots a graph using matplotlib. Next, we attempt to read in two CSV files—one containing data on trains, and one containing data on tweets. It then calculates various statistics on the train data, including the average delay time and the most common delay time. It does the same for the tweet data. Later, we detail a code that selects the top ten tweets that have been labeled as non-racist, shown in Figure 17.7.

Out[3]:	id	label	tweet
0	1	0	@user when a father is dysfunctional and is so selfish he drags his kids into his dysfunction. #run
1	2	0	@user @user thanks for #lyft credit i can't use cause they don t offer wheelchair vans in pdx. #disapointed #getthanked
2	3	0	bihday your majesty
3	4	0	#model i love u take with u all the time in ur□□±!!! 😍😍😍😍😍😍😍😍😍
4	5	0	factsguide: society now #motivation
5	6	0	[2/2] huge fan fare and big talking before they leave. chaos and pay disputes when they get there. #allshowandnogo
6	7	0	@user camping tomorrow @user @user @user @user @user @user danny□;
7	8	0	the next school year is the year for exams😭□□" can't think about that 😭□□ #school #exams #hate #imagine #actorslife #revolutionschool #girl
8	9	0	we won!!! love the land!!! #allin #cavs #champions #cleveland #clevelandcavaliers â□
9	10	0	@user @user welcome here ! i'm it s so #gr8 !

FIGURE 17.7 The function returns top ten tweets labeled as non-racist.

Out[4]:	id	label	tweet
13	14	1	@user #cnn calls #michigan middle school 'build the wall' chant " #tcot
14	15	1	no comment! in #australia #opkillingbay #seashepherd #helpcovedolphins #thecove #helpcovedolphins
17	18	1	retweet if you agree!
23	24	1	@user @user lumpy says i am a.. prove it lumpy.
34	35	1	it's unbelievable that in the 21st century we'd need something like this. again. #neverump #xenophobia
56	57	1	@user lets fight against #love #peace
68	69	1	😡©the white establishment can t have blk folk running around loving themselves and promoting our greatness
77	78	1	@user hey, white people; you can call people 'white' by @user #race #identity #medâ□
82	83	1	how the #altright uses & insecurity to lure men into #whitesupremacy
111	112	1	@user i m not interested in a #linguistics that doesn t address #race & .. racism is about #power. #raciolinguistics bringsa□

FIGURE 17.8 The function printing top ten tweets labeled as 1.

On obtaining the output, we next input code to check the tweets with racist tweets and labeling them as 1 (positive). It then prints the first ten tweets that have been labeled as 1 in Figure 17.8.

Next, the code train.shape, test.shape checks if the input and output shapes of a training and test dataset are the same. If they are not, the code prints an error message. Following, the train["label"].value_counts() code calculates the amount of occasions each word comes in a text file. Finally, we use the code for graphing the distribution of the length of tweets, in terms of words, in the training and test datasets in Figure 17.9.

Next task is to write a code which appends the test data to the train data and then splitting the data into train and test sets. Then, we use a code that removes patterns from a text input using a regular expression. Also, it is attempted to remove the user handles of the tweets in Figure 17.10.

Then, we write a python code that removes punctuation, numbers, and special characters except small and large alphabets and # value in Figure 17.11. Along with that, returning short words and splits the text into individual tokens, in Figure 17.12 and Figure 17.13.

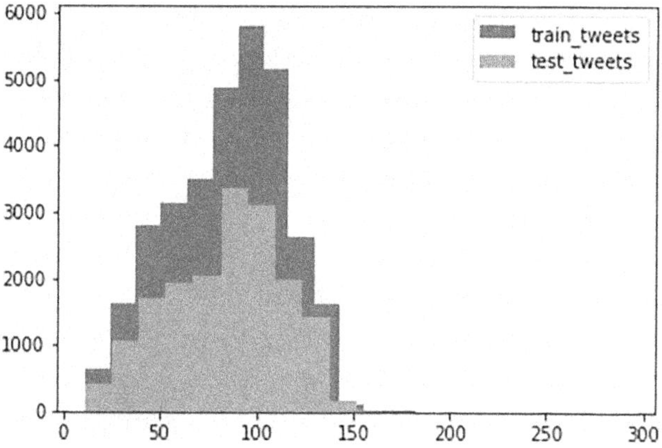

FIGURE 17.9 The graphical distribution of length of tweets.

FIGURE 17.10 Splitting the training dataset.

FIGURE 17.11 Tokenizing the data.

Out[12]:

	id	label	tweet	tidy_tweet
0	1	0.0	@user when a father is dysfunctional and is so selfish he drags his kids into his dysfunction. #run	when father dysfunctional selfish drags kids into dysfunction #run
1	2	0.0	@user @user thanks for #lyft credit i can't use cause they don't offer wheelchair vans in pdx. #disapointed #getthanked	thanks #lyft credit cause they offer wheelchair vans #disapointed #getthanked
2	3	0.0	bihday your majesty	bihday your majesty
3	4	0.0	#model i love u take with u all the time in ur😍😍😍±!!! 😍😍😍😍😍😍😍😍😍😍{😍😍{😍😍{	#model love take with time
4	5	0.0	factsguide: society now #motivation	factsguide society #motivation

FIGURE 17.12 Attempt to remove the user handle.

Out[13]:
```
0                    [when, father, dysfunctional, selfish, drags, kids, into, dysfunction, #run]
1         [thanks, #lyft, credit, cause, they, offer, wheelchair, vans, #disapointed, #getthanked]
2                                                                   [bihday, your, majesty]
3                                                         [#model, love, take, with, time]
4                                                   [factsguide, society, #motivation]
Name: tidy_tweet, dtype: object
```

FIGURE 17.13 Splitting the text to tokens.

Out[16]:
```
0                 when father dysfunct selfish drag kid into dysfunct #run
1         thank #lyft credit caus they offer wheelchair van #disapoint #getthank
2                                                       bihday your majesti
3                                               #model love take with time
4                                               factsguid societi #motiv
Name: tidy_tweet, dtype: object
```

FIGURE 17.14 Stem the words in a tweet.

Next, the Python code in the next section uses the PorterStemmer() function to stem the words in a tweet. Followed by a code that breaks a tweet into individual words and then assigns a "tokenized_tweet" variable to each word. It then calculates a "combined" variable that contains a list of all of the words in the tweet, in the order they occur. Finally, it writes a summary of the tweet in plain English. We next attempt to create a list of tweets that are "tidy"—meaning they are formatted correctly. It then prints the first five tweets from the list in Figure 17.14.

17.6.2 Data Visualization of Tweets

The goal of data exploration is to extract as much knowledge and analysis as possible given Twitter data. It must be emphasized that the information was preprocessed using text. Investigation is done outside for factors which are thought to be worth discussing. For example, consider the variables that were generated. Since we have organized and labeled our data, we can move with downloading packages necessary for implementation of data. We write a code that installs the word cloud package, which is used to create word clouds. It also installs NumPy, Matplotlib, and Pillow, which are all required by word cloud. The code also installs a few packages that are necessary for the word cloud package to work, including matplotlib, word cloud, and kiwisolver packages, and then uses matplotlib to create a word cloud of the text. This will return a list of words from a "tidy tweet" and creates a word cloud using the word cloud library in Figure 17.15.

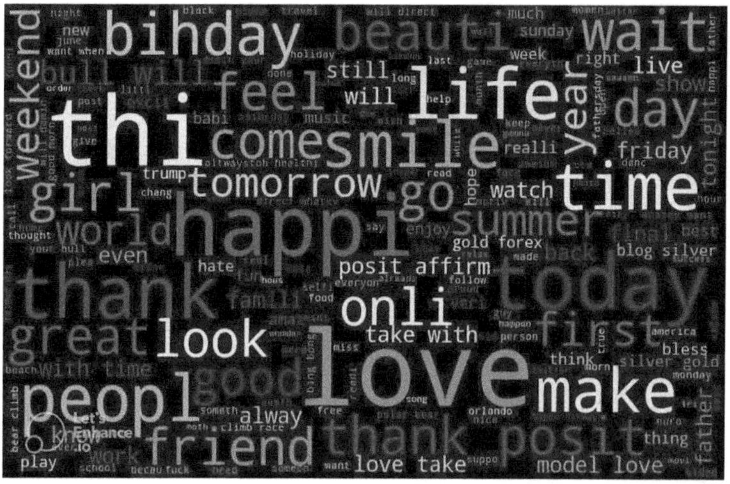

FIGURE 17.15 Word cloud library from a "tidy tweet."

FIGURE 17.16 Word cloud with the most retweeted words.

Next, we can take a list of tweets and sort them by the number of retweets they have received. It then creates a word cloud with the most retweeted words displayed larger than the others, and the output will be shown in Figure 17.16.

Once this is done, we extract hashtags from a tweet moreover from a racist and sexist tweet. It then sums up the totals for both types of tweets. Finally we use nltk. FreqDist() function to create a list of the frequencies of all the hashtags found in the text of a given article. It then uses the pd.DataFrame() function to create a dataframe with two columns, "Hashtag" and "Count." The "nlargest()" function is then used to

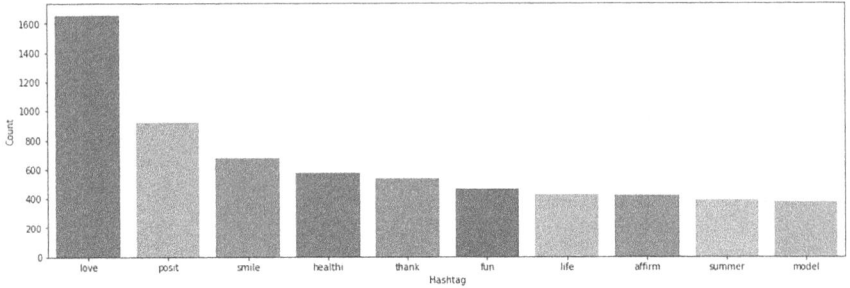

FIGURE 17.17   Frequency of hashtags.

select the ten most frequent hashtags, based on their "Count" column. These hashtags are then plotted as a bar chart. Overall, we use a code to calculate the frequency of hashtags used in tweets that are negative in sentiment and then creates a dataframe of the top ten most frequent hashtags. It then plots a bar chart of the frequency of the hashtags in Figure 17.17 and Figure 17.18.

17.7   RESULTS AND DISCUSSION

After receiving the data visualization for the tweets, one may create the vector for tweets by following these steps, as shown in Figure 17.19.

All algorithms' productivity is assessed using precision, accuracy, F1 score, recall, number of correctly predicted, and number of incorrect predictions. Next, we apply Word2Vec function, which involves the following:

1. The code generates a dataset of 0s with both the rows equivalent to the total of characters in the tweets and 200 columns.

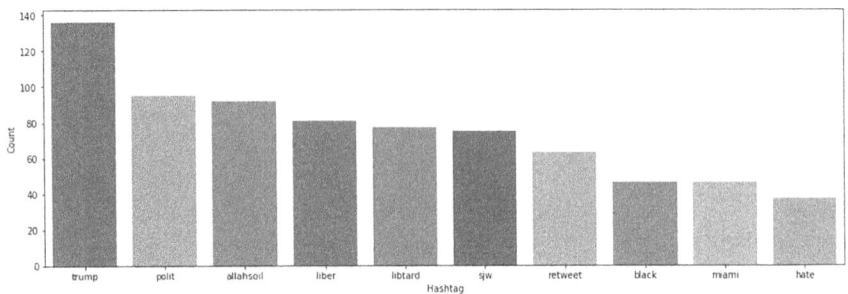

FIGURE 17.18   Top ten of most frequently used hashtags.

```
def word_vector(tokens, size):
    vec = np.zeros(size).reshape((1, size))
    count = 0.
    for word in tokens:
        try:
            vec += model_w2v[word].reshape((1, size))
            count += 1.
        except KeyError: # handling the case where the token is not in
vocabulary continue
            if count != 0:
                vec /= count
    return vec
```

FIGURE 17.19 Code for wordvector.

2. For each token in the tweet, the code creates a column in the dataframe that contains the vector of word vectors for that token.
3. The code trains a LogisticRegression model on the dataframe and measures the F1-score of the model.

Following, one must train the test bag of word features using the train_ bow=bowl[:value",:] to obtain a result for splitting data into training and validation dataset. After which one must train the model.

The F-score, recall, and accuracy were used to calculate the inter annotator agreements (IAA), which were based on the gold standard. In comparison to the first annotator's explanation, which was regarded as the gold standard, precision (P) is the proportion of the apt positive annotated corpus that has been annotated by a second annotator. The number of entities marked up by the second annotator in total as well as the ratio of TP entities, or true positives, were utilized to determine the exactness:

$$P = TP/TP + FP \tag{17.1}$$

FP stands for false positives.

Recall (R) is a proportion of annotated entities that is favorable, according to the second annotator. It is calculated using the following formula as the ratio between the total number of TP and annotations in the gold standard:

$$R = TP/TP + FN. \tag{17.2}$$

TABLE 17.3
Accuracy of the Naïve Bayes Classifier

Accuracy	0.5375
95% CI	0.563, 0.657
Positive Class	1

The F-score is the precision and recall's harmonic mean and is calculated in accordance with the formula as follows:

$$F\text{-}score = 2 * (Precision \times R)/Precision + R. \tag{17.3}$$

17.7.1 Interpretation of the Naïve Bayes Classifier Indicator

Based on Table 17.3, a confidence level of 53%, accuracy was computed as the number of words found in the positive class. A machine learning approach called the Naive Bayes classifier is used to categorize or filter data. It is particularly utilized in natural language processing applications including sentiment analysis, topical categorization of news items, and spam mail screening. Use the naive bayes classifier to categorize, filter, or classify any text data. Examples of use cases are shown here.

- Sorting junk mail
- Categorizing news items
- Analyzing positive and negative opinion

Three separate algorithms make up the Naive Bayes classifier: Guassian, multinomial, and Bernoulli. It is really difficult to fully comprehend any algorithm. To classify texts, you simply need to comprehend multinomial Naive Bayes fits. A multinomial event model, such as bag-of-words, which represents documents as vector spaces by counting words, frequently employs multinomial Naive Bayes.

Based on a confidence level of 53%, accuracy was computed as the number of words found in the positive class. The evaluation that belongs in the category of offensive remarks is positive class. The probability that 56.3% to 65.7% of hate speech will be accurately recognized is 53% based on this degree of confidence. The 65.75% accuracy might be considered to be unsatisfactory, nevertheless.

17.8 ACCURACY OF NEURAL ALGORITHM AND CONFUSION MATRIX

A confusion matrix may be used to correctly gauge a classifier's potential. Each diagonal piece stands for a result that was correctly categorized. The misclassified results are displayed on the confusion matrix's off diagonals. As a result, the ideal classifier will have a confusion matrix that only has diagonal elements, with all other values set to 0. A confusion matrix that results from the categorizing process gives real values and predicted values.

The accuracy gap was attempted to be closed using a neural algorithm, and a confusion matrix was developed as a result of the results. According to the aforementioned matrix, the accuracy percentage for the predicted hate comments category is 53%. As a consequence, this method could be trustworthy enough for users like administrators or government authorities to recognize hate speech online.

The F1 Score was 0.5303408146300915, in addition. Thus, the algorithm chooses the best settings for the twitter data based on the F1 score acquired on a test set. The

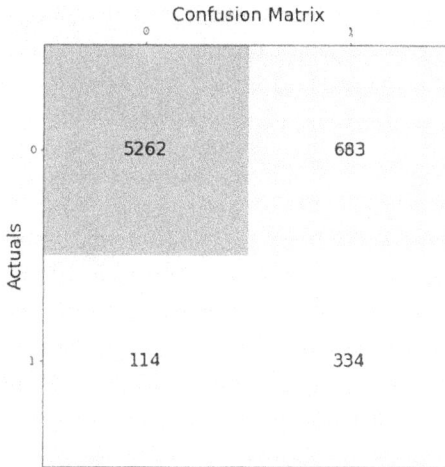

FIGURE 17.20 Confusion matrix of wordvector.

ideal parameters in the confusion matrix of TP TN values are those that offer the highest F1 score in Figure 17.20.

17.9 CONCLUSION

Racist remarks are more widespread on SMS like Twitter, but they will be recognized and deleted immediately to stop it propagating further. This study looks at racism detection via the lens of sentiment analysis, detecting racist tweets by recognizing negative feelings. Many natural language processing (NLP) techniques appear to provide governments with new methods to communicate with their constituents (Figure 17.20). Sentiment analysis methods could be used to track public sentiment and see whether people are reacting positively or negatively to national policy. This may be combined with ML algorithms for better accuracy that is able to detect sentiments well. As an alternative to waiting to talk with a human in a contact center, on official websites, conversational bots can provide quick answers to relevant comments. If NLP technologies are discovered to also be ethnically biased, the goal of rendering government better responsive to people's problems is endangered. Racial bias in NLP has the potential to exacerbate current conflicts by confirming most people's faith in their state doesn't really reflect them, particularly in light of racial disparity in several countries. Although language may appear to be a minor factor in rising racial tensions, it is critical to our capacity to communicate with one another, and in a democracy, everyone is treated equally. Any knowledge which restricts people of color's capacity to freely express sentiment analysis and conversation bots are two possible uses of NLP in government discussed in this research. It looks at three sources of racial prejudice in each of these systems: word embeddings pick up on preexisting biases and assumptions in speech, and programs that use these may reinforce biases towards people of color. Systems must address harsh words or hateful speech, including racial epithets. Predicated on research methods in the established scholarly papers on partiality in NLP and the prevailing techniques used among governments, as well as discussions with academicians and practitioners in NLP. Though NLP may make administrations more responsive to the interests of white residents, it is possible that the needs and views of people of color will be disregarded. We applied the analysis on a Kaggle-based dataset available on GitHub as well. The results show that data visualization helps in categorizing the topics of viral opinion clearly with F1 score to the max accuracy.

17.10   FUTURE WORK

The advent of data visualization greatly accelerates and increases performance by enabling visualizations that may be translated to valuable insights. Nothing beats being able to share real-time information through engaging images. In this study, we looked at the identification of hierarchical feature sentiment for requirement evolution prediction and sentiment analysis and an unsupervised topic model. The main objective is to simplify the recording and understanding of needs changes amongst the various granularities of product feature requirements by product developers. The findings of this study significantly advance efforts for automated text mining analysis for product requirements engineering. The method makes use of hierarchical topic modelling and sentiment analysis to give a comprehensive view of user opinion. The approach was able to locate emotion features that were mentioned in the user emotion text at different granularities using a sentiment orientation. The work presented here demonstrates the potential of topic modelling and sentiment analysis powered by deep neural networks for predicting requirement change from user evaluations. Our collaborative approach is based on the hypothesis that compositional word embedding representations of the relevant characteristics may be used to learn context-dependent sentiment to visualize the data. The adopted model and training objectives may provide a range of distributed word embedding outputs. As a result, the efficiency of the results of the emotion classification may depend on how well the words are embedded. Due to the large number of implicit expressions for characteristics and emotion in the user's review, the text analysis-based approach for user opinion evolution identification needs to be changed to the implicit context. We ought to consider this issue for our forthcoming efforts.

One might also use multidimensional data processing to help you generate an intelligent mapping. The number of analytical tools is rapidly increasing as data visualization becomes the technical forefront. Quick Prototyping is one set of analytical approaches, whereas Charting Libraries is another. By giving our method a voice, we may make further progress in developing more tailored charts to aid the media business in framing stories more efficiently and manipulating statistics if publications appear to be receiving racist comments using ML based model.

17.11 KEYWORDS

Data visualization sentiment analysis
Natural language processing
Twitter
Social media platforms
SARS-CoV-2
Machine learning
Journalism 4.0
Virtual reality
Parts of speech
XGBoost

REFERENCES

1. El Rahman, S. A., AlOtaibi, F. A., & AlShehri, W. A. (2019). Sentiment analysis of twitter data. 2019 International Conference on Computer and Information Science.
2. Kennedy, H. (2012). Perspectives on sentiment analysis. *Journal of Broadcasting & Electronic Media*, 56(4), 435–450. https://doi.org/10.1080/08838151.2012.732141.
3. Medhat, W., Hassan, A., & Korashy, H. (2014). Sentiment analysis algorithms and applications: A survey. *Ain Shams Engineering Journal*, 5(4), 1093–1113. https://doi.org/10.1016/j.asej.2014.04.011.
4. Kaiser, K. R., Kaiser, D. M., Kaiser, R. M., & Rackham, A. M. (2018). Using social media to understand and guide the treatment of racist ideology. *Global Journal of Guidance and Counseling in Schools Current Perspectives*, 8(1), 38–49. https://doi.org/10.18844/gjgc.v8i1.3579.
5. El Rahman, S. A., AlOtaibi, F. A., & AlShehri, W. A. (2019). Sentiment analysis of twitter data. 2019 International Conference on Computer and Information Science.
6. Kennedy, H. (2012). Perspectives on sentiment analysis. *Journal of Broadcasting & Electronic Media*, 56(4), 435–450. https://doi.org/10.1080/08838151.2012.732141.
7. Medhat, W., Hassan, A., & Korashy, H. (2014). Sentiment analysis algorithms and applications: A survey. *Ain Shams Engineering Journal*, 5(4), 1093–1113. https://doi.org/10.1016/j.asej.2014.04.011.
8. Kaiser, K. R., Kaiser, D. M., Kaiser, R. M., & Rackham, A. M. (2018). Using social media to understand and guide the treatment of racist ideology. *Global Journal of Guidance and Counseling in Schools Current Perspectives*, 8(1), 38–49. https://doi.org/10.18844/gjgc.v8i1.3579.
9. Perrin, A., & Anderson, M. (2019, April 10). Share of U.S. adults using social media, including Facebook, is mostly unchanged since 2018. *Pew Research Center*. www.pewresearch.org/fact-tank/2019/04/10/share-of-u-s-adults-using-social-media-including-facebook-is-mostly-unchanged-since-2018/.
10. Twitter—statistics & facts. (n.d.). Statista. Retrieved January 2, 2023, from www.statista.com/topics/737/twitter/
11. Arigo, D., Pagoto, S., Carter-Harris, L., Lillie, S. E., & Nebeker, C. (2018). Using social media for health research: Methodological and ethical considerations for recruitment and intervention delivery. *Digital Health*, 4, 2055207618771757. https://doi.org/10.1177/2055207618771757.

12. Bliuc, A.-M., Faulkner, N., Jakubowicz, A., & McGarty, C. (2018). Online networks of racial hate: A systematic review of 10 years of research on cyber-racism. *Computers in Human Behavior*, 87, 75–86. https://doi.org/10.1016/j.chb.2018.05.026.

13. Price, M. A., Weisz, J. R., McKetta, S., Hollinsaid, N. L., Lattanner, M. R., Reid, A. E., & Hatzenbuehler, M. L. (2022). Meta-analysis: Are psychotherapies less effective for Black youth in communities with higher levels of anti-Black racism? *Journal of the American Academy of Child and Adolescent Psychiatry*, 61(6), 754–763. https://doi.org/10.1016/j.jaac.2021.07.808.10.

14. Williams, D. R., & Cooper, L. A. (2019). Reducing racial inequities in health: Using what we already know to take action. *International Journal of Environmental Research and Public Health*, 16(4), 606. https://doi.org/10.3390/ijerph16040606.

15. Paradies, Y., Ben, J., Denson, N., Elias, A., Priest, N., Pieterse, A., Gupta, A., Kelaher, M., & Gee, G. (2015). Racism as a determinant of health: A systematic review and meta-analysis. *PLoS One*, 10(9), e0138511. https://doi.org/10.1371/journal.pone.0138511.

16. Phelan, J. C., & Link, B. G. (2015). Is racism a fundamental cause of inequalities in health? *Annual Review of Sociology*, 41(1), 311–330. https://doi.org/10.1146/annurev-soc-073014-112305.

17. Devi, B., Shankar, V. G., Srivastava, S., Nigam, K., & Narang, L. (2021). Racist tweets-based sentiment analysis using individual and ensemble classifiers. In *Micro-Electronics and Telecommunication Engineering* (pp. 555–567). Springer Singapore.

18. Bohl, F. (2018, October 6). The future of data visualisation. *Towards Data Science*. https://towardsdatascience.com/the-future-of-data-visualization-2f976b90b93d

19. E Pasquarelli, W. (2019, August 27). Report on "Racial Bias in Natural Language Processing." *FUTURIUM—European Commission*. https://ec.europa.eu/futurium/en/european-ai-alliance/report-racial-bias-natural-language-processing.html

20. Abernathy, P. M. (n.d.). Will local news survive? *Usnewsdeserts.com*. Retrieved May 19, 2023, from www.usnewsdeserts.com/wp-content/uploads/2020/06/2020_News_Deserts_and_Ghost_Newspapers.pdf

21. Stray, J. (2019). Making artificial intelligence work for investigative journalism. *Digital Journalism*, 7(8), 1076–1097. https://doi.org/10.1080/21670811.2019.1630289

22. BibBase. (n.d.). Spacy 2: Natural language understanding with Bloom embeddings, convolutional neural networks and incremental parsing. *Bibbase.org*. Retrieved May 19, 2023, from https://bibbase.org/network/publication/honnibal-montani-spacy2naturallanguageunderstandingwithbloomembeddingsconvolutionalneuralnetworksandincrementalparsing-2018

23. Moody, C. E. (2016). Mixing Dirichlet topic models and word embeddings to make lda2vec. In arXiv [cs.CL]. http://arxiv.org/abs/1605.02019

24. Nikolenko, S. I. (2016). Topic quality metrics based on distributed word representations. Proceedings of the 39th International ACM SIGIR Conference on Research and Development in Information Retrieval.

25. Shah, S. (2017). Reuters tracer: Toward automated news production using large scale social media data. 2017 IEEE International Conference on Big Data (Big Data). www.academia.edu/84267208/Reuters_tracer_Toward_automated_news_production_using_large_scale_social_media_data

26. Kherwa, P., & Bansal, P. (2018). Topic modeling: A comprehensive review. *ICST Transactions on Scalable Information Systems*, 159623. https://doi.org/10.4108/eai.13-7-2018.159623

27. Deerwester, S., Dumais, S. T., Furnas, G. W., Landauer, T. K., & Harshman, R. (1990). Indexing by latent semantic analysis. *Journal of the American Society for Information Science. American Society for Information Science*, 41(6), 391–407. https://doi.org/10.1002/(sici)1097-4571(199009)41:6<391::aid-asi1>3.0.co;2-9

28. Paatero, P., & Tapper, U. (1994). Positive matrix factorization: A non-negative factor model with optimal utilization of error estimates of data values. *Environmetrics*, *5*(2), 111–126. https://doi.org/10.1002/env.3170050203

29. Paatero, P. (1997). Least squares formulation of robust non-negative factor analysis. *Chemometrics and Intelligent Laboratory Systems: An International Journal Sponsored by the Chemometrics Society*, *37*(1), 23–35. https://doi.org/10.1016/s0169-7439(96)00044-5

18 A Data-Driven Model-Based Application to Interpret Anomalies Utilizing Mutual Information in Satellite Data

*Zakea Il Agure, Shaikha Al Qassemi,
and Hicham Itani*

18.1 INTRODUCTION

Satellites have become a part of critical infrastructure in the modern world, widely implemented in commercial, defense, and civil sectors. Satellites provide inevitable services such as communication, remote sensing, and navigation. They further play inevitable roles in different departments to facilitate decision-making. Government agencies implement satellites to conduct scientific research to advance their knowledge of space and the terrestrial environment (Münz et al. 2007). The governments also get to understand the earth's surface, solar system, and atmosphere to understand the possible changes in weather and climate (Münz et al. 2007). Such knowledge informs their policy development to improve human conditions and develop programs that enable them to monitor weather, identify natural and man-made hazards, agricultural development, and the global impact of human activities. One of the applications of satellites has been in the security sector. Global governments have invested in satellites as a part of security surveillance for their armed forces (Azevedo et al. 2012). The defense sectors also engage satellite services to gain intelligence and monitor the environment to improve national security and strategic advantages (Münz et al. 2007). The commercial sector is overly reliant on satellite services because they are the cornerstone of all global communications. The satellites set on the earth's orbit have leveraged efficient business operations globally through electronic monetary transactions and quick information distribution.

Damaged or malfunctioning satellites cause loss of essential strategic services and financial loss. Therefore, early detection, diagnosis, and prevention of these anomalies and faults help enhance the reliability and durability of satellite systems (Azevedo et al. 2012). It should be known that satellites are a composite technique

DOI: 10.1201/9781003343332-18

that has many components that are interconnected and that are restricted simultaneously. Satellites have paved the way for humans to research more into the world and expand their knowledge of what they already have. According to human knowledge or the ability to perceive things that a satellite is in a harsh outer space environment, such as solar radiation, unforeseen anomalies or failures may occur during satellite revolution activities. It combines multidisciplinary technologies such as telemetry sensing, wireless communication, and navigation control. Taking caution to detest these unexpected anomalies or failures will ensure long-term stability in satellite revolution activities. In a nutshell, anomaly detection plays a vital role in satellite fault detection and current health monitoring. The detection and prevention of anomalies in the satellite system help them endure continuity of services. According to Agrawal and Agrawal (2015), satellite owners monitor the health of their systems regularly to track and identify unusual problems in the entire system. The owners track these anomalies through sensors that measure the state of the satellite and its conditions. The sensors then transmit the data on every satellite health to ground stations. The satellite damages have different severity levels based on the impact. The anomalies negatively impact satellites' operations (Agrawal and Agrawal 2015). The severity of satellite damages ranges from temporary errors within bob-critical components to complete mission failures, which are costly to fix (Azevedo et al. 2012). This chapter utilizes link mining to detect satellite data anomalies and their causes. The link mining approach interprets the anomalies in satellite data semantically and then characterizes them in categories for ease of identification and deployment of mitigation measures.

To further delve into the research topic and the mining approach, it is critical to understand what is exactly meant by anomalies and its respective causes. As such, anomalies refer to patterns in data that do not observe normal behavior. There are different causes of anomalies, including malicious activity or some kind of intrusion. This abnormal behavior found in the dataset is interesting to the analyst, and this is the most crucial element for anomaly detection. Khraisat et al. (2019) surveyed the most relevant works in the field of automatic network intrusion, and from their research, there are different types of anomalies, with the main types and subtypes of known fraud. The scholars also presented the nature of data evidence collected within affected industries by providing a detailed survey of data mining applications and their feature scope. Overall, anomaly detection is an application of data mining where various techniques can be applied. Padhy et al. (2012) described ready-made data mining techniques that can be applied directly to detect intrusion, as cited by Agrawal and Agrawal (2015). A wide perspective of the techniques can be practically deployed by viewing the possible causes for the lack of acceptance of the proposed novel approaches.

Satellites experience failures due to malfunctions in their hardware and software components. Such unprecedented failures compromise satellites' efficiency and general performance (Chandola et al. 2009). Therefore, satellites need regular checkups to ensure uninterrupted satellite services. A failure on either of the components, whether the hardware or the software, lends the satellite non-functional. Satellite anomalies occur due to various causes like human error, faulty equipment, and the hazardous natural space environment (Il-Agure 2015). Furthermore, the anomalies may result from the impacts of falling debris from the orbit, malicious actors, and unintended interference from other satellite transmitters (Chandola et al. 2009). The anomalies can result from one of these factors or a combination of these factors. However, they may not immediately be detected by the satellite operators at the time of the event. Anomalies and its respective subtypes further occur when satellite sensors have been damaged or the transmission equipment is interfered with. Changes are detected when monitoring data or when measuring raw data. These significantly affect the condition assessment of measured structures. Detecting abnormal monitoring data is generally difficult and poses serious challenges in arriving at accuracy. However, with the knowledge of the correlation analysis, the hourly segmented measured data is transformed into the matrix form as the input of the deep learning model for training. The ability and success of the proposed method are validated with two datasets of acceleration data, one from an arch bridge and the second from a cable-stayed bridge. The model is trained with the dataset of the arch bridge, and it identifies most anomaly data types in the test dataset with perfect classification performance. The satellite operators need more databases to investigate and catalog the anomalies. These limitations arise from the evolving nature of anomalies. The few databases are limited to specific anomalies affiliated with scientific satellites rather than the broader satellite community Chandola et al. (2009). However, the available databases are openly shared. They are limited to historical anomalies encountered when the operators and sponsoring agencies have limited resources but are willing to share the information openly, a move that poses more security risks (Il-Agure 2015). The databases are sensitive and are designed for particular purposes. Hence, they leverage empirical records about the hardware and software with high vulnerability and specific regions within the most hazardous space under varying solar-terrestrial conditions Chandola et al. (2009).

No centralized, accurate, and up-to-date anomaly database is usable by the broader community. This implies that satellite operators have limited access to information about anomalies. They need more helpful information for diagnosing the causes of different anomalies. Anomalies arising from similar sources occurring at the same time and same regions of space point to environmental hazards as the cause (Il-Agure 2015). The environmental hazards likely to cause anomalies simultaneously and region include solar flares, solar proton, and magnetosphere storms. However, anomalies on a single satellite among other satellites within the same region imply it developed unique problems in its specific hardware and software (Getoor and Diehl 2005). The recurrent anomalies in specific hardware or software components of satellites may create awareness among the operators and learn how to overcome the challenge when it arises again. This saves on the costs of investigating and recovering from the anomalies. It is important to develop a central anomaly database to leverage long-lasting solutions to anomalies in community satellites. A centralized anomaly database outlines the latest information regarding different anomalies on the causes, the components they affect and the regions with hazardous or solar-terrestrial conditions that may affect the satellites.

At the same time, considering the unbearable environment that satellites work in and the complicated design of the system, detecting anomaly performance is not possible in the outer space environment. Currently, the most commonly used method is to directly collect in-orbit operation information of each satellite element by installing many sensors on every part of the satellite. The next step is for these sensors to transmit the in-orbit operation data to the ground telemetry center. The third and last step is that ground personnel can analyze these satellite telemetry data to conduct satellite anomaly detection. Evidently, there has been a growing interest in leveraging satellite data to better understand the environment and its associated phenomena. An important aspect of this is identifying and interpreting anomalies in the data. To do this effectively, it is important to use an approach that can accurately identify and explain the anomalous behavior that may be present in the data. As such, this research will employ the link mining methodology to explore the anomalies and correlation between satellite data.

18.2 METHODOLOGY

This section interprets satellite datasets to leverage valuable insights into establishing the nature of links and their future relationships. It shows how mutual information is useful in exploring and interpreting the detected anomalies in satellite data. However, this approach cannot interpret real-time datasets. Link mining is a preferred methodology for anomaly detection in satellites. In the past few years, link mining has emerged as the most effective method for discovering hidden patterns in datasets. Therefore, it can be utilized for explicit links between objects. Cluster analysis has been utilized in this study to help in grouping objects. The grouping is done based on similarity so that similar objects are grouped. The grouping of objects based on similarity will help identify any anomalies in similar objects Getoor and Diehl (2005).

Several machine learning techniques and detection systems were developed to facilitate the identification of anomalies. Clustering analysis was implemented to help group data based on their behavior and structure without previous assessment of the data (Il-Agure 2016). Using this clustering approach, any data that did not conform to these groups were identified as having anomalies. This mode of anomaly detection approach is achievable through unsupervised machine learning. Data mining

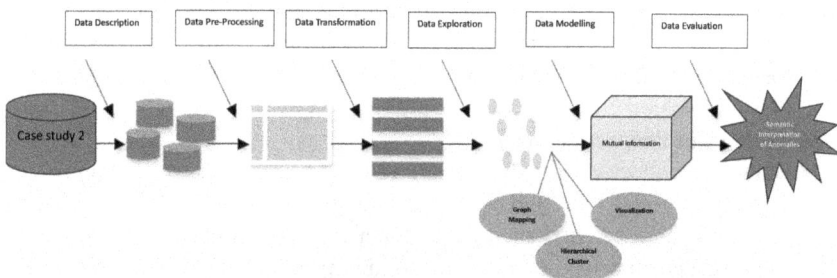

FIGURE 18.1 Link mining methodology.

analyzes satellite datasets to detect anomalies. It simplifies data to help analysts and scientists discover satellites' hidden performance (Getoor and Diehl 2005). Data mining works in seven stages that are sequential and follow each other, namely, the description of data, preprocessing, data transformation, exploration, data modelling, visualization, and finally, data evaluation, as demonstrated in the next figure 18.1.

Stage 1: Data Description

Satellite operators store databases of satellite anomalies, though some departments, such as defense and commercial sectors, do not share their database with the public or other organizations. The satellite databases all conform to different standards for collecting and cataloguing data and in which format; therefore, they need to be identified. Databases with anomaly information from multiple satellite owners have been analyzed in this section. This section also examines the concept of a potential future database helpful in diagnosing anomalies through mutual information. The available databases address specific satellite operation communities. However, the future database should be comprehensive, broadly available, and centralized. This ensures they enhance the ability of the entire space community to identify and diagnose spacecraft problems efficiently. There is also nonparametric reliability of satellite subsystems that is referred to in the database as a Class I failure, which is a failure of a subsystem resulting in the retirement of a satellite, to compute the reliability of the considered subsystem. The nonparametric analysis provides powerful results since the reliability calculation cannot fit any predefined lifetime distribution.

However, this flexibility makes nonparametric results neither easy nor convenient for various purposes often encountered in engineering design, such as reliability-based design optimization. Additionally, the parametric analysis makes some failure trends and patterns more clearly identified and recognizable. Several possible methods are the relative contribution of each subsystem to satellite failure, comparative analysis of subsystem failure, and identify how the culprit subsystems driving satellite unreliability are made. To be specific, there is a quantified relative contribution of each subsystem to the failure of the satellites in a sample. There is also a time dimension to this analysis for analyzing the evolution over time of the relative contribution of each subsystem to the satellite loss. The anomaly database that has been compiled needs to be more comprehensive and is limited to specific satellites from a given timeframe. Figure 18.2 describes the existing anomaly databases to determine how a centralized database may add value.

Stage 2: Data Preprocessing

The raw data collected from the satellites have massive errors, including misspellings, missing data, and noisy and incomplete or wrong data. Therefore, analysis cannot be conducted on raw data collected directly from the data sources. They have to be preprocessed to improve quality and increase the accuracy of the analysis process (Figure 18.3). Preprocessing adequately prepares the data and occurs in the following phases.

FIGURE 18.2 Anomaly database.

* Testing the methodology impeded zero and unknown (x) values to the dataset, to ensure that the system will detect the anomalies or not.

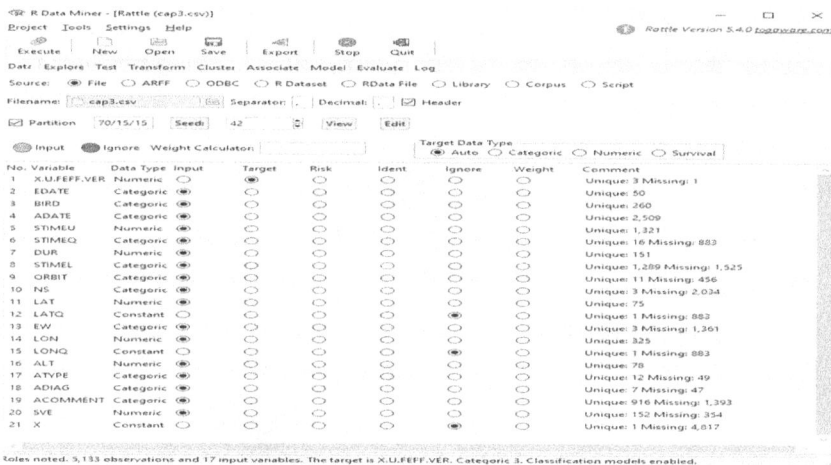

FIGURE 18.3 Data preprocessing.

No.	Variable	Data Type	Input	Target	Risk	Ident	Ignore	Weight	Comment
1	X.U.FEFF.VER	Numeric	○	●	○	○	○	○	Unique: 3 Missing: 1
2	EDATE	Categoric	●	○	○	○	○	○	Unique: 50
3	BIRD	Categoric	●	○	○	○	○	○	Unique: 260
4	ADATE	Categoric	●	○	○	○	○	○	Unique: 2,509
5	STIMEU	Numeric	●	○	○	○	○	○	Unique: 1,321
6	STIMEQ	Categoric	●	○	○	○	○	○	Unique: 16 Missing: 883
7	DUR	Numeric	●	○	○	○	○	○	Unique: 151
8	STIMEL	Categoric	●	○	○	○	○	○	Unique: 1,289 Missing: 1,525
9	ORBIT	Categoric	●	○	○	○	○	○	Unique: 11 Missing: 456
10	NS	Categoric	●	○	○	○	○	○	Unique: 3 Missing: 2,034
11	LAT	Numeric	●	○	○	○	○	○	Unique: 75
12	LATQ	Constant	○	○	○	○	●	○	Unique: 1 Missing: 883
13	EW	Categoric	●	○	○	○	○	○	Unique: 3 Missing: 1,361
14	LON	Numeric	●	○	○	○	○	○	Unique: 325
15	LONQ	Constant	○	○	○	○	●	○	Unique: 1 Missing: 883
16	ALT	Numeric	●	○	○	○	○	○	Unique: 78
17	ATYPE	Categoric	●	○	○	○	○	○	Unique: 12 Missing: 49
18	ADIAG	Categoric	●	○	○	○	○	○	Unique: 7 Missing: 47
19	ACOMMENT	Categoric	●	○	○	○	○	○	Unique: 916 Missing: 1,393
20	SVE	Numeric	●	○	○	○	○	○	Unique: 152 Missing: 354
21	X	Constant	○	○	○	○	●	○	Unique: 1 Missing: 4,817

Roles noted. 5,133 observations and 17 input variables. The target is X.U.FEFF.VER. Categoric 3. Classification models enabled.

1. Data reduction is a critical process that helps select the most important data from the high quantity of data obtained. This is because, with a high quantity of data, it can take time to achieve clear results in the relationship. Data reduction is performed using a portion of the data collected.

2. Detecting duplicate and misspelled items: this phase involves detecting and eliminating duplicate data and those with errors like misspellings. Here, data with the same acronyms and lexical forms are considered the same; thus, the decision has to be made on the items to eliminate and retain.

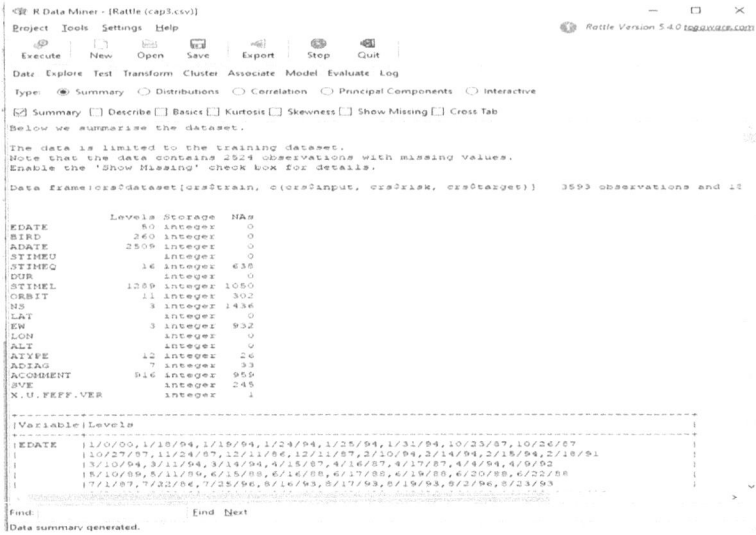

FIGURE 18.4　Data transformation.

Stage 3: Data Transformation

In this phase, the relationships between the nodes are identified, and similarities in their analysis are registered in the clusters. The process of data transformation relies on the use of Rattle. This functions by establishing the existing and prevailing relationships between nodes in data. The data transformation is summarized in Figure 18.4.

Stage 4: Data Exploration

This phase is undertaken when the relationships between the selected nodes are developed. The exploration is done on the data to detect the similarities. The data is explored once a network of relationships between the selected nodes is constructed. When clusters are detected from the analysis, a label is selected and set to each, using the most important cluster terms. Data exploration is gathered in huge numbers, unstructured or disarranged from various sources that align with the analyst's interest. The first is to understand and develop a comprehensive view of the data before having any data for analysis. The exploration process is demonstrated in the diagram in the Figure 18.5.

Stage 5: Data Modelling

In this phase, the data analysts develop a method for anomaly detection on satellite data using data mining and visualization. Testing and modelling are the key processes taking place at this stage. Data mining and visualization help uncover the underlying pattern from the anomalies, a useful step for various fields. The data modelling stage also covers using the R tool for clustering. Cluster analysis is a critical step in evaluating whether mutual information-based measures can be applied to determine the strength of

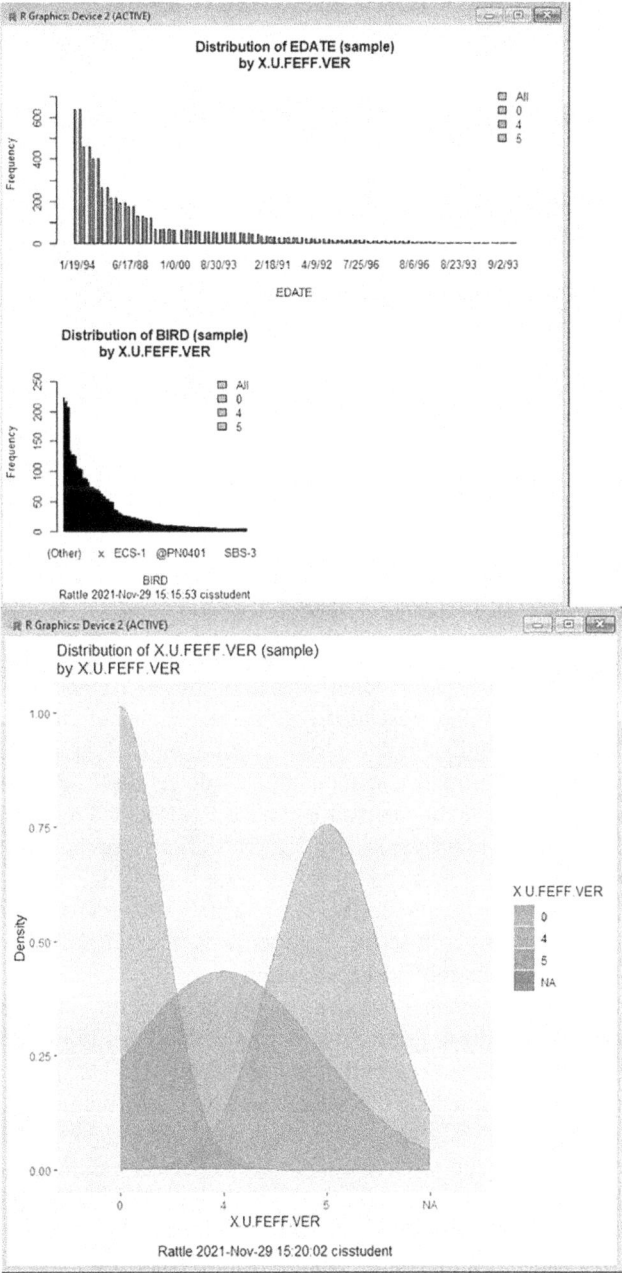

FIGURE 18.5 Data exploration.

FIGURE 18.6   Data modelling.

group ties. The clusters can be introduced purposely for network structure inference. Figure 18.6 represents the ten clusters created in the data modelling phase. Any significant deviations from the normal structural patterns of underlying graphs represent anomalies. To detect these deviations, many processes occur, including preparation, measurement, and results to express the differences between the groups. The deviations can be detected using inference procedures developed from the statistic tests. The primary aim of detecting the anomalies in clusters is to identify the anomalies. These are then highlighted and represented as edges in the presented data.

Stage 6: Data Visualization and Evaluation

This section helps understand the relationships between different entities in a network and the behavior of a network as a whole. This is a result of interactions between entities within the network.

To develop mathematical models to understand and predict the behavior of networks (Newman 2003), researchers have used graphic visualization to develop experiments with several network analysis techniques, including statistical inference and computational algorithms. Satellite data is critical in analyzing data relations in anomalies. Still, it can be used to create maps that give a visual representation of satellite data structure.

All transformations are applied to data to derive similarities from the data. The hierarchical clustering approach is often used by researchers who study satellite data relationships, for example. When doing this, visualization leverages a clear

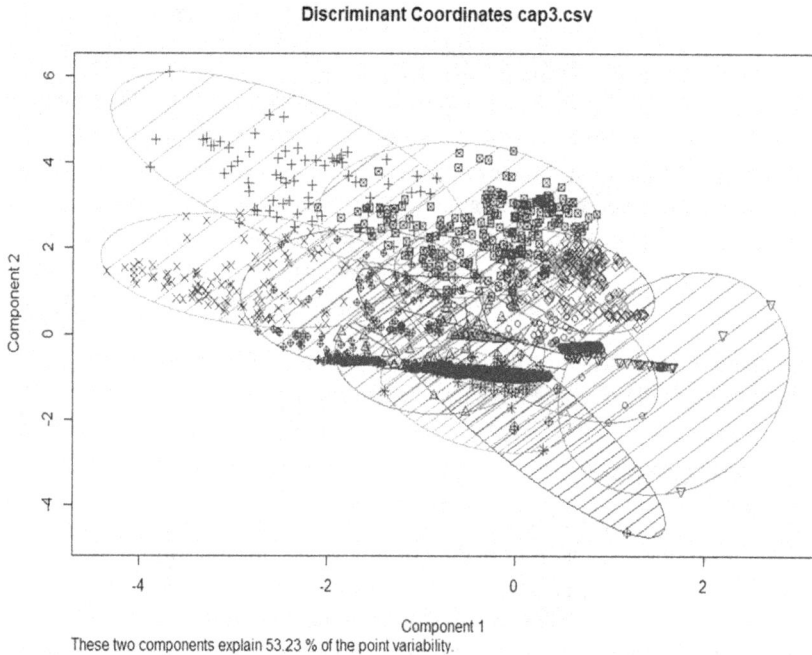

Discriminant Coordinates cap3.csv

These two components explain 53.23 % of the point variability.

FIGURE 18.7 Data modelling (clustering).

understanding and better representation of the output maps represented in satellite data, as demonstrated in the diagram in the next section. The final output map displays several objects and their relations among the objects. Various visualizations can be used, including distance- and graph-based visualizations. In distance-based visualizations, the distance between two clusters reflects the relationship between the clusters. The smaller the distance between two clusters, the stronger the relationship between the nodes. Meanwhile, in the graph-based visualizations in the case study, the distance between two nodes does not reflect the relation between the nodes.

To visually group nodes, lines are typically drawn between them; colors are the simplest way to do this. The circle colors determine the cluster to which an item belongs. The item clusters can be calculated and translated into colors using a color scheme. In this case, the relationship between nodes in clusters is shown by color and size. The red color below the image corresponds to the highest item density in cluster 6 and cluster 8, while blue corresponds to the lowest item density in a cluster. This can give a great insight into the relations inside a group and between different cluster groups (Figure 18.7).

18.2.1 SUMMARY OF THE METHODOLOGY

The link mining methodology is a technique that enables the exploration of correlations and anomalies between satellite data. This approach involves identifying connections between variables and analyzing their mutual information. By mapping the variables to nodes, a network of connections and relationships is created across the various stages that include data description, data preprocessing, data transformation, exploration, modelling, and finally visualization and processing. In a nutshell, using this approach, the complex relationships between variables and identified anomalies across data will be examined. Most importantly, the potential causes of the anomalies will also be examined. In addition, this methodology can use same metrics for interpreting and predicting future anomalies. In order to employ this approach, there will be a need to first gather satellite data. Once it is gathered, various algorithms and techniques will be used to find correlations between different variables and identify anomalous clusters. Afterward, data will be modelled and evaluated to determine the strength of the connections between different variables, and identify any unusual patterns in our data. Overall, by utilizing the link mining methodology, we can interpret and predict anomalies more accurately and efficiently.

18.3   DISCUSSION

Anomalies show a pattern of behavior that does not conform with the defined normal behavior. They may result from malicious activities and other forms of intrusion. Anomaly detection's core is based on analyzing the dataset collected from the satellites through the six stages described in the methodology (Getoor and Diehl 2005). The first stage is critical for successful anomaly detection because it ensures datasets conform to standards for accurate detection. The data description stage leverages more responsibilities to satellite owners and operators to diagnose anomalies in time using mutual information (Münz et al. 2007). For instance, satellite operators in civil, defense, and commercial sectors would develop databases that detect their anomalies to have long-term solutions to ever-evolving anomalies from different causative factors. The data description helps the experts to extract usable data from the large datasets collected from the satellites. It describes the patterns and extracts new and significant information from the datasets to identify anomalies and develop effective mitigation measures (Chandola et al. 2009). The description of the different satellite may take a long time to ensure it is comprehensive and include a wide range of satellites. The compiled anomaly databases developed from standardized anomaly information adds more value to the entire process. This process also involves the selection of data that is restricted to a certain period or season. This type of raw data varies, or there are different types. These data include header information and/or payload and statistical data about packet flows or connections, including start and end timestamps, number of exchanged packets, and bytes. The data mining process and the achievable result depends very heavily on the type of raw data.

The preprocessing data stage facilitates the transformation of raw data collected from satellites into useful and efficient formats. The data preprocessing involves data cleaning, transformation, and data reduction. Data cleaning is critical to eliminate irrelevant and incomplete parts (Münz et al. 2007). The missing data can be handled by ignoring the tuples and filling in missing values; however, in the event of noisy data that might be difficult to interpret, the binning method can be employed to smooth it (Chandola et al. 2009). Regression has also proved to be an effective approach to handling noisy data. When fitted into a regression function, the noisy data become

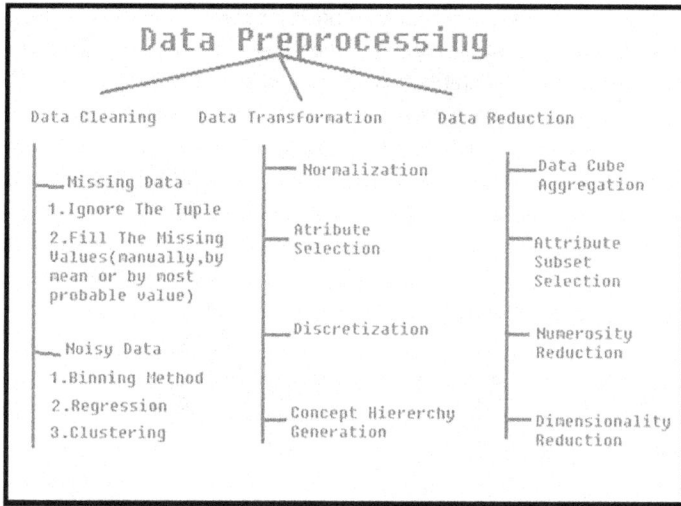

FIGURE 18.8   Data preprocessing breakdown.

smooth. Data transformation can be implemented through different steps to make the data usable by the mining methodology (Chandola et al. 2009). Transformation changes the data into appropriate forms suitable for the mining process. The transformation can be implemented using normalization, attribute selection, discretization, and concept hierarchy generation (Chandola et al. 2009). The monitoring data and the data mining algorithm are the key determinants. It is necessary or important to clean and filter the data to avoid generating misleading or wrong information or trends. Disturbances are the interruptions of known changes, for example, the dependencies on the time of day and missing data fields in some of the data records.

Data reduction leverages more benefits to the mining methodology. It enables the system to handle vast amounts of data by increasing storage efficiency and reducing data storage and analysis costs (Münz et al. 2007). The data cube aggregation is the initial step to data reduction; it is applied to data to develop the data cube. The attribute subset selection leverages relevant attributes that can be utilized, and the rest are discarded (Chandola et al. 2009). During the attribute selection, the level of p-value is critical to set the significance threshold. Any attribute greater than the p-value is discarded (Chandola et al. 2009). Meanwhile, the numerosity reduction during the data reduction process would facilitate the storage of the model of data instead of the entire data (Il-Agure 2015). The dimensionality reduction reduces the size of data through different encoding mechanisms. This can give a great insight into the relations inside a group and between different cluster groups as shown in Figure 18.8. The diagram in the next section is a summary of the preprocessing data stage. Data transformation is creating a new dataset that only contains binary data. The advantage of transformation data representation is the ability to process it with various algorithms for association rules creation. Avoiding this problem with large dimension numbers is using the binary representation of log records. Reduction is achieved by anomaly detection from log files and inserting values into categories and using an interval representative instead of a scalar or time value.

A binary string contains a numerical value of 1 for values present in the record and a numerical value of 0 otherwise. In contrast to this, some of the values are unable to reduce, for example, the IP addresses or ports. On the other hand, when applying data mining methods to raw monitoring data, this does not always lead to useful results because many attributes within the initial or original records are not relevant to the goal of data mining duty. This means that a transformation has to be made that will convert the raw data into a database with a small number of relevant elements. In addition to this, it is important to aggregate the data to decrease the memory and processing of the data mining algorithm.

During data exploration, the relationships between the nodes are established to derive similarities from the data, while data modelling helps the analysts to detect the anomalies in satellite data through clustering and visualization (Il-Agure 2015). The clustering and visualization uncover the underlying patterns from the anomalies detected. This helps various fields evaluate whether measures based on mutual information can be used to show the strength of cluster analysis (Münz et al. 2007). In addition, it helps to examine the extent to which cluster structures derived from mutual information-based measures resemble actual cluster structures. The anomaly pattern recorded from the anomaly deviations helps establish good and bad patterns. In addition, clusters of the data objects can be constructed using different ways, for example, the K-Means, K-Medoids for small datasets and CLARA, CLARANS for large datasets, and BIRCH, Chameleon for performing macro clustering on micro clusters. Cluster-based methods have a procedure in that the assumption is anomalies either belong to a small, sparse cluster or do not belong to any cluster, while on the other hand, the normal objects are part of large and dense clusters.

Data visualization and evaluation are performed on the satellite data anomalies to analyze the similarities and give a visual representation of the output map. It gives study relations among anomalies within different nodes to reflect cluster relations (Il-Agure 2015). The smaller the distance between two clusters, the stronger the relationship between the nodes. On the other hand, in graph-based visualizations in the case study, the distance between two nodes does not reflect the relationship between the nodes (Il-Agure 2015). Instead, drawing lines between nodes from the cluster typically indicates the relationships between nodes; the simplest way to group nodes visually is to use colors. When items have been assigned to clusters, the color of an item's circle can be determined by the cluster to which the item belongs (Münz et al. 2007). Item clusters are calculated and translated into colors using a color scheme. In the case study, the relationship between nodes in clusters is shown by color and size. In this color scheme, red corresponds with the highest item density in clusters, while blue corresponds with the lowest item density in the cluster.

There are other ways of getting these data for analysis; one is using an autoencoder. An autoencoder is a special deep neural network often used to de-noise input data. Generally, the idea is to give some input data. The first step is to enlarge the input into a higher dimensional (Pilastre et al. 2020) sparse representation. Secondly, continue reducing the dimension lower in an iterative fashion until you obtain a compressed data representation. This step is known as encoding. Following this encoding, the opposite operation is performed, now the decoding, where you enlarge this compressed data back to its original size and compare this new result with the original data input into the neural network. With this idea, the network will learn only what is most important to represent the core attributes/features of the data and automatically ignore such things as anomalies and noise. The first half of the network encodes the data into a lower dimensional vector space; the second half decodes the data by enlarging this compressed data representation to reconstruct the original input data. Using an autoencoder neural network instantly is to have two separate applications to ATHMoS; first, the extraction of automatically learned features representing time intervals of the data which can be used to augment the existing human-engineered features currently used. Secondly, directly run the anomaly detection. The key to obtaining the features is to train an autoencoder neural network to correctly recreate the original input training data, which is done up to some accuracy measure such as root-mean-squared. For each telemetry parameter that now has its trained network model, the output is extracted at the smallest hidden layer or, in other words, the compressed representation of the data. Considering users' actual communication activities and interactions, the resulting graph, usually called an activity graph, will be drawn. This activity graph can be categorized as a basic or weighted activity graph. A graph containing similar kinds of edges in every pair of nodes, irrespective of strong or weak ties between them, is called a basic activity graph, but a weighted activity graph represents a graph structure in which the strength of the activation link is also taken into account.

This case study was designed to use mutual information to validate the visualization graph. We used a real dataset where the anomalies were not known in advance, and the data had to be preprocessed. Il-Agure (2015) has shown that the developed approach allows us to work with noisy and inconsistent data when scaled to large datasets and combined with semantic preprocessing. Mutual information supported a semantic interpretation of the clusters, as shown in the discussion of the clusters. Mutual information is a quantitative measure of how much one random variable (B) tells us about another random variable (A). In this case, information is considered a reduction in the uncertainty of a variable; high mutual information indicates a large reduction in uncertainty. High mutual information indicates a large reduction in uncertainty, while low mutual information indicates a small reduction and zero mutual information between variables. This approach can help identify anomalies in the data, characterize them, and understand their properties (Il-Agure 2015).

Most anomalies originate from satellite hardware and software manufacturing and design flaws. They can also result from extreme space weather that interferes with the intensity of electromagnetic radiation and the density of charged particles in the satellite's environment (Münz et al. 2007). The satellite anomalies are also caused by micrometeoroids or space debris, operator error, regular wear and tear from exposure to the plasma environment of space, or interference by human technological activities, either intentional or unintentional.

Satellite charging is another cause of anomalies, especially on spacecraft. This is because of recent observations, charging on high-inclination earth orbiting HBO satellite (Fennell 2022). The HBO charging data showed that the occurrence patterns or trends were consistent with the expected motion of substorm-injected plasma electrons. This HBO data was taken on a scale of altitudes in a 630 inclination orbit. When HBO energetic particle data is combined with CRRES, GOES, and GPS, it estimates some of the worst-case levels of internal charging fluxes. The other form of data is the SCATHA data. This also shows the occurrence trend accordant with the anticipated motions of substorm-injected plasma electrons. This data was taken in the near geosynchronous orbit. These SCATHA data concluded that the internal discharging rates were related to the intensities of energetic electrons are 100 kiloelectronvolt. This also statistically indicated that their occurrence peaked near local noon. This data was collected by the SCATHA spacecraft Charging at High Altitude satellite. It was frowned upon to rule the condition and also document the existence of satellite charging. This satellite was instrumented to measure charging and detect electrostatic discharging that might have occurred. Discharges collected were observed and characterized as either surface or internal charging related (Fennell 2022). These anomalies have caused severe feminine, loss of life to both humans and livestock, and large-scale migration in the Sahel. This is because the network of rain gauge stations in the Sahel has decreased significantly in recent years (Eklund et al. 2016). This added considerable uncertainty to the database based on the station data, for example, the climate research unit rainfall database, which resulted in the wrong analysis (Eklund et al. 2016). This has affected the studies of rainfall regime changes. Telemetry data are some sensor values, for example, temperature, voltage, angular velocity, and temperatures that have low and high limit values. This makes limit checking (Machida 2006) of importance. It is one of the alternatives for anomaly detection. However, this is for some cases when limit values are normal; anomalies can exist. This means that some class anomalies occur without violating the limits on the variables. To take measures before the situation occurs, these anomalies should be predicted.

Moreover, domain experts or operators should always monitor and review out-of-limit values. If the limit values are inappropriate, an anomaly will be detected. This follows that false alarms will be generated, and real anomalies may be missed. Also, Machida (2006) have developed a new method for limit checking. They have combined limit checking and sparse Bayesian learning or the relevance vector machine (RVM). RVM is used to learn a high and low limit values model from old normal telemetry. The telemetry data is usually analyzed and monitored by human operators and analysts. This helps access if the values are not out of a predefined range or pattern. Values out of limits mostly communicate a potential anomaly in a given satellite subsystem. However, this large amount of telemetry data makes it close to impossible to perform a careful and detailed analysis in the current world. Several satellites' mission threats begin as unexplained anomalies in the telemetry. The operators track these anomalies in their databases of detailed satellite status information, which many consider proprietary or classified. They may investigate the cause of these anomalies if they are repetitive or significant enough to threaten the satellite's mission (Münz et al. 2007). In such investigations, a centralized and up-to-date shared database of anomalies experienced by many different satellites in a variety of orbital configurations could help provide context and narrow down the potential causes of the anomalies. The investigators have a critical role in determining unique patterns in satellite operations at any given time. Through applying this method, they can detect the origin of the anomalies, whether hardware defect, accidental interference, purposeful attack, or a space weather event. Still, some challenges may hinder satellites from different sectors from developing their databases or sharing information about their satellite systems with other groups. However, some methods are used to detect these anomalies. Three methods are usually implemented for anomaly detection with monitoring data. These methods include threshold-based, expert experience, and data-driven methods. First is the data-driven method, which we are currently discussing. It is independent of area knowledge, is easy to extend, and has strong learning ability. Most intensive research has been carried out based on a data-driven model. The second threshold-based anomaly detection is the easiest and fastest method used in different sectors. Nevertheless, the threshold should be set by experience and design requirements. It is also relatively large, which may miss some anomalies. Third is the experience-based method, which is a bit more effective than the threshold-based one. This is because it has more injected prior knowledge. However, knowledge limits the completeness of experience, which may restrict the detection ability.

18.4   CONCLUSION

Data mining is a complex or difficult process, and the resulting cluster centroids can detect anomalies in new online monitoring data with a small number of distance calculations. This helps in allowing deploying the detection method for scalable real-time detection, for example, as part of an intrusion detection system. In addition to the aforementioned observations, limited on-orbit data and statistical analyses of spacecraft reliability exist in the technical literature. Reliability has long been recognized as a critical design attribute for space systems. Spacecraft are high-value assets whose cost often exceeds hundreds of millions of dollars and whose "location" or operational environment renders them physically inaccessible for maintenance to compensate for substandard reliability. Satellite anomalies adversely affect on-orbit operational spacecraft. Satellites experience anomalies of some kind throughout their lifetime at any time. The anomalies vary in severity from temporary errors in noncritical subsystems to loss-of-contact and complete mission failure. Anomalies result from various factors, including malicious activities, operator error, and environmental hazards. Satellite operators conduct investigations to establish the cause of a specific anomaly. These investigations are usually conducted at significant expense.

This study has analyzed the nature and causes of satellite anomalies and the potential benefits of a shared and centralized satellite anomaly database. The findings demonstrate that a shared satellite anomaly database significantly benefits the commercial community. However, the main challenges to this include reluctance to share detailed information with the broader community. There is also a lack of dedicated resources accessible to third parties that can help them develop and manage such a database. However, trusted third parties and cryptographic methods such as secure multiparty computing or differential privacy must provide long-term solutions to the challenges. Instead, they have the potential to resolve the issues related to the secure sharing of anomaly data. This study has also demonstrated how a satellite anomaly database could benefit the community of spacecraft operators and how the obstacles inhibiting the development of a shared database could be partially overcome using mutual information. Therefore, those tasked with investigating the causes of severe or recurrent anomalies would benefit from greater awareness of whether other satellites are also experiencing anomalies under similar conditions. There are some challenges posed during anomaly detection. The first challenge is that the presence of noise in the data collected from the sensor makes anomaly detection more challenging since it now has to distinguish between interesting anomalies and unwanted noise/missing values. Anomaly detection techniques are required to operate in an online approach. Due to severe resource constraints, anomaly detection techniques need to be light-weight. The second challenge is that data is collected and distributed; hence, pattern matching is required to analyze the data (Chatzigiannakis 2006). Due to abnormal pattern matching, this will help to detect orbits from the online approach.

Using Bayesian modelling, the data can detect the uncertainties between the data input and the data output under the study. This is because it helps discover the trends from detected anomalies. With this in mind, it is possible to know whether the results can show the strength of feed-forward in single or multiple perceptions of the data being analyzed. The kind of information is very important to know the level of uncertainty of the information. The information should have high mutuality, indicating reduced uncertainty at par. With this, it is possible to identify the anomalies, be able to characterize them, and be able to identify their elements. This model will be of help in detecting the origin of these amenities. This will help identify the rate of occurrence, whether it is decreasing, if the trend persists or occurs more often, and whether it is a threat to humans. This will now be easier due to the centralization of data from different satellites. The previous study shows that the actual interactions or communication is a vital key of the elements. This is because no matter what resources are available within a structure, without communication activity, those resources will remain dormant, and no benefits will be provided for individuals.

REFERENCES

Agrawal, S., & Agrawal, J. (2015). Survey on anomaly detection using data mining techniques. *Procedia Computer Science*, 60, 708–713.

Azevedo, D. R., Ambr´osio, A. M., & Vieira, M. (2012). Applying data mining for detecting anomalies in satellites. In *Ninth European Dependable Computing Conference* (pp. 212–217). IEEE.

Chandola, V., Banerjee, A., & Kumar, V. (2009). Anomaly detection: A survey. *ACM Computing Survey*, 41(3), 1–58.

Chatzigiannakis, L. C. (2006). Sink mobility protocols for data collection in wireless sensor networks. Retrieved from https://dl.acm.org/

Eklund, A., Nichols, T. E., & Knutsson, H. (2016). Cluster failure: Why fMRI inferences for spatial extent have inflated false-positive rates. *Proceedings of the National Academy of Sciences*, 113(28), 7900–7905.

Fennell, J. F. (2022). Spacecraft charging: Observations and relationship to satellite anomalies. Retrieved from https://apps.dtic.mil/sti/citations/ADA394826

Getoor, L., & Diehl, C. (2005). Link mining: A survey. *SIGKDD Explorations*, 7(2), 3–12.

Il-Agure, Z. I. A. (2015). Anomalies in link mining based on mutual information. In *2015 SAI Intelligent Systems Conference (IntelliSys)* (pp. 1057–1059). IEEE.

Il-Agure, Z. I. A. (2016). *Anomalies in Link Mining Based on Mutual Information*. Staffordshire University.

Khraisat, A., Gondal, I., Vamplew, P., & Kamruzzaman, J. (2019). Survey of intrusion detection systems: Techniques, datasets and challenges. *Cybersecurity*, 2(1), 1–22.

Machida, K. M. (2006). Telemetry mining. Retrieved from IEEE Xplore.

Münz, G., Li, S., & Carle, G. (2007). Traffic anomaly detection using k-means clustering. In *GI/ITG Workshop MMBnet*, Vol. 7, September 2007. Hochschule Mannheim, Fakultät für Informatik.

Newman, M. E. J. (2003). The structure and function of complex networks. *SIAM Review*, 45(2), 167–256. doi:10.1137/S003614450342480.

Padhy, N., Mishra, P., & Panigrahi, R. (2012). The survey of data mining applications and feature scope. *International Journal of Computer Science, Engineering and Information Technology (IJCSEIT)*, 2(3), 43–58.

Pilastre, B., Boussouf, L., D'Escrivan, S., & Tourneret, J.-Y. (2020). Anomaly detection in mixed telemetry data using a sparse representation and dictionary learning. *Signal Processing*, 168, 107320. doi:10.1016/j.sigpro.2019.107320.

19 A Review of Reinforcement Learning Approaches for Autonomous Systems in Industry 4.0

*BH Krishna Mohan, Pulicherla Padmaja,
M Durairaj, P Nagamalleswrarao,
K Srinivasarao, and Sagaya Aurelia*

19.1 INTRODUCTION

19.1.1 RESEARCH BACKGROUND

Due to its multifarious relevance across several disciplines, the study of autonomous systems has been rising to extraordinary prominence in recent years [1]. Industries including health care, transportation, manufacturing, as well as agriculture are being transformed by autonomous systems, which include robots, self-driving cars, including unmanned aerial aircraft [2]. These systems claim to be more effective, require less human interaction, and have the capacity to do jobs in risky or difficult circumstances. Reinforcement learning is one of the key elements that gives these systems their autonomy. A kind of artificial intelligence called reinforcement learning enables autonomous entities to gain knowledge from their interactions with the outside world and make wise judgments [3]. It is fundamental in determining how autonomous systems make decisions as well as enabling them to adjust and improve their behavior over time [4]. In order to advance technology and solve safety, ethical, and regulatory problems in the implementation of these systems, it must be maintained to fully comprehend the difficulties and opportunities presented by reinforcement learning in the context of autonomous systems.

19.1.2 RESEARCH AIM AND OBJECTIVES

Aim

The aim of this study is to thoroughly examine the landscape of reinforcement learning in autonomous systems, highlighting both the challenges and opportunities it offers, and providing insights into prospective uses and ramifications.

 DOI: 10.1201/9781003343332-19

Objectives

- To assess the body of knowledge on reinforcement learning in autonomous systems, highlighting major issues, including finding knowledge gaps.
- To investigate the technological difficulties as well as constraints involved in implementing reinforcement learning in autonomous systems.
- To examine the case studies as well as real-world applications where reinforcement learning has shown effectiveness in improving the performance of autonomous systems.
- To offer suggestions to enhance algorithms and methods for reinforcement learning in autonomous systems.

19.1.3 RESEARCH RATIONALE

This study was motivated by the growing significance of autonomous systems in a variety of sectors. It is crucial to comprehend and address the difficulties these systems confront as they are more integrated into society, especially in terms of decision-making and flexibility [5]. Although it has great promise as a key component, reinforcement learning also raises serious technological and moral questions. By exploring these complexities, this research aims to offer beneficial insights for academics and practitioners, encouraging a better understanding of how reinforcement learning can be effectively utilized while at the same time highlighting the crucial issues that demand attention in the pursuit of secure, efficient, and responsibly operated autonomous systems.

19.2 LITERATURE REVIEW

19.2.1 CRITICAL ANALYSIS OF EXISTING LITERATURE

A careful analysis of the reinforcement learning (RL) integration in the context of autonomous systems shows two key aspects. First off, the study of Key Trends in Reinforcement Learning for Autonomous Systems highlights how transformational RL is in this field [6]. Notably, there has been a noticeable move in favor of deep reinforcement learning, which makes use of neural networks to solve difficult decision-making problems. Techniques for transfer learning are becoming more popular, enabling autonomous systems to incorporate information from one area to improve performance in another [7]. This analysis also sheds light on the increasing synergy between RL as well as other machine learning paradigms, allowing for the development of more resilient and flexible autonomous systems. Real-world applications in a variety of industries, like as robots, and self-driving cars, alongside health care, illustrate RL's relevance in the real world along with its potential for transformation [8]. On the other hand, evaluating the identified knowledge gaps in the current literature reveals key areas that require more research. These gaps include a range of technical difficulties, including addressing sample inefficiencies as well as handling the exploration-exploitation trade-offs inherent in RL [9]. The use of RL in autonomous systems raises several ethical issues which have to be taken into account. Furthermore, RL algorithms' interpretability and generalization problems have

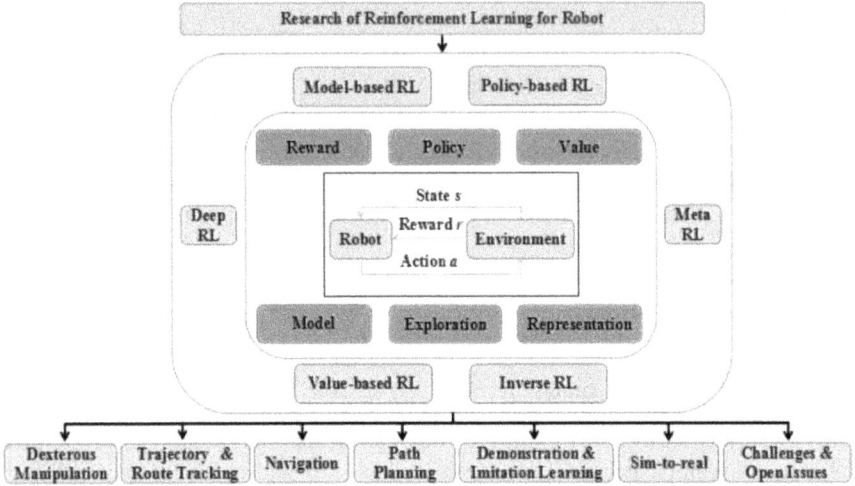

FIGURE 19.1 Reinforcement learning for robots.

not yet been resolved. Understanding these limitations is of the utmost importance because it emphasizes how urgently creative research projects are needed to develop adaptable, safe, and morally good RL solutions [10]. Utilizing the full potential of autonomous systems while tackling urgent research limits requires closing these crucial knowledge gaps. Figure 19.1 provides a broad taxonomy of the primary RL research papers for robotics for ease of viewing.

19.2.2 TECHNICAL CHALLENGES AND LIMITATIONS

Reinforcement learning (RL) integration into autonomous systems is a complex endeavor full of technical difficulties that need careful investigation. In its examination of the complexities of this integration process, Challenges in Reinforcement Learning Integration sheds light on problems like sample inefficiency, which occurs when RL algorithms frequently need a lot of input and interaction with the environment to get desired results [11]. The trade-offs between exploring novel behaviors as well as utilizing well-known ones that are inherent in RL additionally present a substantial challenge for autonomous systems trying to maximize decision-making [12]. Additionally, the computationally demanding nature of RL and the real-time demands of autonomous systems necessitate creative methods to assure effectiveness and responsiveness.

Additionally, Limitations of Existing Approaches examines the shortcomings and limitations of the present approaches and tools employed in RL for autonomous systems [13]. This evaluation takes into account the intrinsic constraints of RL algorithms, including the challenge of generalizing knowledge across various tasks and contexts [14]. The part additionally looks at the difficulties of model-based vs. model-free RL techniques and provides insights into their distinct drawbacks [15]. This research is aimed at contributing to the development of more effective and efficient RL solutions suited to the particular requirements of autonomous systems by giving light on these technological difficulties as well as constraints [16]. Figure 19.2

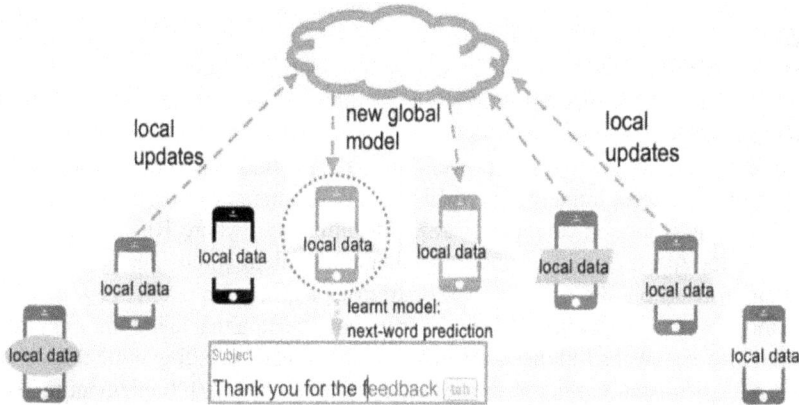

FIGURE 19.2 Technical challenges and limitations.

illustrates a distributed device paradigm in which the computation is carried to the data rather than the data being brought to the computation.

19.2.3 Real-World Applications and Case Studies

In "Real-World Applications and Case Studies," the practical ramifications of reinforcement learning (RL) in the context of autonomous systems are strikingly presented [17]. This part includes a thorough examination of real-world applications where RL has proven to be very effective in strengthening the performance of autonomous systems across a variety of fields [18]. It explores into the field of robotics, where RL-powered robots have been shown to be remarkably adaptable and capable of tackling complex problems. Robotic arms, for example, have been trained to carry out complex tasks like gripping things of various sizes as well as forms, demonstrating the adaptability of RL in practical settings [19].

This section also examines the field of self-driving cars, where RL algorithms have been crucial in autonomous navigation as well as decision-making [20]. Case studies show examples of how RL-equipped cars successfully and effectively negotiated challenging urban situations while making crucial choices to protect passengers [21]. Additionally, the healthcare industry offers interesting use cases, such as medical robots driven by RL which assist with operations and medication development. These real-world applications highlight the significance of RL for numerous industries and provide insightful information about its potential to revolutionize them, paving the path for improved and competent autonomous systems [22].

19.2.4 Recommendations for Optimization

The work optimizing reinforcement learning algorithms explores methods for boosting the efficacy and effectiveness of reinforcement learning (RL) methods when used with autonomous systems [23]. It includes suggestions for dealing with RL's sample inefficiency, a significant problem that frequently obstructs real-world applications. Techniques like experience replay and off-policy procedures have the potential to be used to make better use of gathered data, expediting the learning process, in order to alleviate this difficulty [24]. Additionally, developments in deep RL architectures—such as deep Q-networks as well as policy gradient approaches—offer intriguing ways to improve the functionality of RL algorithms in autonomous systems, making them more adaptive and data-efficient.

In addition, Techniques for Improved Performance makes recommendations that are specific to the requirements and difficulties encountered by autonomous systems [25]. The significance of domain-specific customization, where RL algorithms are tailored to the specifics of the application domain, is highlighted. The application of information from one field to another, or transfer learning, is emerging as a viable tactic for raising performance [26]. In order to make certain that autonomous systems function responsibly, techniques for tackling safety issues including ethical considerations in RL decision-making processes are also researched [27]. Implementing these optimization suggestions would enable more robust and dependable autonomous decision-making while recognizing the nuances unique to this field, allowing academics and practitioners to fully realize the potential of RL in autonomous systems [28].

19.2.5 LITERATURE GAP

The inadequate examination of the sociological and ethical implications of reinforcement learning (RL) in autonomous systems constitutes a significant gap in the research to date. Studies frequently ignore the wider effects of RL-driven choices on privacy, and safety, including human interaction despite a strong focus on technical elements [29]. It is crucial to comprehend the moral and societal ramifications of RL in autonomous systems since this encourages the appropriate use of AI. Closing this gap would encourage thorough research that not only improves technology but also tackles the moral and legal issues raised by autonomy driven by RL.

19.3 METHODOLOGY

To get a thorough understanding of the incorporation of reinforcement learning (RL) in autonomous systems from a human-centric perspective, this research employs an interpretivism philosophical approach. The significance of context and subjectivity is emphasized by interpretivism, which fits well with the complex design of autonomous systems [30]. The implementation of a deductive methodology facilitates the testing of current ideas and hypotheses in relation to the integration of RL in autonomous systems. This method involves coming up with hypotheses on the basis of the body of literature already in existence and then testing them using empirical evidence [31]. The study design uses a descriptive methodology with the goal of offering a thorough investigation of RL integration in autonomous systems. The goal of descriptive research is to offer a thorough explanation of the event being investigated, including the adoption of RL at the moment, the difficulties encountered, and proposed solutions [32]. Secondary data collection is one of the primary methods used to obtain data. There will be a thorough study of the literature as well as analysis of reports, academic papers, books, as well as publications from the business world [33]. This information will come from a variety of sources, which include academic papers, technical manuals, along with case studies, and it will allow for a comprehensive knowledge of RL's function in autonomous systems [34].

To find recurrent themes, patterns, and trends in the literature, the secondary data will be systematically analyzed using thematic analysis. To extract pertinent information on RL integration issues, trends, and potential uses in autonomous systems, the data will be coded as well as categorized [35]. Throughout the study process, ethical concerns will be followed, and appropriate citation, privacy, and confidentiality procedures will be in place. By comparing data from many sources, triangulation will increase the validity and dependability of the results [36]. Additionally, a peer review procedure is going to be used to guarantee the validity and reliability of the study. This technological technique addresses the unique issues as well as developments in this field while providing a solid framework for the thorough investigation of reinforcement learning integration in autonomous systems.

19.4 RESULTS

19.4.1 KEY TRENDS IN RL INTEGRATION

Substantial trends in the incorporation of reinforcement learning (RL) inside autonomous systems were discovered through study of the body of current research. The discipline saw a paradigm change when deep reinforcement learning (DRL) emerged as the dominating trend [37]. DRL employs more complex neural network designs to help autonomous systems manage challenging decision-making problems. These neural networks are capable of processing a lot of input efficiently and build intricate, hierarchical representations of the environment, enabling more sophisticated as well as adaptable decision-making. Transfer learning techniques were additionally essential for RL integration [38]. With the use of these techniques, autonomous systems may use their expertise and experience from one area to perform better in another. Autonomous systems could adjust to new tasks and surroundings more quickly by transferring learned policies, value functions, or features, which eliminates the requirement for costly retraining [39].

It was also clear that RL alongside additional machine learning paradigms like supervised learning and imitation learning work well together [40]. By combining the advantages of several strategies, this partnership increases the flexibility of autonomous systems. For example, although RL fine-tunes and refines these rules through interaction with the environment, supervised learning could provide early policy advice [41]. Collectively, these patterns demonstrate the dynamic progress of RL integration, permitting autonomous systems to tackle a wide range of challenging and varied problems in a variety of fields.

19.4.2 TECHNICAL CHALLENGES AND LIMITATIONS

The technological difficulties in integrating reinforcement learning (RL) into autonomous systems were investigated, and the results revealed important issues that need rapid attention. One of these issues is sample inefficiency, which is a major worry [42]. In order to achieve desired levels of performance, RL algorithms frequently

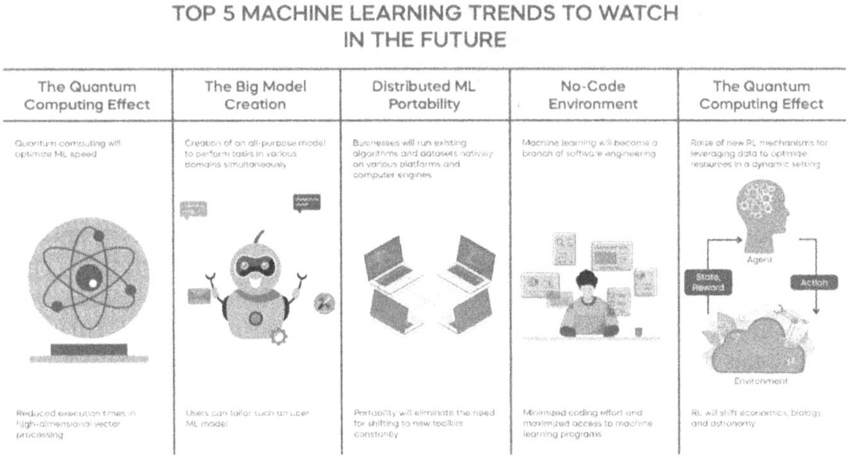

TOP 5 MACHINE LEARNING TRENDS TO WATCH
IN THE FUTURE

The Quantum Computing Effect	The Big Model Creation	Distributed ML Portability	No-Code Environment	The Quantum Computing Effect
Quantum computing will optimize ML speed	Creation of an all-purpose model to perform tasks in various domains simultaneously.	Businesses will run existing algorithms and datasets natively on various platforms and computer engines	Machine learning will become a branch of software engineering	Raise of new RL mechanisms for leveraging data to optimize resources in a dynamic setting
Reduced execution times in high-dimensional vector processing	Users can tailor such an user ML model	Portability will eliminate the need for shifting to new toolkits constantly	Minimized coding effort and maximized access to machine learning programs	RL will shift economics, biology, and astronomy

FIGURE 19.3   Key trends in RL integration.

display a voracious appetite for data and necessitate a significant number of interactions with the environment [43]. This requirement for a lot of data might make real-world applications problematic as well as expensive, which inhibits the use of RL in autonomous systems on a large scale. The exploration-exploitation trade-off that is a part of RL decision-making also presented a significant challenge. Autonomous systems need to strike a fine balance between using known actions to maximize short-term rewards along with experimenting with novel actions to find superior methods. Finding this balance is particularly difficult in constantly changing and unpredictable situations when making the wrong choice might have expensive repercussions [44]. Five machine learning models are shown in Figure 19.3, and they are all based on recent advances and continuing issues facing the sector.

Furthermore, RL techniques' high computational complexity along with associated real-time constraints faced challenges. In real-time applications like autonomous cars or robotic systems, many RL techniques can sometimes be less practicable or affordable since they need significant processing resources [45]. In order to overcome these obstacles, creative approaches must be developed to increase the effectiveness, data utilization, and responsiveness of RL algorithms in autonomous systems. This will ensure their practical applicability along with dependability in challenging, dynamic contexts.

19.4.3   REAL-WORLD APPLICATIONS AND CASE STUDIES

The practical relevance as well as the transformational potential of reinforcement learning (RL) inside autonomous systems have been clearly illustrated by its real-world applications [46]. Case studies in the field of robotics have shown conclusively that RL has been successful in permitting robots to adapt to dynamic and unexpected settings. These applications illustrate RL's adaptability and versatility in real-world circumstances through allowing robotic devices to carry out complex tasks like grabbing items of various shapes and sizes [47]. RL algorithms have produced major advantages for self-driving cars as well. Case studies have shown the ability

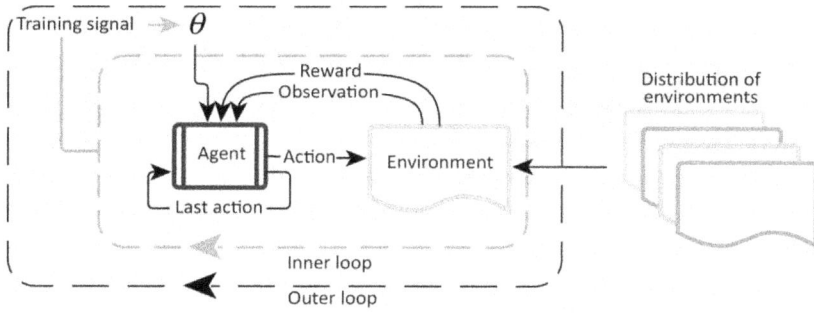

FIGURE 19.4 Reinforcement learning process.

of RL-equipped cars to maneuver in urban areas that are complicated and full of unknowns [48]. These RL-powered cars demonstrate the potential of RL for enhancing autonomous transportation systems, including increase road safety by making critical choices in real-time to protect the safety of passengers and pedestrians.

Figure 19.4 shows the Inner and outer loops of training in the schematic of meta-reinforcement learning. The inner-loop learner, 'An agent', which is instantiated by a recurrent neural network, is determined by the parameter weights θ that are trained by the outer loop and interact with the environment during the episode.

RL has been employed in the healthcare industry in applications like medical robots [49]. These case studies highlight the role that RL plays in supporting surgical operations as well as the drug development process, emphasizing its capacity to streamline medical procedures and enhance medical technology [50]. Together, these real-world examples show the manner in which RL has the potential to revolutionize the way autonomous systems work in a variety of fields, from robotics and transportation to health care and beyond.

19.4.4 RECOMMENDATIONS FOR OPTIMIZATION

A number of optimization techniques were put out to improve the efficacy and efficiency of reinforcement learning (RL) algorithms in autonomous systems. These suggestions address a number of important RL integration issues with the ultimate goal of strengthening its functionality as well as usefulness in real-world situations.

The issue of sample inefficiency, a problem that has long prevented RL's practical usage, was one of the main areas of study. To maximize the use of gathered data, techniques including experience replay as well as off-policy procedures were recommended [51]. In order to gain knowledge more effectively from limited material, experience replay includes saving along with reusing previous experiences. Off-policy approaches provide RL agents the ability to learn from data gathered under several rules, improving data efficiency and hastening learning. Deep RL architectural developments, such as deep Q-networks (DQN) and policy gradient techniques, were cited as potential directions. These methods leverage deep neural networks to address difficult decision-making problems, increasing data flexibility and efficiency as well as enabling improved RL integration inside autonomous systems.

Additionally, it was suggested to use transfer learning and domain-specific customization approaches for adapting RL algorithms to the particular needs as well as difficulties of autonomous systems. These methods eliminate the need for substantial retraining by allowing the reapplication of information acquired in one area to improve performance in another.

The importance of ethical and safe decision-making in autonomous systems was highlighted throughout the exploration of these issues. To make sure that autonomous operations are carried out responsibly and in accordance with ethical norms, strategies for tackling safety issues including moral conundrums were put forth. Together, these optimization recommendations open the path for RL integration in autonomous systems that are more effective, flexible, as well as morally acceptable, encouraging their wider adoption and safer deployment.

19.5   EVALUATION AND CONCLUSION

19.5.1   CONCLUSION

The integration of reinforcement learning (RL) in autonomous systems has been thoroughly examined in this study. There have been recognized and discussed key trends, technical difficulties, real-world applications, as well as optimization suggestions. The research emphasizes the revolutionary potential of RL, especially in industries like robotics, and self-driving cars, alongside health care. Strategies to reduce sample inefficiency, advancements in deep RL architectures, domain-specific customization, and safety and ethical concerns have all been suggested as approaches to improve RL's efficacy. This study contributes to a better understanding of RL's function in autonomous systems alongside acts as a useful resource for academics, practitioners, and policymakers by directing future deployments in an effective and accountable way.

19.5.2   RESEARCH RECOMMENDATION

There are various directions that reinforcement learning (RL) inside autonomous systems can go in the future. Researchers should first look more deeply into the moral and security issues underlying RL integration. As these systems proliferate, it is crucial to look at methods to guarantee that autonomous agents make ethical decisions and deal with ethical conundrums [52]. The applicability of suggested optimization solutions should be confirmed by longitudinal research and real-world applications. This involves evaluating the deep RL algorithms' scalability and flexibility in changing contexts [53]. To create complete frameworks as well as rules that direct the appropriate deployment of RL-powered autonomous systems, multi-disciplinary collaboration involving AI researchers, ethicists, legislators, alongside domain specialists is also essential [54]. The potential for autonomous systems in upcoming sectors like quantum computing, edge computing, and decentralized networks should additionally be explored in further study.

19.5.3   FUTURE WORK

The development of multidisciplinary frameworks for responsible AI deployment, and long-term field experiments, including improving RL's capacity to adapt to changing contexts should be the main goals of future research [55]. Emerging fields like decentralized networks and quantum computing present promising opportunities for RL-powered autonomous systems that demand investigation.

REFERENCES

1. Muhammad, K., Ullah, A., Lloret, J., Del Ser, J. and de Albuquerque, V.H.C., 2020. Deep learning for safe autonomous driving: Current challenges and future directions. *IEEE Transactions on Intelligent Transportation Systems*, 22(7), pp. 4316–4336.
2. Chowdhury, M.Z., Shahjalal, M., Ahmed, S. and Jang, Y.M., 2020.6G wireless communication systems: Applications, requirements, technologies, challenges, and research directions. *IEEE Open Journal of the Communications Society*, 1, pp. 957–975.
3. Sarker, I.H., 2021. Machine learning: Algorithms, real-world applications and research directions. *SN Computer Science*, 2(3), p. 160.
4. Kunduru, A.R., 2023. Machine learning in drug discovery: A comprehensive analysis of applications, challenges, and future directions. *International Journal on Orange Technologies*, 5(8), pp. 29–37.
5. Morocho-Cayamcela, M.E., Lee, H. and Lim, W., 2019. Machine learning for 5G/B5G mobile and wireless communications: Potential, limitations, and future directions. *IEEE Access*, 7, pp. 137184–137206.
6. Wahab, O.A., Mourad, A., Otrok, H. and Taleb, T., 2021. Federated machine learning: Survey, multi-level classification, desirable criteria and future directions in communication and networking systems. *IEEE Communications Surveys & Tutorials*, 23(2), pp. 1342–1397.
7. Sun, A.Y. and Scanlon, B.R., 2019. How can Big Data and machine learning benefit environment and water management: A survey of methods, applications, and future directions. *Environmental Research Letters*, 14(7), p. 073001.
8. Chataut, R. and Akl, R., 2020. Massive MIMO systems for 5G and beyond networks—overview, recent trends, challenges, and future research direction. *Sensors*, 20(10), p. 2753.
9. Sarker, I.H., Khan, A.I., Abushark, Y.B. and Alsolami, F., 2023. Internet of things (iot) security intelligence: A comprehensive overview, machine learning solutions and research directions. *Mobile Networks and Applications*, 28(1), pp. 296–312.
10. Nguyen, T.T. and Reddi, V.J., 2021. Deep reinforcement learning for cyber security. *IEEE Transactions on Neural Networks and Learning Systems*, 34(8), 3779–3795, Aug. 2023, doi: 10.1109/TNNLS.2021.3121870.
11. Alzubaidi, L., Zhang, J., Humaidi, A.J., Al-Dujaili, A., Duan, Y., Al-Shamma, O., Santamaría, J., Fadhel, M.A., Al-Amidie, M. and Farhan, L., 2021. Review of deep learning: Concepts, CNN architectures, challenges, applications, future directions. *Journal of big Data*, 8, pp. 1–74.
12. Luong, N.C., Hoang, D.T., Gong, S., Niyato, D., Wang, P., Liang, Y.C. and Kim, D.I., 2019. Applications of deep reinforcement learning in communications and networking: A survey. *IEEE Communications Surveys & Tutorials*, 21(4), pp. 3133–3174.
13. Zhang, J.M., Harman, M., Ma, L. and Liu, Y., 2020. Machine learning testing: Survey, landscapes and horizons. *IEEE Transactions on Software Engineering*, 48(1), pp. 1–36.
14. Wang, F. and Preininger, A., 2019. AI in health: State of the art, challenges, and future directions. *Yearbook of medical informatics*, 28(01), pp. 016–026.
15. Singh, B., Kumar, R. and Singh, V.P., 2022. Reinforcement learning in robotic applications: A comprehensive survey. *Artificial Intelligence Review*, pp. 1–46.
16. Helm, J.M., Swiergosz, A.M., Haeberle, H.S., Karnuta, J.M., Schaffer, J.L., Krebs, V.E., Spitzer, A.I. and Ramkumar, P.N., 2020. Machine learning and artificial intelligence: Definitions, applications, and future directions. *Current Reviews in Musculoskeletal Medicine*, 13, pp. 69–76.

17. Benzaid, C. and Taleb, T., 2020. AI-driven zero touch network and service management in 5G and beyond: Challenges and research directions. *IEEE Network*, *34*(2), pp. 186–194.

18. Brunke, L., Greeff, M., Hall, A.W., Yuan, Z., Zhou, S., Panerati, J. and Schoellig, A.P., 2022. Safe learning in robotics: From learning-based control to safe reinforcement learning. *Annual Review of Control, Robotics, and Autonomous Systems*, *5*, pp. 411–444.

19. Liu, Y., Yuan, X., Xiong, Z., Kang, J., Wang, X. and Niyato, D., 2020. Federated learning for 6G communications: Challenges, methods, and future directions. *China Communications*, *17*(9), pp. 105–118.

20. Tan, D.K.P., He, J., Li, Y., Bayesteh, A., Chen, Y., Zhu, P. and Tong, W., 2021, February. Integrated sensing and communication in 6G: Motivations, use cases, requirements, challenges and future directions. In *2021 1st IEEE International Online Symposium on Joint Communications & Sensing (JC&S)* (pp. 1–6). IEEE.

21. Li, T., Sahu, A.K., Talwalkar, A. and Smith, V., 2020. Federated learning: Challenges, methods, and future directions. *IEEE Signal Processing Magazine*, *37*(3), pp. 50–60.

22. Haydari, A. and Yılmaz, Y., 2020. Deep reinforcement learning for intelligent transportation systems: A survey. *IEEE Transactions on Intelligent Transportation Systems*, *23*(1), pp. 11–32.

23. Tsurumine, Y., Cui, Y., Uchibe, E. and Matsubara, T., 2019. Deep reinforcement learning with smooth policy update: Application to robotic cloth manipulation. *Robotics and Autonomous Systems*, *112*, pp. 72–83.

24. Gill, S.S., Xu, M., Ottaviani, C., Patros, P., Bahsoon, R., Shaghaghi, A., Golec, M., Stankovski, V., Wu, H., Abraham, A. and Singh, M., 2022. AI for next generation computing: Emerging trends and future directions. *Internet of Things*, *19*, p. 100514.

25. Carlucho, I., De Paula, M., Wang, S., Petillot, Y. and Acosta, G.G., 2018. Adaptive low-level control of autonomous underwater vehicles using deep reinforcement learning. *Robotics and Autonomous Systems*, *107*, pp. 71–86.

26. Lei, L., Tan, Y., Zheng, K., Liu, S., Zhang, K. and Shen, X., 2020. Deep reinforcement learning for autonomous internet of things: Model, applications and challenges. *IEEE Communications Surveys & Tutorials*, *22*(3), pp. 1722–1760.

27. Asghar, A., Farooq, H. and Imran, A., 2018. Self-healing in emerging cellular networks: Review, challenges, and research directions. *IEEE Communications Surveys & Tutorials*, *20*(3), pp. 1682–1709.

28. El Zaatari, S., Marei, M., Li, W. and Usman, Z., 2019. Cobot programming for collaborative industrial tasks: An overview. *Robotics and Autonomous Systems*, *116*, pp. 162–180.

29. Khan, L.U., Yaqoob, I., Imran, M., Han, Z. and Hong, C.S., 2020. 6G wireless systems: A vision, architectural elements, and future directions. *IEEE Access*, *8*, pp. 147029–147044.

30. Recht, B., 2019. A tour of reinforcement learning: The view from continuous control. *Annual Review of Control, Robotics, and Autonomous Systems*, *2*, pp. 253–279.

31. Gandhi, A., Adhvaryu, K., Poria, S., Cambria, E. and Hussain, A., 2023. Multimodal sentiment analysis: A systematic review of history, datasets, multimodal fusion methods, applications, challenges and future directions. *Information Fusion*, *91*, pp. 424–444.

32. Shi, Z., Yao, W., Li, Z., Zeng, L., Zhao, Y., Zhang, R., Tang, Y. and Wen, J., 2020. Artificial intelligence techniques for stability analysis and control in smart grids: Methodologies, applications, challenges and future directions. *Applied Energy*, *278*, p. 115733.

33. Ullah, Z., Al-Turjman, F., Mostarda, L. and Gagliardi, R., 2020. Applications of artificial intelligence and machine learning in smart cities. *Computer Communications*, *154*, pp. 313–323.

34. Nguyen, H. and La, H., 2019, February. Review of deep reinforcement learning for robot manipulation. In *2019 Third IEEE International Conference on Robotic Computing (IRC)* (pp. 590–595). IEEE.

35. Fink, O., Wang, Q., Svensen, M., Dersin, P., Lee, W.J. and Ducoffe, M., 2020. Potential, challenges and future directions for deep learning in prognostics and health management applications. *Engineering Applications of Artificial Intelligence*, *92*, p. 103678.
36. Fallah, S.N., Deo, R.C., Shojafar, M., Conti, M. and Shamshirband, S., 2018. Computational intelligence approaches for energy load forecasting in smart energy management grids: State of the art, future challenges, and research directions. *Energies*, *11*(3), p. 596.
37. Saeed, W. and Omlin, C., 2023. Explainable AI (XAI): A systematic meta-survey of current challenges and future opportunities. *Knowledge-Based Systems*, *263*, p. 110273.
38. Yaacoub, J.P.A., Salman, O., Noura, H.N., Kaaniche, N., Chehab, A. and Malli, M., 2020. Cyber-physical systems security: Limitations, issues and future trends. *Microprocessors and Microsystems*, *77*, p. 103201.
39. Qi, X., Chen, G., Li, Y., Cheng, X. and Li, C., 2019. Applying neural-network-based machine learning to additive manufacturing: Current applications, challenges, and future perspectives. *Engineering*, *5*(4), pp. 721–729.
40. Kiran, B.R., Sobh, I., Talpaert, V., Mannion, P., Al Sallab, A.A., Yogamani, S. and Pérez, P., 2021. Deep reinforcement learning for autonomous driving: A survey. *IEEE Transactions on Intelligent Transportation Systems*, *23*(6), pp. 4909–4926.
41. Zhang, K., Yang, Z. and Başar, T., 2021. Multi-agent reinforcement learning: A selective overview of theories and algorithms. *Handbook of Reinforcement Learning and Control*, pp. 321–384.
42. Olowononi, F.O., Rawat, D.B. and Liu, C., 2020. Resilient machine learning for networked cyber physical systems: A survey for machine learning security to securing machine learning for CPS. *IEEE Communications Surveys & Tutorials*, *23*(1), pp. 524–552.
43. Furstenau, L.B., Rodrigues, Y.P.R., Sott, M.K., Leivas, P., Dohan, M.S., López-Robles, J.R., Cobo, M.J., Bragazzi, N.L. and Choo, K.K.R., 2023. Internet of things: Conceptual network structure, main challenges and future directions. *Digital Communications and Networks*, *9*(3), pp. 677–687.
44. Mason, K. and Grijalva, S., 2019. A review of reinforcement learning for autonomous building energy management. *Computers & Electrical Engineering*, *78*, pp. 300–312.
45. Asharf, J., Moustafa, N., Khurshid, H., Debie, E., Haider, W. and Wahab, A., 2020. A review of intrusion detection systems using machine and deep learning in internet of things: Challenges, solutions and future directions. *Electronics*, *9*(7), p. 1177.
46. Padala, V.S., Gandhi, K. and Dasari, P., 2019. Machine learning: The new language for applications. *IAES International Journal of Artificial Intelligence*, *8*(4), p. 411.
47. Isele, D., Rahimi, R., Cosgun, A., Subramanian, K. and Fujimura, K., 2018, May. Navigating occluded intersections with autonomous vehicles using deep reinforcement learning. In *2018 IEEE International Conference on Robotics and Automation (ICRA)* (pp. 2034–2039). IEEE.
48. Ni, D., Xiao, Z. and Lim, M.K., 2020. A systematic review of the research trends of machine learning in supply chain management. *International Journal of Machine Learning and Cybernetics*, *11*, pp. 1463–1482.
49. Ding, R.X., Palomares, I., Wang, X., Yang, G.R., Liu, B., Dong, Y., Herrera-Viedma, E. and Herrera, F., 2020. Large-Scale decision-making: Characterization, taxonomy, challenges and future directions from an Artificial Intelligence and applications perspective. *Information Fusion*, *59*, pp. 84–102.
50. Sarker, I.H., 2021. Deep learning: A comprehensive overview on techniques, taxonomy, applications and research directions. *SN Computer Science*, *2*(6), p. 420.
51. Shang, C. and You, F., 2019. Data analytics and machine learning for smart process manufacturing: Recent advances and perspectives in the big data era. *Engineering*, *5*(6), pp. 1010–1016.

52. Theissler, A., Pérez-Velázquez, J., Kettelgerdes, M. and Elger, G., 2021. Predictive maintenance enabled by machine learning: Use cases and challenges in the automotive industry. *Reliability Engineering & System Safety*, *215*, p. 107864.

53. Alazab, M., RM, S.P., Parimala, M., Maddikunta, P.K.R., Gadekallu, T.R. and Pham, Q.V., 2021. Federated learning for cybersecurity: Concepts, challenges, and future directions. *IEEE Transactions on Industrial Informatics*, *18*(5), pp. 3501–3509.

54. Feriani, A. and Hossain, E., 2021. Single and multi-agent deep reinforcement learning for AI-enabled wireless networks: A tutorial. *IEEE Communications Surveys & Tutorials*, *23*(2), pp. 1226–1252.

55. Sharma, S.K., Woungang, I., Anpalagan, A. and Chatzinotas, S., 2020. Toward tactile internet in beyond 5G era: Recent advances, current issues, and future directions. *IEEE Access*, *8*, pp. 56948–56991.

Index

For Product Safety Concerns and Information please contact our EU
representative GPSR@taylorandfrancis.com
Taylor & Francis Verlag GmbH, Kaufingerstraße 24, 80331 München, Germany

9 7 8 1 0 3 2 3 8 0 6 4 3